RADAR SIGNALS

RADAR SIGNALS

NADAV LEVANON
ELI MOZESON

IEEE PRESS

A JOHN WILEY & SONS, INC., PUBLICATION

Copyright © 2004 by John Wiley & Sons, Inc. All rights reserved.

Published by John Wiley & Sons, Inc., Hoboken, New Jersey.
Published simultaneously in Canada.

No part of this publication may be reproduced, stored in a retrieval system, or transmitted in any form or by any means, electronic, mechanical, photocopying, recording, scanning, or otherwise, except as permitted under Section 107 or 108 of the 1976 United States Copyright Act, without either the prior written permission of the Publisher, or authorization through payment of the appropriate per-copy fee to the Copyright Clearance Center, Inc., 222 Rosewood Drive, Danvers, MA 01923, 978-750-8400, fax 978-646-8600, or on the web at www.copyright.com. Requests to the Publisher for permission should be addressed to the Permissions Department, John Wiley & Sons, Inc., 111 River Street, Hoboken, NJ 07030, (201) 748-6011, fax (201) 748-6008.

Limit of Liability/Disclaimer of Warranty: While the publisher and author have used their best efforts in preparing this book, they make no representations or warranties with respect to the accuracy or completeness of the contents of this book and specifically disclaim any implied warranties of merchantability or fitness for a particular purpose. No warranty may be created or extended by sales representatives or written sales materials. The advice and strategies contained herein may not be suitable for your situation. You should consult with a professional where appropriate. Neither the publisher nor author shall be liable for any loss of profit or any other commercial damages, including but not limited to special, incidental, consequential, or other damages.

For general information on our other products and services please contact our Customer Care Department within the U.S. at 877-762-2974, outside the U.S. at 317-572-3993 or fax 317-572-4002.

Wiley also publishes its books in a variety of electronic formats. Some content that appears in print, however, may not be available in electronic format.

Library of Congress Cataloging-in-Publication Data:

Levanon, Nadav.
    Radar signals / Nadav Levanon, Eli Mozeson.
       p. cm.
    Includes bibliographical references and index.
    ISBN 0-471-47378-2 (cloth)
    ISBN 13: 978-0-471-47378-7
    1. Radar.   I. Mozeson, Eli II. Title.

TK6575.L478 2004
621.3848–dc22
                              2003056882

10 9 8 7 6 5 4 3 2 1

*To the memory of my father Chaim Levanon,
who gave me two homes: our family and Tel Aviv University,
which he founded in 1953.*

N.L.

To Dganit, Noam, and Nadav.
E.M.

CONTENTS

Preface   xiii

1 Introduction   1

    1.1 Basic Relationships: Range–Delay and Velocity–Doppler   2
         Box 1A: Doppler Effect   3
    1.2 Accuracy, Resolution, and Ambiguity   7
    1.3 Environmental Diagram   13
    1.4 Other Trade-Offs and Penalties in Waveform Design   15
    1.5 Concluding Comments   17
         Problems   18
         References   19

2 Matched Filter   20

    2.1 Complex Representation of Bandpass Signals   20
         Box 2A: **I** and **Q** Components of Narrow Bandpass Signal   22
    2.2 Matched Filter   24
         Box 2B: Filter Matched to a Baseband Rectangular Pulse   27
    2.3 Matched Filter for a Narrow Bandpass Signal   29
    2.4 Matched-Filter Response to Its Doppler-Shifted Signal   31
         Problems   32
         References   33

3 Ambiguity Function — 34

- 3.1 Main Properties of the Ambiguity Function — 34
- 3.2 Proofs of the AF Properties — 36
- 3.3 Interpretation of Property 4 — 38
- 3.4 Cuts Through the Ambiguity Function — 40
- 3.5 Additional Volume Distribution Relationships — 42
- 3.6 Periodic Ambiguity Function — 42
  - Box 3A: Variants of the Periodic Ambiguity Function — 44
- 3.7 Discussion — 46
  - Appendix 3A: MATLAB Code for Plotting Ambiguity Functions — 47
  - Problems — 51
  - References — 52

4 Basic Radar Signals — 53

- 4.1 Constant-Frequency Pulse — 53
- 4.2 Linear Frequency-Modulated Pulse — 57
  - 4.2.1 Range Sidelobe Reduction — 61
  - 4.2.2 Mismatch Loss — 66
- 4.3 Coherent Train of Identical Unmodulated Pulses — 67
  - Problems — 72
  - References — 73

5 Frequency-Modulated Pulse — 74

- 5.1 Costas Frequency Coding — 74
  - 5.1.1 Costas Signal Definition and Ambiguity Function — 75
  - 5.1.2 On the Number of Costas Arrays and Their Construction — 80
  - 5.1.3 Longer Costas Signals — 83
- 5.2 Nonlinear Frequency Modulation — 86
  - Appendix 5A: MATLAB Code for Welch Construction of Costas Arrays — 96
  - Problems — 97
  - References — 99

6 Phase-Coded Pulse — 100

- Box 6A: Aperiodic Correlation Function of a Phase-Coded Pulse — 101

		Box 6B: Properties of the Cross-Correlation Function of a Phase Code	104
	6.1	Barker Codes	105
		6.1.1 Minimum Peak Sidelobe Codes	106
		6.1.2 Nested Codes	107
		6.1.3 Polyphase Barker Codes	109
	6.2	Chirplike Phase Codes	113
		6.2.1 Frank Code	115
		Box 6C: Perfectness of the Frank Code	117
		6.2.2 P1, P2, and Px Codes	118
		6.2.3 Zadoff–Chu Code	122
		Box 6D: Perfectness of the Zadoff–Chu Code	124
		Box 6E: Rotational Invariance of the Zadoff–Chu Code Aperiodic ACF Magnitude	125
		6.2.4 P3, P4, and Golomb Polyphase Codes	126
		6.2.5 Phase Codes Based on a Nonlinear FM Pulse	128
	6.3	Asymptotically Perfect Codes	132
	6.4	Golomb's Codes with Ideal Periodic Correlation	134
		Box 6F: Deriving the Perfect Golomb Biphase Code	135
		Box 6G: Deriving the Golomb Two-Valued Code with Ideal Periodic Cross-Correlation	136
	6.5	Ipatov Code	137
	6.6	Optimal Filters for Sidelobe Suppression	140
	6.7	Huffman Code	142
	6.8	Bandwidth Considerations in Phase-Coded Signals	145
	6.9	Concluding Comments	155
		Appendix 6A: Galois Fields	156
		Appendix 6B: Quadriphase Barker 13	158
		Appendix 6C: Gaussian-Windowed Sinc	159
		Problems	160
		References	164
7	**Coherent Train of LFM Pulses**		**168**
	7.1	Coherent Train of Identical LFM Pulses	169
	7.2	Filters Matched to Higher Doppler Shifts	173
	7.3	Interpulse Weighting	176
	7.4	Intra- and Interpulse Weighting	179
	7.5	Analytic Expressions of the Delay–Doppler Response of an LFM Pulse Train with Intra- and Interpulse Weighting	180
		7.5.1 Ambiguity Function of N LFM Pulses	181
		7.5.2 Delay–Doppler Response of a Mismatched Receiver	182

	7.5.3 Adding Intrapulse Weighting	183
	7.5.4 Examples	185
	Problems	189
	References	190

8 Diverse PRI Pulse Trains — 191

8.1 Introduction to MTI Radar — 191
    8.1.1 Single Canceler — 192
    8.1.2 Double Canceler — 193
8.2 Blind Speed and Staggered PRF for an MTI Radar — 195
    8.2.1 Staggered-PRF Concept — 195
    8.2.2 Actual Frequency Response of Staggered-PRF MTI Radar — 199
    8.2.3 MTI Radar Performance Analysis — 202
    Box 8A: Improvement Factor Introduced through the Autocorrelation Function — 204
    Box 8B: Optimal MTI Weights — 206
8.3 Diversifying the PRI on a Dwell-to-Dwell Basis — 210
    8.3.1 Single-PRF Pulse Train Blind Zones and Ambiguities — 210
    8.3.2 Solving Range–Doppler Ambiguities — 212
    8.3.3 Selection of Medium-PRF Sets — 214
    Box 8C: Binary Integration — 220
    Problems — 222
    References — 225

9 Coherent Train of Diverse Pulses — 226

9.1 Diversity for Recurrent Lobes Reduction — 226
9.2 Diversity for Bandwidth Increase: Stepped Frequency — 228
    9.2.1 Ambiguity Function of a Stepped-Frequency Train of LFM Pulses — 229
    9.2.2 Stepped-Frequency Train of Unmodulated Pulses — 231
    9.2.3 Stretch-Processing a Stepped-Frequency Train of Unmodulated Pulses — 236
    9.2.4 Stepped-Frequency Train of LFM Pulses — 245
9.3 Train of Complementary Pulses — 262
    Box 9A: Operations That Yield Equivalent Complementary Sets — 265
    9.3.1 Generating Complementary Sets Using Recursion — 266
    9.3.2 Complementary Sets Generated Using the PONS Construction — 267
    9.3.3 Complementary Sets Based on an Orthogonal Matrix — 269
9.4 Train of Subcomplementary Pulses — 270
9.5 Train of Orthogonal Pulses — 273

Box 9B: Autocorrelation Function of Orthogonal-Coded
Pulse Trains 274
9.5.1 Orthogonal-Coded LFM Pulse Train 277
9.5.2 Orthogonal-Coded LFM–LFM Pulse Train 279
9.5.3 Orthogonal-Coded LFM–NLFM Pulse Train 281
9.5.4 Frequency Spectra of Orthogonal-Coded Pulse Trains 284
Appendix 9A: Generating a Numerical Stepped-Frequency
Train of LFM Pulses 284
Problems 286
References 291

10 Continuous-Wave Signals 294

10.1 Revisiting the Periodic Ambiguity Function 295
10.2 PAF of Ideal Phase-Coded Signals 297
10.3 Doppler Sidelobe Reduction Using Weight Windows 301
10.4 Creating a Shifted Response in Doppler and Delay 305
10.5 Frequency-Modulated CW Signals 306
    10.5.1 Sawtooth Modulation 309
    10.5.2 Sinusoidal Modulation 311
    10.5.3 Triangular Modulation 315
10.6 Mixer Implementation of an FM CW Radar Receiver 318
Appendix 10A: Test for Ideal PACF 323
Problems 324
References 326

11 Multicarrier Phase-Coded Signals 327

Box 11A: Orthogonal Frequency-Division Multiplexing 330
11.1 Multicarrier Phase-Coded Signals with Low PMEPR 332
    11.1.1 PMEPR of an IS MCPC Signal 333
Box 11B: Closed-Form Multicarrier Bit Phasing with Low
PMEPR 335
    11.1.2 PMEPR of an MCPC Signal Based on COCS of a
    CLS 339
11.2 Single MCPC Pulse 341
    11.2.1 Identical Sequence 342
    11.2.2 MCPC Pulse Based on COCS of a CLS 345
11.3 CW (Periodic) Multicarrier Signal 350
11.4 Train of Diverse Multicarrier Pulses 358
    11.4.1 ICS MCPC Diverse Pulse Train 358
    11.4.2 COCS of a CLS MCPC Diverse Pulse Train 360
    11.4.3 MOCS MCPC Pulse Train 361
    11.4.4 Frequency Spectra of MCPC Diverse Pulse Trains 364
11.5 Summary 365

	Problems	367
	References	372

Appendix: Advanced MATLAB Programs — **373**

A.1	Ambiguity Function Plot with a GUI	373
A.2	Creating Complex Signals for Use with ambfn1.m or ambfn7.m	390
A.3	Cross-Ambiguity Function Plot	394
A.4	Generating a CW Periodic Signal with Weighting on Receive	400

Index — **403**

PREFACE

This book is devoted to the design and analysis of radar signals. The last comprehensive book dedicated to this subject was written in 1967 (Cook and Bernfeld; see Chapter 1 references). Since then, many journal and conference papers on radar signals have been published, as well as some good book chapters. Furthermore, the incredible progress in digital signal processing removed many of the constraints that squelched new signal ideas. Thus a new book on radar signals seems long overdue, and we believe that this book will fill the void.

Classical, enduring concepts such as the matched filter (Chapter 2) and the ambiguity function (Chapter 3) are discussed in detail, and useful related software is provided. Basic and advanced radar signals are described and analyzed. Many, if not most of the radar signals described in the open literature are presented and their performance analyzed. The knowledge gathered from these signals is used to suggest several new or modified signals that provide improved performance in resolution, ambiguity, spectral efficiency, and diversity.

The book contains tables of preferred signals and MATLAB codes for generating coded signals and for calculating and plotting ambiguity functions and other signal features. Each chapter is followed by a set of problems. Thus, the book can serve as a general reference as well as a textbook in an advanced radar course or a supplemental text in a basic radar course. In style and methodology it follows the approach used by Levanon in his 1988 text *Radar Principles* (see Chapter 1 references).

Following the chapters on matched filters and the ambiguity function, basic radar signals are discussed in Chapter 4 in terms of both analytical and numerical analysis. These include a constant-frequency pulse, a linear-FM pulse (with a weight window), and a coherent pulse train. Use of the MATLAB software provided is demonstrated using these simple signals.

In Chapter 5 we expand on other frequency-modulated schemes with detailed analysis of Costas coding and various nonlinear-FM signals. Chapter 6 provides a very comprehensive presentation of phase-coded signals, starting from Barker codes (binary and polyphase), minimum peak sidelobe codes, noiselike codes (PRNs), chirplike codes (Frank, Zadoff–Chu, P-codes), and codes suggested by Golomb, Ipatov, Huffman, and others. The chapter contains an important section on bandlimiting schemes in which the rectangular code element is replaced by a Gaussian windowed sinc or by a quadriphase waveform.

In Chapter 7 we expand on the most popular radar signal: a coherent train of linear-FM pulses. We offer a detailed analysis of its delay–Doppler performance, including intra- and interpulse weighting (matched or on-receive) for range and Doppler sidelobe reduction.

Diversity is widely used in radar signals, beginning with diversity of the pulse repetition interval (Chapter 8), which mitigates blind speeds. More elaborate diversity schemes are presented in Chapter 9, with emphasis on stepped-frequency pulses (both unmodulated and linear FM). The stepped-frequency signal is also used to demonstrate the important stretch-processing concept. The chapter ends with sections on complementary and orthogonal pulses.

Continuous-wave radar signals have experienced a revival in both military and civilian applications and are the subject of Chapter 10. Their analysis tool is the periodic ambiguity function, revisited in this chapter. Both analog and digital coded continuous-wave signals are presented and analyzed. Both matched-filter and simple mixer processing are studied.

Chapter 11 is devoted to multicarrier radar signals. Multicarrier is well known in communications but is a relatively new signal concept in radar. It offers the designer more degrees of freedom and more dimensions through which to introduce diversity. However, it entails variable amplitudes. Presently, this is a major hindrance in high-power amplifiers. But the aim of this book is to cover all worthy signals, including those that presently suffer from implementation difficulties.

In addition to the MATLAB programs embedded throughout the book, more elaborate MATLAB programs are provided in an appendix at the end of the book. Some of these programs include graphic user interfaces, which simplifies their use and allows quick change of parameters. Mastering the use of these programs is well worth the effort.

NADAV LEVANON
ELI MOZESON

1

INTRODUCTION

This book is devoted to the design and analysis of radar signals. Basic concepts such as the matched filter and the ambiguity function are discussed in detail, and useful related software is provided. Basic and advanced radar signals are presented and analyzed. The various chapters include many, if not most, of the radar signals described in the open literature. The knowledge obtained is utilized to suggest several new or modified signals with improved performance.

In the history of radar, signal ideas usually preceded implementation by many years, because of processing complexity and hardware limitations. In the introduction to their classical book *Radar Signals*, Cook and Bernfeld (1967) describe how the concept of pulse compression, developed and patented during World War II, was buried as a curiosity in the patent files. Only when the necessary transmitter components (e.g., high-power klystrons) became available did pulse compression win renewed interest.

In the first chapter of *Radar Design Principles*, Nathanson (1991) presents a checklist of possible constraints that can limit radar design. Among them are such questions as:

1. Can the transmitter support complex waveforms?
2. Is the transmitter suitable for pulsed- or continuous-wave transmission?
3. Are there unavoidable bandwidth limitations in the transmitter, receiver, or antenna?
4. Is frequency shifting from pulse to pulse practical?

Radar Signals, By Nadav Levanon and Eli Mozeson
ISBN 0-471-47378-2 Copyright © 2004 John Wiley & Sons, Inc.

Lack of coherent signal generation and amplification, which hampered pulse compression during World War II, seems naive today. So does the first item in Nathanson's checklist, dealing with complex waveforms. Extrapolating to the future, present limitations, such as the linear power amplifiers required for variable-amplitude radar signals, may raise eyebrows several years from now. In an attempt to extend its relevance, in this book we allow considerable freedom in the various characteristics of waveforms.

The word *Radar* (derived from "radio detection and ranging") summarizes the two main tasks of radar: detecting a target and determining its range. Fairly early range has expanded to include direction to the target and radial velocity between the radar and the target. Presently, more information on the target can be sought, such as its shape, size, and trajectory.

In most cases the reliability of detection, including the statistics of hits, misses, and false alarms, depends mostly on the signal's energy compared to the receiver's thermal noise level, and much less on the waveform. Determining the spatial direction to the target depends (in a stationary radar) on the antenna and its tracking system. The signal's waveform is responsible for the accuracy, resolution, and ambiguity of determining the range and radial velocity (range rate) of the target. Range is associated with the *delay* of the signal received. Range rate is associated with the *Doppler* shift of the signal received. These relationships are discussed next.

1.1 BASIC RELATIONSHIPS: RANGE–DELAY AND VELOCITY–DOPPLER

When the target can be approximated by a small point and the environment is free space, the relationship between range R and delay τ is simply

$$R = \tfrac{1}{2} C_p \tau \tag{1.1}$$

where C_p is the velocity of propagation. The factor $\tfrac{1}{2}$ is due to the fact that the radar signal traverses the distance R twice (round trip). Equation (1.1) is only an approximation. In the lower atmosphere, C_p is not a constant but changes with altitude; hence the radar signal propagates a slightly longer distance along a bent path. Since the effect is minor and not very related to radar signals, it will be ignored.

The Doppler shift is developed in Box 1A with the help of Fig. 1.1, in which the propagation of two peaks (A and B) of a sinusoidal signal are followed as they propagate toward a target moving at a constant radial velocity v. It is shown that when the signal bandwidth is narrow compared with the carrier frequency (as in the case of a pure sinusoidal) and where the target radial velocity v is much smaller than the propagation velocity C_p, the *Doppler shift*, defined as the difference between the frequency received, f_R, and the frequency transmitted,

BASIC RELATIONSHIPS: RANGE–DELAY AND VELOCITY–DOPPLER

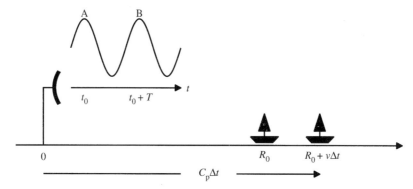

FIGURE 1.1 Timing in a Doppler scene.

BOX 1A: Doppler Effect

The Doppler shift is developed with the help of Fig. 1.1. Peak A departs at time $t = t_0$ when the target is at R_0 and reaches the target after travel time Δt, during which the target advanced an additional distance; hence,

$$C_p \, \Delta t = R_0 + v \, \Delta t \qquad (1A.1)$$

where R_0 is the target location when peak A leaves the radar ($t = t_0$), Δt the travel time of peak A to reach the target, and $v \, \Delta t$ the distance advanced by target during Δt. Rewriting (1A.1), the signal's travel time is given by

$$\Delta t = \frac{R_0}{C_p - v} \qquad (1A.2)$$

The moment t_1 in which peak A returns to the radar is given by

$$t_1 = t_0 + 2 \, \Delta t = t_0 + \frac{2 R_0}{C_p - v} \qquad (1A.3)$$

Similar expressions can be worked out for the second peak B, which left the radar T seconds after peak A and returned at t_2:

$$t_2 = t_0 + T + \frac{2 R_1}{C_p - v} \qquad (1A.4)$$

where R_1 is the target location when peak B leaves the radar (at $t = t_0 + T$), t_2 the time of return of peak B to radar, and T the period of transmitted sinusoidal waveform. Note that R_1 in (1A.4) can be replaced by

$$R_1 = R_0 + vT \qquad (1A.5)$$

The period of the received waveform T_R is equal to the difference between the arrival times of the two peaks:

$$T_R = t_2 - t_1 = t_0 + T + \frac{2(R_0 + vT)}{C_p - v} - \left(t_0 + \frac{2R_0}{C_p - v}\right) = T\frac{C_p + v}{C_p - v} \quad (1A.6)$$

The ratio between the received and transmitted periods is therefore

$$\frac{T_R}{T} = \frac{C_p + v}{C_p - v} \quad (1A.7)$$

and the ratio between the corresponding frequencies is

$$\frac{f_R}{f_0} = \frac{C_p - v}{C_p + v} = \frac{1 - v/C_p}{1 + v/C_p} \quad (1A.8)$$

yielding the received frequency,

$$f_R = f_0 \frac{1 - v/C_p}{1 + v/C_p} \quad (1A.9)$$

In electromagnetic propagation (contrary to acoustic propagation) the expected target velocities are always much smaller than the velocity of propagation, $v \ll C_p$, yielding the approximation

$$\frac{1}{1 + v/C_p} = 1 - \frac{v}{C_p} + \frac{v^2}{C_p^2} - \cdots \quad (1A.10)$$

Using (1A.10) in (1A.9) yields

$$f_R = f_0 \left(1 - \frac{v}{C_p}\right)\left(1 - \frac{v}{C_p} + \frac{v^2}{C_p^2} - \cdots\right)$$

$$= f_0 \left(1 - \frac{2v}{C_p} + \cdots\right) \approx f_0 \left(1 - \frac{2v}{C_p}\right) \quad (1A.11)$$

Rewriting, we get

$$f_R \approx f_0 - \frac{2v}{C_p/f_0} = f_0 - \frac{2v}{\lambda} \quad (1A.12)$$

where λ is the wavelength transmitted. The Doppler shift is defined as

$$f_D = f_R - f_0 \approx -\frac{2v}{\lambda} \quad (1A.13)$$

f_0, is given by

$$f_D = f_R - f_0 \approx -\frac{2v}{\lambda} \qquad (1.2)$$

where λ is the wavelength transmitted.

Figure 1.1 is a special case in which the velocity is exactly in the radial direction, hence equal to the range rate

$$v = \dot{R} \qquad (1.3)$$

The more general approximation to the Doppler shift is therefore

$$f_D \approx -\frac{2\dot{R}}{\lambda} \qquad (1.4)$$

The scenario depicted in Fig. 1.1 was also a special case from another point of view. The signal was a pure sinusoid at a frequency f_0. What happens when the signal contains modulation: that is, other frequencies? In other words, can we talk about a single Doppler shift when the signal has considerable bandwidth?

In wide (or ultrawide) bandwidth signals we have to go back to equation (1A.7) and note that the target's movement created a time scale between the signal transmitted and the signal received:

$$T = \frac{C_p - v}{C_p + v} T_R \underset{v \ll C_p}{\approx} \left(1 - \frac{2v}{C_p}\right) T_R \qquad (1.5)$$

This time scale applies not only to the period of the signal but to the time axis in general. In other words, ignoring attenuation, the signal received, $s_R(t)$, can be written as a time-scaled and delayed version of the signal transmitted, $s(t)$:

$$s_R(t) = s\left[\left(1 - \frac{2v}{C_p}\right)t - \tau\right] \qquad (1.6)$$

The delay τ is twice the one-way signal travel time Δt defined in (1A.2): namely,

$$\tau = 2\,\Delta t = \frac{2R_0}{C_p - v} = \frac{2R_0}{C_p(1 - v/C_p)} \approx \frac{2R_0}{C_p} \qquad (1.7)$$

Errors resulting from the various approximations can be found in Appendix A of DiFranco and Rubin (1968). A popular rule of thumb says that if the signal bandwidth is less than one-tenth of the carrier frequency, the signal is considered a narrowband signal and it is reasonable to assume that the target motion causes only a Doppler shift of the carrier frequency according to (1.4). Otherwise, time scaling should be considered, which also affects the envelope of the signal. In the following chapters we use the narrowband assumption. Numerical simulations with rather complicated signals showed that the difference between the calculated performances was very small, even when the narrowband assumption was used with a signal whose bandwidth reached 40% of the center frequency.

Another assumption made above and used henceforth is lack of radial acceleration: (i.e., $\ddot{R} \approx 0$). In most radar applications it is justified to assume that over the *coherently processed signal duration*, the radial velocity remains constant. In practice, the expected target acceleration is usually limited below some predetermined value characterizing the targets in question. The design of radar signals should take this value into consideration such that the target does not change its Doppler in an amount higher than Doppler resolution during the coherence processing period.

The phrases *coherently processed signal duration* and *Doppler resolution* need clarification. They can be explained with the example of a train of unmodulated pulses. As we will learn later in the book, Doppler resolution of a signal is a function of the total duration of the signal. A common approach to extending the signal is to repeat it periodically. A single pulse has poor Doppler resolution because the Doppler shift creates little change during the pulse duration. On the other hand, a train of pulses exhibits good Doppler resolution because of the changes (due to Doppler) between the pulses. This change is primarily in the Doppler-induced initial phase of each pulse. To extract this Doppler-induced phase change it is necessary for the receiver to know the original initial phase of each pulse. That is what we mean when we refer to the pulse train as a *coherent* pulse train. A simple example of a coherent pulse train is shown in Fig. 1.2.

The simple example in Fig. 1.2 is a special case in which the coherence was obtained by on–off switching of a continuous sinusoidal signal. However, any

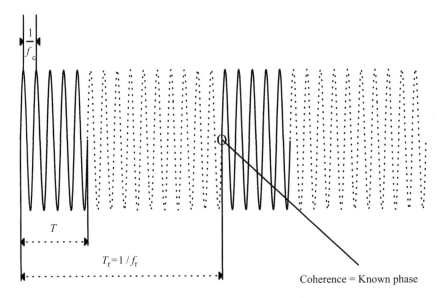

FIGURE 1.2 Coherent pulse train.

other initial phases (of the second and later pulses) are acceptable as long as the receiver knows what they were when transmitted. An example is radar utilizing a noncoherent transmitting device, where each transmitted pulse has a randomly generated initial phase. In such radar systems it is common to lock on the transmitted pulse phase using a dedicated circuit and to use this memorized phase as a reference for the pulse received. Implementations where the phase value is known only one pulse backward are usually referred to as *coherent on receive*.

Having associated range and velocity with two signal parameters, delay and Doppler, we can now discuss how well we can determine them and how the signal design can help.

1.2 ACCURACY, RESOLUTION, AND AMBIGUITY

Let us begin with a simple example. We need to measure the frequency of a sinusoidal signal. If the signal-to-noise-plus-interference ratio (SNIR) is very high (i.e., there are no other sinusoids and negligible random noise), we can measure the frequency with a counter. A counter counts the number of cycles within a given time span or measures the time interval between several zero crossings. A counter will produce an erroneous result when there is additive noise or when other sinusoidal signals are present: that is, when the signal-to-noise ratio (SNR) is low or when there are interferences from other signals. The lower the SNIR, the bigger the measurement error will be. Below a threshold SNIR, the counter will fail completely. What can be done in the low-SNIR scenario is to feed the received signal into many narrow bandpass filters, each centered at a different frequency. We can then find the filter that yields the highest output, or pick those filters whose outputs exceed a predetermined threshold. One way to implement such a bank of filters is to sample the input signal (plus noise) and perform fast Fourier transform (FFT).

The radar scenario is almost always a low-SNIR scenario. In some applications the radar performance is noise limited, while in other applications the performance is interference limited. The reflection from a target is almost always accompanied by reflections from the surrounding environment (ground, ocean), referred to as *clutter*, or by reflections from neighboring targets or targets farther away. For targets at a great distance or closer targets with a lower radar cross section (RCS), the thermal noise becomes a significant background. For this reason, measurements in radar are usually performed by a bank of filters, in delay and in Doppler.

Still, in many applications the target may stand alone and provide a high SNIR value (e.g., as in the case of ground-based antenna direction measurement of airborne targets). Indeed, in these cases it is practical to perform a measurement (e.g., angle measurement using the monopulse method) and not necessarily employ a multibeam array, which is equivalent to a bank of spatial filters. A more detailed discussion of measurement versus filtering in radar may be found in (Levanon, 1988).

The filter used in radar to measure the delay of a returned known signal is usually the matched filter. The matched filter is so important in radar that it is the subject of Chapter 2. The matched filter concentrates the entire energy of the signal into an output peak at a predetermined additional delay. It is therefore optimal for causing the output to cross the threshold and identify a detected reflection at the corresponding delay in the presence of the receiver thermal noise.

The peak of the output of a matched filter, when fed by the signal to which it was matched, is a function of the signal's energy and not of the signal's waveform. However, the output before and after the peak are strongly affected by the waveform. If the output level remains high over an extended delay, the threshold will be crossed in many delay cells, resulting in uncertainty as to which is the true delay. This implies that the measurement *accuracy* is proportional to the shape of the matched filter response close to the peak (actually, the second derivative of the response) and inversely proportional to the SNR.

Furthermore, if there are weaker neighboring targets that should be detected, their matched-filter output peak could be masked by a wide peak (mainlobe) or high sidelobes of a strong target. Thus the *resolution* or minimal separable distance (MSD) is proportional to the ambiguity function mainlobe width (usually measured at the -3 dB point) and inversely proportional to the SNIR. The interference level itself is a function of the nature of the interference. In a case where the interference is caused by a point target, the interference level is proportional to the interfering target RCS and the matched-filter sidelobe level expected at a given separation. The matched-filter peak sidelobe level ratio (PSLR) is often used to characterize the level of interference expected from point targets. For volume or surface clutter the interference level is characterized by the matched-filter integrated sidelobe level ratio (ISLR).

The science (or art) of designing radar signals is based on finding signals that yield a matched-filter response that matches a given application. For example, if closely separated targets are to be detected and distinguished in a low-SNR scenario, a radar signal having a matched-filter response that exhibits a narrow mainlobe (the peak) and low sidelobes is required. The mainlobe width and sidelobe level requirements are a function of the expected target separation and expected target RCS difference.

Two targets can be near each other in range (e.g., an aircraft flying over a patch of land) but way apart in radial velocity. For this reason radar receivers create filters matched not only to the signal transmitted but also to several different Doppler-shifted versions of it. Here again it is important to achieve a narrow response in Doppler, so that, for example, the moving target could be distinguished from the stationary background. Each one will cause a peak at a different Doppler-shifted matched filter. So the response of a matched filter needs to be studied in two dimensions: delay τ and Doppler ν. The tool for that is the ambiguity function (Woodward, 1953; Rihaczek, 1969), which describes that two-dimensional response. The ambiguity function $|\chi(\tau, \nu)|$ is the subject of Chapter 3. A basic question when designing a radar signal is: What is a good ambiguity function (AF), and can it be obtained? Intuitively, one could think that the ideal AF should exhibit a single sharp peak at the origin (which is the nominal delay and Doppler for that matched filter), and near-zero level everywhere else (thumbtack shape). Even if it is the ideal AF, it cannot be produced completely. We learn in Chapter 3 that the AF peak at the origin cannot exceed a value of 1 and that the volume underneath the ambiguity function squared is a constant. If the AF is lowered in one area of the delay–Doppler plane, it must rise somewhere else.

Several AF shapes are presented in Figs. 1.3 to 1.6. Only two quadrants of the AF (positive Doppler) are plotted. Two adjacent quadrants contain all the information because the AF is symmetrical with respect to the origin. The four plots are presented in order to demonstrate different possible distributions of the AF volume over the delay–Doppler plane. The corresponding signals are discussed in more detail in later chapters of the book.

Figure 1.3 shows the AF of the most basic signal—an unmodulated pulse of width T (see Section 4.1). The delay axis is normalized with respect to T. The Doppler axis is normalized with respect to $1/T$. (The same type of normalization is used in Figs. 1.4 and 1.5.) The 0.5, 0.25, and 0.1 contour lines are also shown on top of the AF contour. The AF demonstrates the expected resolutions in delay and Doppler. The ambiguity function is zero for delays higher than the pulse width; thus no interference is expected with targets having range separation higher than the pulse duration. The ambiguity function shows relatively large sidelobes

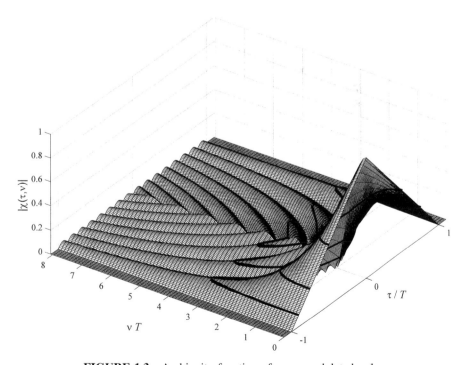

FIGURE 1.3 Ambiguity function of an unmodulated pulse.

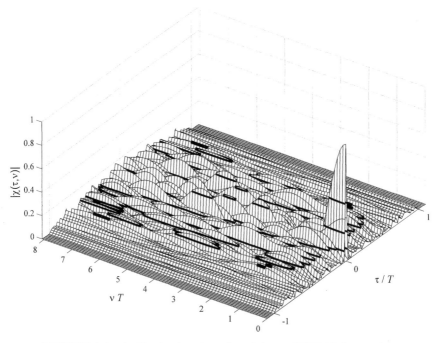

FIGURE 1.4 Ambiguity function of a minimum PSLR biphase pulse.

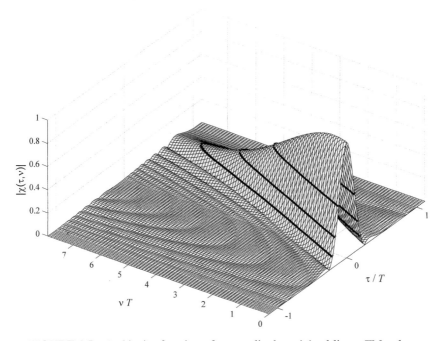

FIGURE 1.5 Ambiguity function of an amplitude-weighted linear-FM pulse.

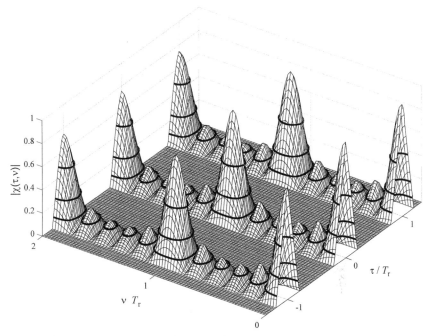

FIGURE 1.6 Ambiguity function of a train of six unmodulated pulses.

in Doppler (PSLR of -13 dB) and Doppler resolution of $1/T$ (both the -3 dB point separation and the first null are at $1/T$). Figure 1.3 is a poor approximation of a thumbtack shape using a signal with a time–bandwidth product of 1. (The time in the time–bandwidth product is the signal duration; the bandwidth is either the separation between the center frequency and the first spectral null or that between the two -3 dB frequencies.)

Figure 1.4 shows a better approximation of the thumbtack shape obtained with a very large time–bandwidth product by binary-phase modulating the pulse in a "random" way (a MPS 48-element biphase code was used; see Section 6.1.1). Similar ambiguity is also obtained by "randomly" frequency stepping the original pulse (using Costas codes; see Section 5.1). The time–bandwidth product used for the plot is 48 (the number of chips in the phase code). The normalized Doppler axis extends to only one-sixth of the time–bandwidth product, but a low sidelobe pedestal extends, in normalized Doppler, as far as the time–bandwidth product.

Signals with a large time–bandwidth product such as the one described here and the one used for Fig. 1.5 are also called *pulse compression waveforms*. The central spike of the AF (the waveform resolution cell size) has time duration $1/B$ and Doppler width $1/T$. The average height of the pedestal of the thumbtack function is $1/TB$, where TB is the time–bandwidth product (or *compression ratio*) of the signal. The AF is spread over a time interval T and a Doppler interval B. The total volume within the pedestal thus is 1, compared to a total volume of $1/TB$ in the central spike.

Figure 1.5 is the AF of a linear-FM pulse (see Section 4.2). The time–bandwidth product is 8. The normalized Doppler axis extends as far as that time–bandwidth product. The figure shows how increasing the bandwidth through modulation narrows (improves) the delay resolution, moving the AF volume from the vicinity of zero Doppler to higher Doppler shifts, yielding a ridge-shaped ambiguity function. Similar AF shapes are obtained by using ordered frequency or phase coding (e.g., P4 or Frank phase coding; see Section 6.2). The result of using "ordered" (in this case, linear) modulation is range–Doppler coupling, easily observed in the AF in the form of a diagonal ridge. Instead of a uniform sidelobe pedestal, the response energy is essentially concentrated at the area of the diagonal ridge, causing lower sidelobes outside the ridge.

Figure 1.6 is a periodic AF of a train of $N = 6$ unmodulated pulses with a duty cycle of $T/T_r = \frac{1}{5}$. (The *periodic* ambiguity function implies that the signal received is an infinitely long train of pulses, while the receiver coherently processes six pulses; see Section 4.3.) The plotted delay axis is normalized with respect to the pulse repetition interval T_r and extends over more than two periods. The Doppler axis plotted is normalized with respect to $1/T_r$ and extends as far as twice the repetition frequency. The figure shows a completely different AF shape in which the mainlobe at the origin is narrow along both the delay and the Doppler axis. The AF volume is now spread at many recurrent lobes, almost as high as the mainlobe at the origin. This shape is referred to as a *bed of nails*. This type of signal achieves good resolution in both delay and Doppler but creates both range and Doppler ambiguities. For example, a target at delay of $\tau + T_r$ will produce a response peak at exactly the same delay as a target at delay τ (usually referred to as *second-time-around echo* or *range folding*). Such an ambiguity is difficult to resolve.

Rihaczek (1971) identified and classified the four classes of ambiguity function described in Figs. 1.3 to 1.6. Table 1.1 summarizes the properties of the various classes, as can be observed from the corresponding AF plots. The constant volume under the ambiguity function squared puzzled many researchers and yielded several unfounded variations of the ambiguity function, which are no more than

TABLE 1.1 Waveform Classification and Ambiguity Functions

	Class			
	A	B1	B2	C
Figure	1.3	1.4	1.5	1.6
Time–bandwidth product	Unity	Large compared with unity		
Ambiguity function	Unsheared ridge	Thumbtack	Sheared ridge	Bed of nails
Resolution cell size	Unity	$1/TB$	Unity	$1/TB$
Ambiguities	No	No	Range–Doppler coupling	Spikes
Sidelobes	Low	High	Low	Low

mathematical fiction. Fourteen years after his book was published, Woodward attempted to dispose of some "grandiose clearance schemes." In a technical note of the Royal Radar Establishment (Woodward, 1967; Nathanson et al., 1991), Woodward wrote: "There is continued speculation on the subject of ambiguity clearance. Like slums, ambiguity has a way of appearing in one place as fast as it is made to disappear in another. That it must be conserved is completely accepted but the thought remains that ambiguity might be segregated in some unwanted part of the time–frequency plane where it will cease to be a practical embarrassment." Efficient (matched-filter) coherent processing of radar signals must obey the reality of the constant volume of AF squared. All the signal designer can do is to manipulate the AF volume so that it will best fit the expected radar target and its surrounding radar environment.

1.3 ENVIRONMENTAL DIAGRAM

The AF is an important tool in characterizing waveforms in terms of the resolution, sidelobe level, and ambiguity. The question of which ambiguity function should be preferred depends not only on the desired delay and Doppler resolution, or on the complexity of the required processor, but also on where the clutter or competing targets are located in the delay–Doppler plane (i.e., in the radar environment). Cook and Bernfeld (1967) stated that "in the extreme case, all signals (waveforms) are equally good (or bad) as long as they are not compared against a specific radar environment." In other words, when coming to the point of selecting a waveform (or waveform class) for a given radar application, the AF should be tested against the environmental parameters that characterize the application.

The environment that the radar encounters may consist of a variety of clutter conditions, countermeasure interference (such as chaff or deliberate electronic emissions), and interference from neighboring radars. The environmental diagram details spectral, spatial, and amplitude characteristics of the radar environment and is used as the basis against which the ambiguity diagram is played in selecting a waveform design. Nathanson et al. (1969) presented a basic model of a radar environmental diagram. An example of an environmental diagram of surveillance radar located at a coastal site is illustrated in Fig. 1.7. Radial velocity (Doppler) is given on the ordinate and the target extent (delay) is indicated along the abscissa. Diagrams such as Fig. 1.7 are sometimes referred to as *R–V diagrams* or *target space*. This environmental diagram shows only the regions in which land or sea clutter, rainstorm, and high-altitude chaff can be expected. The figure does not show their relative power level as seen by the radar. This additional information could be presented using a three-dimensional plot similar to an AF plot, taking into account the radar antenna pattern and direction of interest. For example, looking in the direction of the sea, the land clutter is received only through the antenna sidelobes (or even backlobes), whereas the sea clutter is in the mainlobe area.

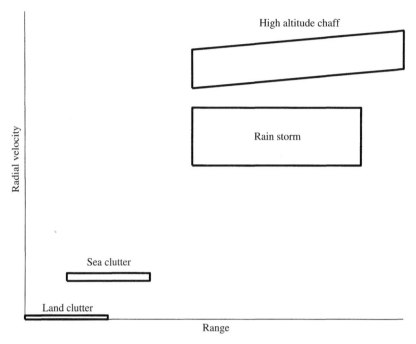

FIGURE 1.7 Basic environmental diagram of a coastal-based radar.

The basic environmental diagram gives a pictorial view of the clutter in range and velocity that the radar must contend with. By selecting the target trajectories expected within the R–V diagram and superimposing the ambiguity diagram of a particular waveform, it is possible to evaluate certain desirable characteristics inherent in the waveform. In Fig. 1.8 the AF contour of a pulse burst waveform is overlaid on the basic environmental diagram. As the target follows a particular trajectory, the ambiguity diagram will move accordingly, and AF ambiguous peaks will enter and exit the chaff and rainstorm space.

Aasen (1976) extended the concept of environmental diagrams to include other electromagnetic radiations from transmitters within the general locality of a radar site. Interference signals will appear to the radar receiver as signals with particular velocity and range characteristics. For example, a stable CW signal within the receiver bandwidth would appear as a horizontal straight line in an environmental diagram. If the CW signal were frequency modulated, the width of the line would increase according to the modulation bandwidth.

A different and far more complicated environmental diagram is of airborne radar. In this case the ground clutter is amplitude, range, and Doppler modulated due to the platform velocity, altitude, and antenna direction. The ground clutter appears in the environmental diagram in the form of a strong mainlobe clutter (MLC) caused by the antenna mainlobe illumination on a specific spot on the ground (limited in Doppler and delay) and a much weaker sidelobe clutter (SLC). The SLC extends in range from the minimal range determined by the platform

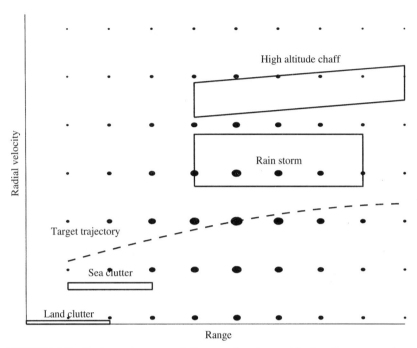

FIGURE 1.8 Basic environmental diagram showing ambiguity diagram overlay.

altitude (minimal distance to the ground) to a maximal range determined by the radar horizon at the given altitude. The Doppler spread of the SLC is determined by the platform velocity vector direction relative to the ground plane. A third dominant ground clutter contribution is the altitude line caused by strong reflection of the ground area underneath the platform. The altitude line return is concentrated in range close to the platform altitude. When the platform is flying straight and level, the SLC Doppler will extend, in corresponding velocities, from $-V$ to $+V$ (where V is the platform velocity) and the altitude line will be positioned at zero Doppler.

1.4 OTHER TRADE-OFFS AND PENALTIES IN WAVEFORM DESIGN

The constant volume underneath the ambiguity function squared implies that signal design involves trade-offs. If we want narrow response in one dimension, we have to accept either poor response in the other dimension or additional ambiguous peaks. We will learn that if we want the ambiguous AF peaks to be well spaced in delay, we have to accept them as closely spaced in Doppler (and vice versa). If we want high range resolution, we should expect a penalty of wide spectral width. If we want good resolution in Doppler, the penalty is long coherent signal duration.

One of the trade-offs in radar signal design is between constant amplitude and AF sidelobes. Efficient RF power amplifiers are presently operating at saturation and do not allow linear changes in amplitude. On the other hand, sidelobe reductions in range or Doppler usually require amplitude variations (weighting). To maintain optimal matched filter performance, the amplitude variations should be split between the transmitter and the receiver. If they cannot be implemented in the high-power transmitter, the full weight window will have to be implemented in the receiver, causing mismatch filtering. We will learn that such a mismatch carries a penalty in the output SNR.

Another conflict involving a linear power amplifier relates variable amplitude and spectrum. In Section 6.8 two variations of a P4 phase-coded signal are compared: one in which a code element (bit) has a rectangular shape and a second in which the shape is Gaussian-windowed sinc. The resulting real amplitudes of a 25-bit signal are plotted in Fig. 1.9. The rectangular bit shape results in a constant-amplitude pulse (dotted line), while the Gaussian-windowed sinc results in a variable-amplitude pulse (solid line). The phase evolution is quite similar in both signals. The resulting AF and autocorrelation function are also very similar. What differ dramatically are the spectrums, plotted in Fig. 1.10. The bottom subplot presents the spectrum of the constant-amplitude pulse. The spectral sidelobes decay very slowly, at a rate of 6 dB/octave. Such a long-tailed spectrum may violate spectral emission regulations, can cause interference to neighboring radars, and may be too wide for the next radio-frequency (RF) stage—the antenna. On the other hand, the spectrum of the variable-amplitude pulse (top subplot) exhibits practically no sidelobes and reaches a -60 dB level at a frequency of $1.2/t_b$, where t_b is the bit length.

Examples of other radar signals with variable real amplitude are the Huffman-coded signal, discussed in Section 6.8, and multicarrier signals, discussed in Chapter 11. The few examples above suggest that removing the constant-amplitude restriction provides radar signals with an additional degree of freedom, which can be used to improve performance. So building an efficient low-cost linear power RF amplifier joins the list of hardware challenges that radar designers faced during radar's development.

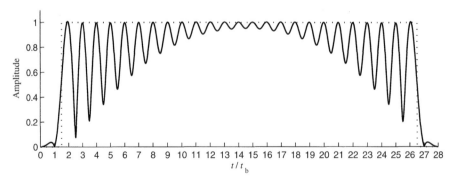

FIGURE 1.9 Amplitude of P4 (25-bit) with Gaussian-windowed sinc (solid) and rectangular (dotted) bit shapes.

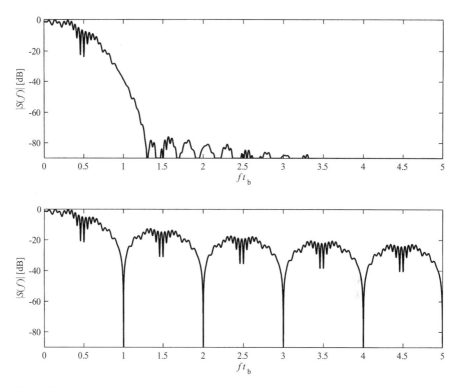

FIGURE 1.10 Spectrum of P4 (25-bit) with Gaussian-windowed sinc (top) and rectangular (bottom) bit shapes.

1.5 CONCLUDING COMMENTS

This was an opening chapter for a text on radar signals, not one on radar principles. In it and the remaining chapters it is assumed that the reader has some radar background, from courses, general radar texts, or experience. In the introduction we presented the basics of radar measurements together with the conflicts in designing suitable waveforms for different applications. In the following chapters we provide the basic mathematical tools required for analyzing and comparing different radar signals. Among these tools are the matched filter and the ambiguity function, and their relation to the signal frequency spectrum. Also discussed are useful building blocks and concepts (e.g., linear FM, weight windows, orthogonality, or complementarity) used in operational radar systems. Use of these building blocks in designing specific waveforms and their parameters is also demonstrated for many types of signals. Chapter 11 is an example of utilizing many of the building blocks to design more complex waveforms.

As already stated, the work (or art) of designing radar waveforms is based mostly on experience and expertise obtained through successive designs. This experience is gained by manipulating signal parameters while using special building blocks with desirable mathematical properties. To get more experience with the various signals, it is recommended that the MATLAB codes presented in the book be used together with the ambiguity function plotting tool presented in Chapter 3, to change signal parameters and observe the changes in waveform performance.

Finally, it is worth mentioning that the book does not cover at least three subjects related to radar waveforms. Selection of the radar center frequency (or band) and polarization are both well covered in other textbooks. We also limited our list of signals to deterministic, fully specified waveforms. Noise radars, or truly random waveforms, are not covered. Clearly specified pseudorandom signals are covered.

PROBLEMS

1.1 Doppler effect in acoustic radar
The transmitted acoustic radar signal is

$$s_T(t) = A_1 \sin[2\pi(f_0 - \Delta f)t] + A_2 \sin[2\pi(f_0 + \Delta f)t + \pi/3]$$

where $A_1 = 1$, $A_2 = 0.5$, and $f_0 = 150\,\text{Hz}$. The point target's radial velocity and the sound propagation velocity are $v = 5\,\text{m/s}$ and $C_p = 340\,\text{m/s}$. Assume that the target is initially at zero delay (i.e., $\tau = 0$). For **(a)** $\Delta f = 5\,\text{Hz}$ (narrowband case) and **(b)** $\Delta f = 50\,\text{Hz}$ (ultrawideband case), using $0 \le t \le 0.05\,\text{s}$, plot:
1. The signal transmitted
2. The signal received using the time compression equation (1.6).
3. The signal received using Doppler shift calculated at the center frequency f_0.

Interpret the results.

1.2 Time compression
For electromagnetic propagation, at what radial velocity (m/s) of the target will the time scale be 1:0.9999?

1.3 Spectrum of radar waveforms
(a) Plot the following two baseband signals:

$$s_1(t) = \begin{cases} 1, & |t| \le \dfrac{t_b}{2} \\ 0, & \text{elsewhere} \end{cases} \qquad \text{rectangular pulse}$$

$$s_2(t) = \begin{cases} \dfrac{\sin(\pi t/t_b)}{\pi t/t_b} \exp\left(\dfrac{-t^2}{t_b^2}\right), & |t| \le 2t_b \\ 0, & \text{elsewhere} \end{cases} \qquad \text{Gaussian-windowed sinc}$$

(b) For part (a), calculate (analytically or numerically) and plot the normalized energy spectral densities $|S(f)/S(0)|$ over the frequency span of $0 \leq f \leq 15/t_b$ and over the vertical scale of -60 to $0\,\text{dB}$.

REFERENCES

Aasen, M. D., Improvement of EMC by applying ambiguity and environmental diagram to the design of radar waveforms, *IEEE Transactions on Electromagnetic Compatibility*, vol. EMC-18, no. 2, May 1976, pp. 74–79.

Cook, C. E., and M. Bernfeld, *Radar Signals: An Introduction to Theory and Application*, Academic Press, New York, 1967.

DiFranco, J. V., and W. L. Rubin, *Radar Detection*, Prentice-Hall, Englewood Cliffs, NJ, 1968.

Levanon, N., *Radar Principles*, Wiley, New York, 1988.

Nathanson, F. E., J. P. Reilly, and M. N. Cohen, *Radar Design Principles: Signal Processing and the Environment*, 2nd ed., McGraw-Hill, New York, 1991.

Rihaczek, A. W., *Principles of High Resolution Radar*, McGraw-Hill, New York, 1969.

Rihaczek, A. W., Radar waveform selection—a simplified approach, *IEEE Transactions on Aerospace and Electronic Systems*, vol. AES-7, no. 6, November 1971, pp. 1078–1086.

Woodward, P. M., *Probability and Information Theory, with Applications to Radar*, Pergamon Press, Oxford, 1953.

Woodward, P. M., *Radar Ambiguity Analysis*, Technical Note 731, Royal Radar Establishment, Malvern, England, 1967.

2

MATCHED FILTER

The *impulse response*, or *transfer function*, of a matched filter is defined by the particular signal to which the filter is matched. Matching will result in the maximum attainable signal-to-noise ratio (SNR) at the output of the filter when the signal to which it was matched, plus white noise, are passed through it. The highest SNR happens at a particular instant, which is a design parameter. In radar applications, SNR is of paramount importance, and matched filters are used extensively. Most practical radar signals can be considered as narrow bandpass signals. Designing a matched filter for a narrow bandpass signal can be made relatively simple by using the complex envelope of the signal. We therefore begin this chapter by describing narrow bandpass signals with the help of their complex envelope.

2.1 COMPLEX REPRESENTATION OF BANDPASS SIGNALS

The majority of radar signals are narrow bandpass signals. Their Fourier transform is limited to an angular-frequency bandwidth of $2W$ centered about a carrier angular frequency $\pm\omega_c$. A narrow bandpass signal can be written in several forms. The basic representation is

$$s(t) = g(t)\cos[\omega_c t + \phi(t)] \qquad (2.1)$$

where $g(t)$ is the *natural envelope* of $s(t)$ and $\phi(t)$ is the instantaneous phase. Another form is the canonical form

$$s(t) = g_c(t)\cos\omega_c t - g_s(t)\sin\omega_c t \qquad (2.2)$$

Radar Signals, By Nadav Levanon and Eli Mozeson
ISBN 0-471-47378-2 Copyright © 2004 John Wiley & Sons, Inc.

COMPLEX REPRESENTATION OF BANDPASS SIGNALS

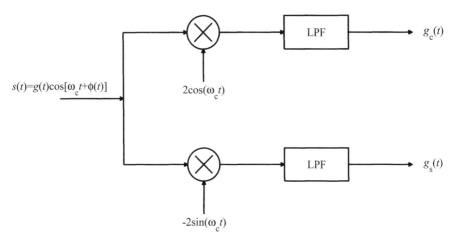

FIGURE 2.1 I/Q detector.

where $g_c(t)$ and $g_s(t)$ are the *in-phase* and *quadrature components*, respectively, given by

$$g_c(t) = g(t) \cos \phi(t) \qquad (2.3a)$$

$$g_s(t) = g(t) \sin \phi(t) \qquad (2.3b)$$

Both $g_c(t)$ and $g_s(t)$ are baseband signals bounded by W.

The in-phase **I** and quadrature **Q** components of a narrow-bandpass signal can be revealed using an I/Q detector, depicted in Fig. 2.1. The cutoff angular frequency of the low-pass filter (LPF) is above W and below $2\omega_c$. Readers are encouraged to prove to themselves that $g_c(t)$ and $g_s(t)$ produced in Fig. 2.1 are indeed those defined in equation (2.3). Box 2A demonstrates an important case that we will meet in a later chapter.

The complex envelope $u(t)$ of the signal $s(t)$ is defined as

$$u(t) = g_c(t) + jg_s(t) \qquad (2.4)$$

The complex envelope provides a third form of the signal:

$$s(t) = \text{Re}\{u(t) \exp(j\omega_c t)\} \qquad (2.5)$$

The carrier angular frequency ω_c is actually an arbitrary mathematical entity. Exactly the same signal $s(t)$ could be described using a different choice of ω_c. This would clearly result in a different $\phi(t)$, different quadrature components, and a different complex envelope $u(t)$.

BOX 2A: I and Q Components of Narrow Bandpass Signal

Consider a signal in which the real envelope is a train $p(t)$ of nearly rectangular pulses ("nearly" because otherwise the signal would not be *narrow* bandpass), and the instantaneous phase was caused by a Doppler frequency shift f_D. Thus,

$$s(t) = p(t) \cos[(\omega_c + 2\pi f_D)t] \qquad (2A.1)$$

The bandpass signal $s(t)$ and its two baseband components $g_c(t)$ and $g_s(t)$ are depicted in Fig. 2.2.

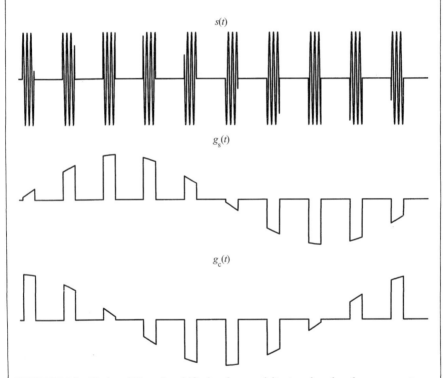

FIGURE 2.2 Train of Doppler-shifted pulses and its two baseband components as detected by an **I/Q** detector.

Of the infinitely many choices of ω_c, the preferred choice, which will make the mathematical analysis simpler, is the one that will result in the following identity:

$$u(t) \exp(j\omega_c t) = s(t) + j\hat{s}(t) \qquad (2.6)$$

where $\hat{s}(t)$ is the Hilbert transform of $s(t)$, namely,

$$\hat{s}(t) = s(t) \otimes \frac{1}{\pi t} = \frac{1}{\pi} \int_{-\infty}^{\infty} \frac{s(\tau)}{t - \tau} d\tau \qquad (2.7)$$

and $\otimes$ represents convolution. The importance of choosing ω_c that satisfies (2.6) increases as the signal's relative bandwidth increases. In most radar applications $s(t)$ is a narrow bandpass signal. Furthermore, in radar the receiver usually knows what the transmitted signal was, and what actual carrier frequency was used in the modulation process.

The complex envelope $u(t)$ is also a baseband signal bounded by W. The relative spectrums are depicted in Fig. 2.3. Observe that the natural envelope of the signal equals the magnitude of the complex envelope:

$$g(t) = |u(t)| \qquad (2.8)$$

Since $u(t)$ is a complex signal, its spectrum is not necessarily symmetrical about the origin, namely $U(\omega) \neq U^*(-\omega)$, where $U^*(\omega)$ is the complex conjugate of $U(\omega)$.

It can easily be checked that a fourth representation of a narrow bandpass signal is

$$s(t) = \tfrac{1}{2} u(t) \exp(j\omega_c t) + \tfrac{1}{2} u^*(t) \exp(-j\omega_c t) \qquad (2.9)$$

The fourth presentation defined in (2.9) will be used in finding the response of a filter matched to a bandpass signal. But first we have to define the matched filter itself.

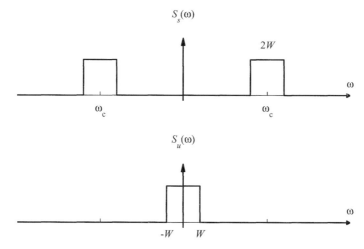

FIGURE 2.3 Spectrums of a narrow bandpass signal (top) and its complex envelope (bottom).

2.2 MATCHED FILTER

In radar applications we generally utilize the reflected "known" signal to detect the existence of a reflecting target. The probability of detection is related to the SNR rather than to the exact waveform of the signal received. Hence we are more interested in maximizing the SNR than in preserving the shape of the signal. A specific matched filter is a linear filter whose impulse response is determined by a specific signal in a way that will result in the maximum attainable SNR at the output of the filter when that particular signal and white noise are passed through the filter. A matched filter is sometimes referred to as a *North filter*, after D. O. North, who first described it in an RCA report (North, 1943).

Matched filters can be derived for baseband as well as for bandpass real signals. For the latter case it will usually suffice to implement a filter matched to the complex envelope of the signal. Hence, we need to be able to design matched filters for complex signals as well. Consider the block diagram in Fig. 2.4. The input to the matched filter is the signal $s(t)$ plus additive white Gaussian noise with a two-sided power spectral density of $N_0/2$. We look for the impulse response $h(t)$ or the frequency response $H(\omega)$ that will yield the maximum output SNR at a predetermined delay t_0. In other words, we look for $h(t)$ or $H(\omega)$ that will maximize

$$\left(\frac{S}{N}\right)_{\text{out}} = \frac{|s_o(t_0)|^2}{\overline{n_o^2(t)}} \qquad (2.10)$$

The matched filter impulse response will be a function only of the waveform $s(t)$ and the predetermined delay t_0.

Let the Fourier transform of $s(t)$ be $S(\omega)$; then the output signal at t_0 is given by

$$s_o(t_0) = \frac{1}{2\pi} \int_{-\infty}^{\infty} H(\omega) S(\omega) \exp(j\omega t_0) \, d\omega \qquad (2.11)$$

The mean-squared value of the noise, which is independent of t, is

$$\overline{n_o^2(t)} = \frac{N_0}{4\pi} \int_{-\infty}^{\infty} |H(\omega)|^2 \, d\omega \qquad (2.12)$$

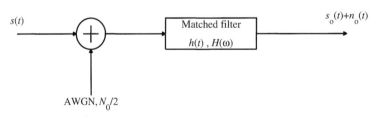

FIGURE 2.4 Matched filter definitions.

Substituting (2.11) and (2.12) in (2.10) yields

$$\left(\frac{S}{N}\right)_{out} = \frac{\left|\int_{-\infty}^{\infty} H(\omega)S(\omega)\exp(j\omega t_0)\,d\omega\right|^2}{\pi N_0 \int_{-\infty}^{\infty} |H(\omega)|^2\,d\omega} \qquad (2.13)$$

We will now use the *Schwarz inequality*, which says that for any two complex signals $A(\omega)$ and $B(\omega)$, the following inequality is true:

$$\left|\int_{-\infty}^{\infty} A(\omega)B(\omega)\,d\omega\right|^2 \leq \int_{-\infty}^{\infty} |A(\omega)|^2\,d\omega \int_{-\infty}^{\infty} |B(\omega)|^2\,d\omega \qquad (2.14)$$

The equality holds if and only if

$$A(\omega) = KB^*(\omega) \qquad (2.15)$$

where K is an arbitrary constant (that can have a dimension). Choosing

$$A(\omega) = H(\omega), \qquad B(\omega) = S(\omega)\exp(j\omega t_0) \qquad (2.16)$$

in (2.14) and using it in (2.13) yields

$$\left(\frac{S}{N}\right)_{out} \leq \frac{1}{\pi N_0} \int_{-\infty}^{\infty} |S(\omega)|^2\,d\omega = \frac{2E}{N_0} \qquad (2.17)$$

where E is the energy of the finite-time signal, namely,

$$E = \int_{-\infty}^{\infty} s^2(t)\,dt = \frac{1}{2\pi}\int_{-\infty}^{\infty} S^2(\omega)\,d\omega \qquad (2.18)$$

Using (2.16) in (2.15), we find out that equality, which means maximum output SNR, is obtained when

$$H(\omega) = KS^*(\omega)\exp(-j\omega t_0) \qquad (2.19)$$

We have thus obtained the frequency response of the matched filter. Equation (2.19) satisfies intuition because it says that the filter weighs its frequency response according to the spectrum of the signal.

Taking the inverse Fourier transform of (2.19) yields the impulse response of the matched filter:

$$h(t) = Ks^*(t_0 - t) \qquad (2.20)$$

This says that the impulse response is a delayed mirror image of the conjugate of the signal. For the filter to be causal, $h(t)$ must be zero for $t < 0$. This can

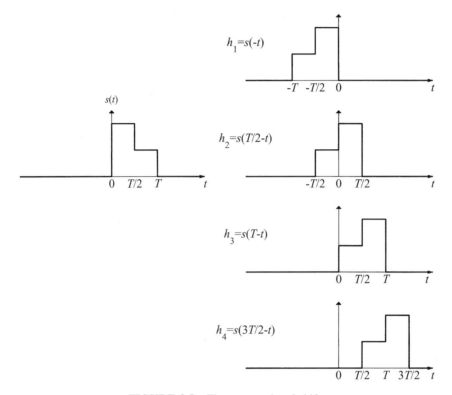

FIGURE 2.5 Time reversal and shift.

happen only if t_0 is equal or longer than the duration of the signal $s(t)$. A few examples are shown in Fig. 2.5.

When a filter is matched to a signal according to (2.20), equality holds in (2.17) and the output SNR (at $t = t_0$) is the highest one attainable, which is $2E/N_0$. It is interesting to note that the maximum SNR at the output of a matched filter (when the input signal is the one the filter is matched to, and the input noise is white) is a function of the signal's energy and not its shape. The same is true for the output signal level at the predetermined delay t_0:

$$s_o(t_0) = F^{-1}\{H(f)S(f)\}_{t=t_0} = \int_{-\infty}^{\infty} H(f)S(f)\exp(j2\pi f t_0)\,df \quad (2.21)$$

Using (2.19) in (2.21) yields

$$s_o(t_0) = \int_{-\infty}^{\infty} KS^*(f)\exp(-j2\pi f t_0)S(f)\exp(j2\pi f t_0)\,df$$

$$= K\int_{-\infty}^{\infty} |S^*(f)|^2\,df \quad \therefore s_o(t_0) = KE \quad (2.22)$$

Equation (2.22) states that regardless of the waveform, at the predetermined delay t_0 the output signal will be related to the energy of the waveform (up to a constant K). From (2.22) we can also deduce that K cannot be dimensionless and its dimension is $(V \cdot s)^{-1}$.

What about the matched filter output at other delays? It can easily be derived using convolution between the signal and the matched filter impulse response,

$$s_o(t) = s(t) \otimes h(t) = \int_{-\infty}^{\infty} s(\tau) h(t-\tau) \, d\tau = \int_{-\infty}^{\infty} s(\tau) K s^*[t_0 - (t-\tau)] \, d\tau$$

$$\underset{K=1, t_0=0}{=} \int_{-\infty}^{\infty} s(\tau) s^*(\tau - t) \, d\tau \qquad (2.23)$$

The right-hand side of (2.23) is recognized as the *autocorrelation function* of $s(t)$.

We can now summarize the main results concerning the matched filter: The impulse response is linearly related to the time-inverted complex-conjugate signal; when the input to the matched filter is the correct signal plus white noise, the peak output response is linearly related to the signal's energy. At that instance the SNR is the highest attainable, which is $2E/N_0$; elsewhere, the response is described by the autocorrelation function of the signal. The design and properties of a matched filter are demonstrated using a rectangular pulse in Box 2B. Next, we will find out how to simplify the design of a filter matched to a narrow bandpass signal by using its complex envelope.

BOX 2B: Filter Matched to a Baseband Rectangular Pulse

The design and properties of a matched filter will be demonstrated using a very simple signal, a rectangular pulse of duration T and height A. Choosing $t_0 = T$, the signal and the filter impulse response appear in Fig. 2.6. The signal and its matched filter are given by

$$s(t) = A, \quad 0 \le t \le T, \quad \text{zero elsewhere;}$$
$$h(t) = KA, \quad 0 \le t \le T, \quad \text{zero elsewhere} \qquad (2B.1)$$

The signal's energy is clearly

$$E = A^2 T \qquad (2B.2)$$

The first half ($0 \le t \le T$) of the output signal is given by

$$s_o(t) = \int_{-\infty}^{\infty} s(t) h(t-\tau) \, d\tau = \int_0^t KA^2 \, d\tau = KA^2 t, \quad 0 \le t \le T \qquad (2B.3)$$

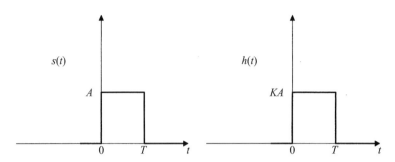

FIGURE 2.6 Rectangular pulse and its matched filter.

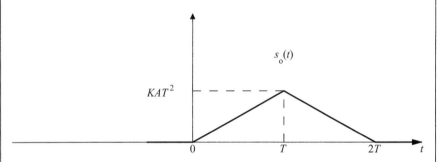

FIGURE 2.7 Matched-filter output.

The second half ($T \leq t \leq 2T$) is symmetrical with respect to the pulse duration T. The complete output signal is shown in Fig. 2.7.

Setting $t = T$ in (2B.3) yields, as expected,

$$s_o(T) = KA^2T = KE \tag{2B.4}$$

To determine the SNR, we need the mean-squared value of the noise. It is given by

$$\overline{n_o^2(t)} = \frac{N_0}{2} \int_{-\infty}^{\infty} |H(f)|^2 \, df = \frac{N_0}{2} \int_{-\infty}^{\infty} |h(t)|^2 \, dt$$

$$= \frac{N_0}{2} \int_0^T |KA|^2 \, dt = \frac{N_0}{2} K^2 A^2 T \tag{2B.5}$$

The SNR at $t = T$ can be obtained from the preceding two equations, yielding the result predicted:

$$\frac{s_o^2(T)}{\overline{n_o^2(t)}} = \frac{(KA^2T)^2}{(N_0/2)K^2A^2T} = \frac{2A^2T}{N_0} = \frac{2E}{N_0} \tag{2B.6}$$

The reader is invited to calculate the highest SNR when the same rectangular pulse is fed to an RC filter (Fig. 2.8), optimized for the given pulse width T.

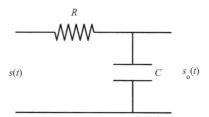

FIGURE 2.8 RC low-pass filter.

The highest maximum SNR is obtained when

$$RC = \frac{T}{0.4\pi} \tag{2B.7}$$

The resulting maximum SNR is

$$\frac{s_o^2(T)}{n_o^2(t)} = \frac{1.62 A^2 T}{N_0} = \frac{1.62 E}{N_0} \tag{2B.8}$$

which is 0.92 dB lower than the SNR obtained with a matched filer.

2.3 MATCHED FILTER FOR A NARROW BANDPASS SIGNAL

In this section we develop a good approximation of a filter matched to a narrow bandpass signal. Using the representation of $s(t)$ as given in (2.9) inside (2.23) yields

$$s_o(t) = \frac{K}{4} \int_{-\infty}^{\infty} [u(\tau) \exp(j\omega_c \tau) + u^*(\tau) \exp(-j\omega_c \tau)]$$
$$\cdot \{u^*(\tau - t + t_0) \exp[-j\omega_c(\tau - t + t_0)]$$
$$+ u(\tau - t + t_0) \exp[j\omega_c(\tau - t + t_0)]\} \, d\tau \tag{2.24}$$

Performing the cross product yields

$$s_o(t) = \frac{K}{4} \exp[j\omega_c(t - t_0)] \int_{-\infty}^{\infty} u(\tau) u^*(\tau - t + t_0) \, d\tau$$
$$+ \frac{K}{4} \exp[-j\omega_c(t - t_0)] \int_{-\infty}^{\infty} u^*(\tau) u(\tau - t + t_0) \, d\tau$$

$$+ \frac{K}{4} \exp[j\omega_c(t-t_0)] \int_{-\infty}^{\infty} u^*(\tau)u^*(\tau-t+t_0) \exp(-j2\omega_c\tau)\, d\tau$$

$$+ \frac{K}{4} \exp[-j\omega_c(t-t_0)] \int_{-\infty}^{\infty} u(\tau)u(\tau-t+t_0) \exp(j2\omega_c\tau)\, d\tau \quad (2.25)$$

Note that the second term on the right-hand side is the complex conjugate of the first term and that the fourth term is the complex conjugate of the third term. Using $(a + jb) + (a - jb) = 2a = 2\operatorname{Re}\{a + jb\}$, we can write

$$s_o(t) = \frac{1}{2} K \operatorname{Re}\left\{\exp[j\omega_c(t-t_0)] \int_{-\infty}^{\infty} u(\tau)u^*(\tau-t+t_0)\, d\tau\right\}$$

$$+ \frac{1}{2} K \operatorname{Re}\left\{\exp[j\omega_c(t-t_0)] \int_{-\infty}^{\infty} u^*(\tau)u^*(\tau-t+t_0) \exp(-j2\omega_c\tau)\, d\tau\right\} \quad (2.26)$$

The integral in the second term on the right-hand side of (2.26) is recognized as the Fourier transform of $u^*(\tau)u^*(\tau-t+t_0)$ evaluated at $\omega = \omega_c$. If, indeed, the signal $s(t)$ is narrowband around ω_c then the spectrum of its complex envelope $u(t)$ is cut off well below ω_c and we will be justified in neglecting the second term, getting

$$s_o(t) \approx \frac{1}{2} K \operatorname{Re}\left\{\exp[j\omega_c(t-t_0)] \int_{-\infty}^{\infty} u(\tau)u^*(\tau-t+t_0)\, d\tau\right\}$$

$$= \operatorname{Re}\left\{\left[\frac{1}{2} K \exp(-j\omega_c t_0) \int_{-\infty}^{\infty} u(\tau)u^*(\tau-t+t_0)\, d\tau\right] \exp(j\omega_c t)\right\} \quad (2.27)$$

In the square brackets of (2.27) we have defined a new complex envelope:

$$u_o(t) = K_u \int_{-\infty}^{\infty} u(\tau)u^*(\tau-t+t_0)\, d\tau, \qquad K_u = \tfrac{1}{2} K \exp(-j\omega_c t_0) \quad (2.28)$$

We can now write the output of the matched filter as

$$s_o(t) \approx \operatorname{Re}\{u_o(t) \exp(j\omega_c t)\} \quad (2.29)$$

Equations (2.28) and (2.29) state that the output of a filter matched to a narrow bandpass (NBP) signal has a complex envelope $u_o(t)$ obtained by passing the complex envelope $u(t)$ of the NBP signal through its own matched filter. The constant K_u differs from K by a fixed phase shift. We can conclude that in NBP radar it is sufficient to study the complex envelope $u(t)$ of the radar signal and its matched filter output $u_o(t)$. Once $u_o(t)$ is obtained, $s_o(t)$ is given through (2.29). There is also a practical justification for dealing only with the complex envelope of the radar signal. Modern radar processing utilizes **I** and **Q** detection (Fig. 2.1) early in the receiver, after which the sampled complex envelope $u(t)$ is available digitally, and matched processing is performed on it (digitally).

2.4 MATCHED-FILTER RESPONSE TO ITS DOPPLER-SHIFTED SIGNAL

A signal reflected from a moving target is Doppler affected. As discussed in Chapter 1, for narrow bandpass signals it is practical to treat the Doppler effect as a change in the carrier frequency. Without exact knowledge of the Doppler shift, the radar receiver cannot modify its matched receiver to the new carrier frequency exactly, and mismatch occurs. In this section we find the output of the matched filter $u_o(t)$ when the input complex envelope contains a Doppler frequency shift, ν. The Doppler-shifted complex envelope $u_D(t)$ is therefore

$$u_D(t) = u(t)\exp(j2\pi\nu t) \tag{2.30}$$

Replacing the first u in (2.28) by u_D and choosing $t_0 = 0$, $K_u = 1$ yields a function of both time and Doppler shift:

$$u_o(t, \nu) = \int_{-\infty}^{\infty} u(\tau)\exp(j2\pi\nu\tau)u^*(\tau - t)\,d\tau \tag{2.31}$$

Reversing the roles of t and τ yields a modified expression,

$$\chi(\tau, \nu) = \int_{-\infty}^{\infty} u(t)u^*(t - \tau)\exp(j2\pi\nu t)\,dt \tag{2.32}$$

which is one of the versions of the very important *ambiguity function*.

As shown above, the ambiguity function (AF) has an important practical meaning—it describes the output of a matched filter when the input signal is delayed by τ and Doppler shifted by ν relative to nominal values for which the matched filter was designed. The AF was introduced by Woodward (1953) and is the main tool in several important radar textbooks (Cook and Bernfeld, 1967; Rihaczek, 1969). Unfortunately, those references differ as to exactly what the signs of τ and ν imply regarding longer or shorter delay, and closing or opening velocities. There are also differences with regard to the function as is, or its magnitude or its square or its magnitude squared. An attempt to standardize the definition (Sinsky and Wang, 1974) proposes the format

$$|\chi(\tau, \nu)|^2 = \left|\int_{-\infty}^{\infty} u(t)u^*(t + \tau)\exp(j2\pi\nu t)\,dt\right|^2 \tag{2.33}$$

where a target farther from the radar than the reference ($\tau = 0$) position will correspond to positive τ and a positive ν implies a target moving toward the radar.

Representation of the AF of various signals is more often done through graphic plots than through analytic expressions. In the plots there is an emphasis on sidelobes relative to the mainlobe. Using the magnitude square will suppress the sidelobes in the graphs, while using logarithmic scale may boost the sidelobe appearance too much. We have therefore elected to use and plot $|\chi(\tau, \nu)|$ rather than $|\chi(\tau, \nu)|^2$. So in this book the term *ambiguity function* will usually refer to

$$|\chi(\tau, \nu)| = \left| \int_{-\infty}^{\infty} u(t) u^*(t+\tau) \exp(j 2\pi \nu t)\, dt \right| \tag{2.34}$$

The properties of this important function are studied in Chapter 3.

PROBLEMS

2.1 Matched-filter response
  (a) Find the impulse response of a filter matched to a signal

  $$f(t) = \begin{cases} A/2, & 0 \le t \le T/2 \\ -A/2, & T/2 \le t \le T \\ 0, & \text{elsewhere} \end{cases}$$

  Choose $t_0 = T$.

  (b) Find the peak SNR at the output of the filter when the input includes the signal plus additive white Gaussian noise (AWGN) with two-sided power spectral density $N_0/2$. Check the agreement with equation (2.17).

2.2 Signal bandwidth effect on matched-filter response
  (a) For each of the two signals

  $$s_1(t) = b, \qquad 0 \le t \le T, \quad \text{zero elsewhere}$$
  $$s_2(t) = b \sin(8\pi t/T), \qquad 0 \le t \le T, \quad \text{zero elsewhere}$$

  find and plot the matched-filter output.

  (b) For $s_2(t)$, try both the straightforward approach [see equation (2.23)] and the low-pass equivalent approach [see equations (2.28) and (2.29)]. Do the two approaches yield the same result? Is $s_2(t)$ a narrow bandpass signal?

2.3 Matched-filter peak output SNR
  Find the peak SNR at the output of a filter matched to the signal

  $$s(t) = A \operatorname{rect}\left(\frac{t}{T}\right) \cos \frac{\pi t}{T} \cos \omega_0 t, \qquad \omega_0 T \gg 1$$

  when the input noise is AWGN with two-sided power spectral density $N_0/2$.

2.4 Matched-filter response to a biphase pulse

(a) Calculate the complex envelope $u_o(t)$ of the matched-filter response of a signal whose complex envelope is given by

$$u(t) = \begin{cases} A\exp(-j\phi), & 0 \le t \le t_b \\ A\exp(+j\phi), & t_b \le t \le 2t_b \end{cases}, \text{ zero elsewhere}$$

(b) Plot the magnitude of the complex envelope $|u_o(t)|$ for $\phi = \pi/2$ and $\phi = \pi/3$.

2.5 Matched-filter response to a two-tone pulse

(a) Calculate the complex envelope $u_o(t)$ of the matched-filter response of a signal whose complex envelope is given by

$$u(t) = \begin{cases} A\exp(-j2\pi f_1 t), & -t_b \le t \le 0 \\ A\exp(+j2\pi f_1 t), & 0 \le t \le t_b \end{cases}, \text{ zero elsewhere}$$

(b) Plot the magnitude of the complex envelope $|u_o(t)|$ for $f_1 = 0.5/t_b$, $f_1 = 0.4545/t_b$, and $f_1 = 1.5/t_b$.

REFERENCES

Cook, C. E., and M. Bernfeld, *Radar Signals: An Introduction to Theory and Application*, Academic Press, New York, 1967.

North, D. O., *Analysis of Factors Which Determine Signal-to-Noise Discrimination in Radar*, Report PTR-6c, RCA Laboratories, Princeton, NJ, June 1943.

Rihaczek A. W., *Principles of High Resolution Radar*, McGraw-Hill, New York, 1969.

Sinsky, A. I., and C. P. Wang, Standardization of the definition of the radar ambiguity function, *IEEE Transactions on Aerospace and Electronic Systems*, vol. AES-10, no. 4, July 1974, pp. 532–533.

Woodward, P. M., *Probability and Information Theory, with Applications to Radar*, Pergamon Press, Oxford, 1953.

3

AMBIGUITY FUNCTION

The ambiguity function (AF) represents the time response of a filter matched to a given finite energy signal when the signal is received with a delay τ and a Doppler shift ν relative to the nominal values (zeros) expected by the filter. As explained in Chapter 2, the AF definition followed in this book is

$$|\chi(\tau, \nu)| = \left| \int_{-\infty}^{\infty} u(t) u^*(t+\tau) \exp(j 2\pi \nu t) \, dt \right| \qquad (3.1)$$

where u is the complex envelope of the signal. A positive ν implies a target moving toward the radar. Positive τ implies a target farther from the radar than the reference ($\tau = 0$) position. The ambiguity function is a major tool for studying and analyzing radar signals. It will serve us extensively in the following chapters, where different signals are described. This chapter presents important properties of the ambiguity function and proves several of them.

3.1 MAIN PROPERTIES OF THE AMBIGUITY FUNCTION

We list the four main properties of the ambiguity function. Proof of the four properties is provided in the next section. The first two properties assume that the energy E of $u(t)$ is normalized to unity.

Radar Signals, By Nadav Levanon and Eli Mozeson
ISBN 0-471-47378-2 Copyright © 2004 John Wiley & Sons, Inc.

Property 1: Maximum at (0,0)

$$|\chi(\tau, \nu)| \leq |\chi(0, 0)| = 1 \qquad (3.2)$$

This property says that the ambiguity function can nowhere be higher than at the origin (where it is normalized to unity by normalizing the signal energy).

Property 2: Constant volume

$$\int_{-\infty}^{\infty} \int_{-\infty}^{\infty} |\chi(\tau, \nu)|^2 \, d\tau \, d\nu = 1 \qquad (3.3)$$

Property 2 states that the total volume under the normalized ambiguity surface (squared) equals unity, independent of the signal waveform.

Properties 1 and 2 imply that if we attempt to squeeze the ambiguity function to a narrow peak at the origin, that peak cannot exceed a value of 1, and the volume squeezed out of that peak must reappear somewhere else. More restrictions on volume dispersion will be discussed later. The next two properties apply to all signals, normalized or not.

Property 3: Symmetry with respect to the origin

$$|\chi(-\tau, -\nu)| = |\chi(\tau, \nu)| \qquad (3.4)$$

Property 3 suggests that it is sufficient to study and plot only two adjacent quadrants of the AF. The remaining two can be deduced from the symmetry property. Our AF plots will usually contain only quadrants 1 and 2 (i.e., positive Doppler values).

Property 4: Linear FM effect
If a given complex envelope $u(t)$ has an ambiguity function $|\chi(\tau, \nu)|$: namely,

$$u(t) \Leftrightarrow |\chi(\tau, \nu)| \qquad (3.5)$$

then adding linear frequency modulation (LFM), which is equivalent to a quadratic-phase modulation, implies that

$$u(t) \exp(j\pi k t^2) \Leftrightarrow |\chi(\tau, \nu - k\tau)| \qquad (3.6)$$

Property 4 says that adding LFM modulation shears the resulting ambiguity function. The meaning of the shear will be demonstrated following the proof of property 4. This important property is the basis for an important pulse compression technique.

3.2 PROOFS OF THE AF PROPERTIES

Proofs of properties 1 to 4 are presented below. Most of the proofs follow Papoulis (1977).

Property 1: To prove this property, we apply the Schwarz inequality to the AF squared:

$$|\chi(\tau, \nu)|^2 = \left| \int_{-\infty}^{\infty} u(t) u^*(t+\tau) \exp(j2\pi\nu t)\, dt \right|^2$$

$$\leq \int_{-\infty}^{\infty} |u(t)|^2\, dt \int_{-\infty}^{\infty} |u^*(t+\tau) \exp(j2\pi\nu t)|^2\, dt \quad (3.7)$$

$$= \int_{-\infty}^{\infty} |u(t)|^2\, dt \int_{-\infty}^{\infty} |u^*(t+\tau)|^2\, dt = E \cdot E = 1 \cdot 1 = 1$$

$$\therefore |\chi(\tau, \nu)|^2 \leq 1, \quad \therefore |\chi(\tau, \nu)| \leq 1$$

Equality [i.e., $|\chi(\tau, \nu)|^2 = 1$] will replace the inequality in (3.7) when the functions in the two integrals [second expression in (3.7)] are conjugates of each other: namely, when

$$u(t) = [u^*(t+\tau) \exp(j2\pi\nu t)]^* = u(t+\tau) \exp(-j2\pi\nu t) \quad (3.8)$$

which obviously happens when $\tau = 0$, $\nu = 0$. Thus, we can conclude that

$$|\chi(\tau, \nu)| \leq |\chi(0, 0)| = 1 \quad (3.9)$$

Property 2: To prove this property, we rewrite $\chi(\tau, \nu)$, replacing ν with $-f$:

$$\chi(\tau, -f) = \int_{-\infty}^{\infty} [u(t) u^*(t+\tau)] \exp(-j2\pi f t)\, dt \quad (3.10)$$

which is recognized as the Fourier transform

$$\chi(\tau, -f) = F[\beta(\tau, t)] \quad (3.11)$$

of the function

$$\beta(\tau, t) = u(t) u^*(t+\tau) \quad (3.12)$$

The energy in the time domain is equal to the energy in the frequency domain (Parseval's theorem):

$$\int_{-\infty}^{\infty} |\beta(\tau, t)|^2\, dt = \int_{-\infty}^{\infty} |\chi(\tau, -f)|^2\, df = \int_{-\infty}^{\infty} |\chi(\tau, \nu)|^2\, d\nu \quad (3.13)$$

Integrating both sides with respect to τ yields the volume V under the ambiguity function squared

$$\int_{-\infty}^{\infty}\int_{-\infty}^{\infty}|\beta(\tau,t)|^2\,dt\,d\tau = \int_{-\infty}^{\infty}\int_{-\infty}^{\infty}|\chi(\tau,-f)|^2\,df\,d\tau$$

$$= \int_{-\infty}^{\infty}\int_{-\infty}^{\infty}|\chi(\tau,\nu)|^2\,d\nu\,d\tau = V \qquad (3.14)$$

We will now evaluate the integral on the left-hand side, starting with a change of variables:

$$t = t_1, \qquad t + \tau = t_2 \qquad (3.15)$$

Using (3.12) and (3.15) in (3.14) yields

$$\int_{-\infty}^{\infty}\int_{-\infty}^{\infty}|u(t)u^*(t+\tau)|^2\,dt\,d\tau = \int_{-\infty}^{\infty}\int_{-\infty}^{\infty}|u(t_1)u^*(t_2)|^2|\mathbf{J}(t_1,t_2)|\,dt_1\,dt_2 = V \qquad (3.16)$$

where the Jacobian is given by

$$\mathbf{J}(t_1,t_2) = \begin{vmatrix} \dfrac{\partial t_1}{\partial t} & \dfrac{\partial t_1}{\partial \tau} \\ \dfrac{\partial t_2}{\partial t} & \dfrac{\partial t_2}{\partial \tau} \end{vmatrix} = \begin{vmatrix} 1 & 0 \\ 1 & 1 \end{vmatrix} = 1 \qquad (3.17)$$

Using (3.17) in (3.16), we get

$$V = \int_{-\infty}^{\infty}\int_{-\infty}^{\infty}|u(t_1)u^*(t_2)|^2|1|\,dt_1\,dt_2$$

$$= \int_{-\infty}^{\infty}|u(t_1)|^2\,dt_1 \int_{-\infty}^{\infty}|u^*(t_2)|^2\,dt_2 = E \cdot E = 1 \cdot 1 = 1 \qquad (3.18)$$

Property 3: To prove this property, we set $-\tau$ and $-\nu$ in the equation for $|\chi(\tau,\nu)|$:

$$\chi(-\tau,-\nu) = \int_{-\infty}^{\infty} u(t)u^*(t-\tau)\exp(-j2\pi\nu t)\,dt \qquad (3.19)$$

and make one change of variable, $t_1 = t - \tau$, which yields

$$\chi(-\tau,-\nu) = \int_{-\infty}^{\infty} u(t_1+\tau)u^*(t_1)\exp[-j2\pi\nu(t_1+\tau)]\,dt_1$$

$$= \exp(-j2\pi\nu\tau)\int_{-\infty}^{\infty} u(t_1+\tau)u^*(t_1)\exp(-j2\pi\nu t_1)\,dt_1 \qquad (3.20)$$

$$= \exp(-j2\pi\nu\tau)\int_{-\infty}^{\infty} u(t+\tau)u^*(t)\exp(-j2\pi\nu t)\,dt$$

Because integration is a linear operation, the integral of a conjugate is equal to the conjugate of the integral; hence,

$$\chi(-\tau, -\nu) = \exp(-j2\pi\nu\tau) \left[\int_{-\infty}^{\infty} u^*(t+\tau) u(t) \exp(j2\pi\nu t) \, dt \right]^*$$
$$= \exp(-j2\pi\nu\tau) \chi^*(\tau, \nu) \qquad (3.21)$$

Taking the absolute value yields property 3,

$$|\chi(-\tau, -\nu)| = |\chi(\tau, \nu)| \qquad (3.22)$$

Property 4: To prove this property, we define a new complex envelope, in which quadratic phase was added to the original envelope $u(t)$:

$$u_1(t) = u(t) \exp(j\pi k t^2) \qquad (3.23)$$

The ambiguity function of $u_1(t)$,

$$u_1(t) \Leftrightarrow |\chi_1(\tau, \nu)| \qquad (3.24)$$

is what we look for. This new ambiguity function (without the absolute value) is

$$\chi_1(\tau, \nu) = \int_{-\infty}^{\infty} u_1(t) u_1^*(t+\tau) \exp(j2\pi\nu t) \, dt$$
$$= \int_{-\infty}^{\infty} u(t) \exp(j\pi k t^2) u^*(t+\tau) \exp[-j\pi k (t+\tau)^2] \exp(j2\pi\nu t) \, dt$$
$$= \exp(-j\pi k \tau^2) \int_{-\infty}^{\infty} u(t) u^*(t+\tau) \exp[j2\pi(\nu - k\tau)t] \, dt$$
$$= \exp(-j\pi k \tau^2) \chi(\tau, \nu - k\tau) \qquad (3.25)$$

Taking the absolute value, we get property 4:

$$|\chi_1(\tau, \nu)| = |\chi(\tau, \nu - k\tau)| \qquad (3.26)$$

3.3 INTERPRETATION OF PROPERTY 4

Before proceeding to other issues concerning the ambiguity function, it may be worthwhile to interpret property 4, the LFM effect. We explain the shearing caused by the LFM effect with the help of Fig. 3.1. Let the horizontal ellipsoid $|\chi(\tau, \nu)| = c$ be the contour of the original ambiguity function having a specific value c. The contour intersects the negative Doppler axis at point A with the coordinates ($\tau = 0$, $\nu = \nu_A$), and it intersects the positive delay axis at point B

INTERPRETATION OF PROPERTY 4

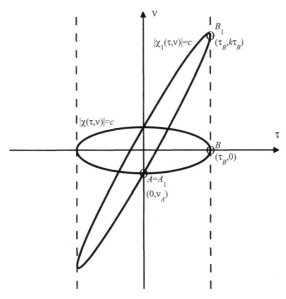

FIGURE 3.1 Linear-FM shearing effect on the ambiguity function.

with the coordinates ($\tau = \tau_B$, $\nu = 0$). The corresponding points A_1 and B_1, of $|\chi_1(\tau, \nu)| = c$, have the same delay coordinates, respectively ($\tau_{A1} = 0$, $\tau_{B1} = \tau_B$). We use (3.26) to find their respective Doppler coordinates, ν_{A1} and ν_{B1}.

Because the delay coordinate of the original intersection of A is zero, we get for A_1

$$c = |\chi_1[0, (\nu_A - k \cdot 0)]| = |\chi_1(0, \nu_A)| \qquad (3.27)$$

implying that the Doppler coordinate of A_1 is identical to the Doppler coordinate of A.

Next we ask where the contour $|\chi_1(\tau, \nu)| = c$ meets the delay τ_B; or what is ν_{B1} so that $|\chi_1(\tau_B, \nu_{B1})| = c$? Using (3.26), we get

$$|\chi_1(\tau_B, \nu_{B_1})| = |\chi(\tau_B, \nu_{B_1} - k\tau_B)| = c \qquad (3.28)$$

However, in Fig. 3.1 we note that $|\chi(\tau_B, 0)| = c$, which implies that $\nu_{B1} - k\tau_B = 0$, or

$$\nu_{B_1} = k\tau_B \qquad (3.29)$$

We thus found that point B at the coordinates ($\tau = \tau_B$, $\nu = 0$), whose delay coordinate is the largest delay through which the contour $|\chi(\tau, \nu)| = c$ passes, was moved, by adding LFM, to point B_1 at the coordinates ($\tau = \tau_B$, $\nu = k\tau_B$). Other points of the contour $|\chi(\tau, \nu)| = c$, at delays $0 < \tau < \tau_B$, were moved in a similar way, thus resulting in the sheared contour $|\chi_1(\tau, \nu)| = c$, also shown

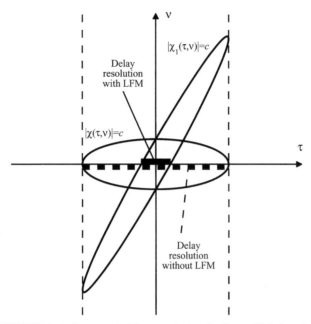

FIGURE 3.2 Improved delay resolution by linear-FM shearing.

in Fig. 3.1. The shearing property of linear FM, which we just studied, reduces (improves) the delay resolution, as pointed out in Fig. 3.2.

Finally, note that for the definition of the ambiguity function used here, the shape of the LFM ridge passing from the third quadrant to the first quadrant of the delay–Doppler space is typical for positive LFM slope ($k > 0$). This implies that for positive LFM slope signal a positive error in estimating target range (the target is assumed to be farther than it really is) will translate to lower closing velocity (negative Doppler).

3.4 CUTS THROUGH THE AMBIGUITY FUNCTION

Some insight into the two-dimensional ambiguity function (AF) can be obtained from its one-dimensional cuts. Consider first the cut along the delay axis. Setting $v = 0$ in (3.1) gives

$$|\chi(\tau, 0)| = \left| \int_{-\infty}^{\infty} u(t) u^*(t + \tau) \, dt \right| = |R(\tau)| \qquad (3.30)$$

where $R(\tau)$ is the autocorrelation function (ACF) of $u(t)$. We got that the zero-Doppler cut of the AF, known as the *range window* for a matched-filter receiver, is the ACF. On the other hand, the ACF equals the inverse Fourier transform of the power spectral density. Thus, we get the relationship

$$\text{range window} \Leftrightarrow \text{autocorrelation} \Leftrightarrow \mathbf{F}^{-1}\{\text{power spectrum}\}$$

This relationship reiterates the importance of LFM. Adding linear frequency modulation broadens the power spectrum, hence narrows the range window, as shown in Fig. 3.3.

The second interesting cut is along the Doppler frequency axis. Setting $\tau = 0$ in (3.1) results in

$$|\chi(0, \nu)| = \left| \int_{-\infty}^{\infty} |u(t)|^2 \exp(j2\pi\nu t)\, dt \right| \qquad (3.31)$$

Equation (3.31) says that the zero-delay cut is the Fourier transform of the *magnitude* squared of the complex envelope $u(t)$. In other words, this cut is indifferent to any phase or frequency modulation in $u(t)$; it is a function only of the amplitude.

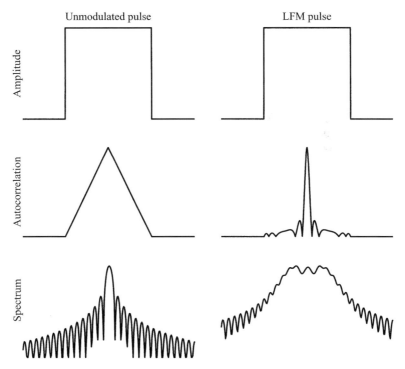

FIGURE 3.3 Comparison between unmodulated pulse and linear-FM pulse.

3.5 ADDITIONAL VOLUME DISTRIBUTION RELATIONSHIPS

The volume distribution of the ambiguity function (squared) in range and in Doppler is constrained by two more refined relationships (Rihaczek, 1969):

$$\int_{-\infty}^{\infty} |\chi(\tau, v)|^2 \, d\tau = \int_{-\infty}^{\infty} |\chi(\tau, 0)|^2 \exp(j2\pi v\tau) \, d\tau \qquad (3.32)$$

$$\int_{-\infty}^{\infty} |\chi(\tau, v)|^2 \, dv = \int_{-\infty}^{\infty} |\chi(0, v)|^2 \exp(j2\pi v\tau) \, dv \qquad (3.33)$$

These two transform relations tell us that if the central peak is squeezed along the delay axis, the volume must spread out in the Doppler domain, and when it is squeezed along the Doppler axis, the volume must spread in delay. Thus, close target separability in one parameter is gained at the expense of spreading volume over a large interval of the other parameter.

3.6 PERIODIC AMBIGUITY FUNCTION

Like the matched filter on which it is based, the ambiguity function is defined for finite-duration signals. However, there are several types of signals that are periodically continuous. Two prominent examples are the periodic continuous-wave (CW) radar signal and a coherent train of *identical* pulses. In both cases, the receiver is usually "matched" to a finite number of periods or pulses, smaller than the number of periods or pulses transmitted. An example of a typical coherent pulse train is shown in Fig. 3.4.

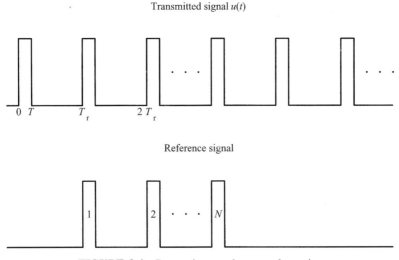

FIGURE 3.4 Processing a coherent pulse train.

To handle this type of signal and processor we define the *periodic ambiguity function* (PAF) and study some of its properties (Freedman and Levanon, 1994; Getz and Levanon, 1995). As shown in Fig. 3.4, a coherent pulse train will help us demonstrate the PAF. We must first explain what the word *coherent* implies. When we introduced the complex envelope of a finite-duration signal (e.g., a pulse), we assumed that the carrier frequency is known. Now, when we consider a train of separated pulses, we need to extend the assumption. We need to assume that the carrier frequency remains the same for all the pulses and that we also know the initial phase of each pulse. A simple variation is to consider the pulses as an interrupted CW signal. In that case the reference signal, constructed from N pulses, can be described as

$$s(t) = \text{Re}[u_N(t)\exp(j2\pi f_c t)] \tag{3.34}$$

and its complex envelope as

$$u_N(t) = \frac{1}{\sqrt{N}} \sum_{n=1}^{N} u_n[t - (n-1)T_r] \tag{3.35}$$

We will usually assume that the pulses are identical: namely, $u_n(t) = u_1(t)$. The complex envelope of the signal received (at zero relative delay and zero Doppler), not limited to N pulses, will be described as an infinite signal:

$$u(t) = \sum_{n=-\infty}^{\infty} u_1[t - (n-1)T_r] \tag{3.36}$$

Doppler-induced phase shifts of the signal received will be handled by the ambiguity function. Other than those phase shifts, (3.36) assumes that the target return exhibits constant phase during the dwell. We now return to Fig. 3.4, where the matched receiver performs a correlation between a received train $u(t)$ of many ($>N$) identical pulses and a reference train $u_N(t)$ of exactly N identical (and same) pulses. The normalized response of this processor, in the presence of Doppler shift, is given by the periodic ambiguity function

$$|\chi_{NT}(\tau, \nu)| = \left| \frac{1}{NT_r} \int_0^{NT_r} u(t) u^*(t+\tau) \exp(j2\pi\nu t)\, dt \right| \tag{3.37}$$

Of special interest is the single-pulse reference signal, which yields

$$|\chi_T(\tau, \nu)| = \left| \frac{1}{T_r} \int_0^{T_r} u(t) u^*(t+\tau) \exp(j2\pi\nu t)\, dt \right| \tag{3.38}$$

Note that as for the aperiodic ambiguity function, a different version of the periodic ambiguity function also exists. In Box 3A some other versions are described and compared to the definition used here.

BOX 3A: Variants of the Periodic Ambiguity Function

The single-period periodic ambiguity function of a signal with complex envelope $u(t)$ was defined in (3.38). The definition in (3.38) represents the straightforward implementation of a filter matched to a signal $u(t)$ delayed by τ and Doppler shifted by ν. Positive values of ν (positive Doppler) represent closing targets, while positive values of τ imply a target farther from the radar than the reference ($\tau = 0$). This definition is not unique. Alternative definitions can be adopted (Freedman and Levanon, 1994). Such definitions are

$$^1|\chi_T(\tau, \nu)| = \left| \frac{1}{T_r} \int_0^{T_r} u\left(t - \frac{\tau}{2}\right) u^*\left(t + \frac{\tau}{2}\right) \exp(j2\pi\nu t) \, dt \right| \quad (3A.1)$$

or

$$^2|\chi_T(\tau, \nu)| = \left| \frac{1}{T_r} \int_{-T_r/2}^{T_r/2} u(t) u^*(t + \tau) \exp(j2\pi\nu t) \, dt \right| \quad (3A.2)$$

These definitions represent an off-line noncasual implementation. For all three definitions the multiperiod ambiguity function relation to the single-period ambiguity function defined in (3.39) holds. Still these definitions are not equivalent, as described below.

Since the signal is periodic with period T_r, the PAF is also periodic in the delay axis direction. The period is T_r for the definitions given in (3.38) or (3A.2). The period of the PAF defined in (3A.1), on the other hand, is $2T_r$.

$$|\chi_T(\tau + nT_r, \nu)| = |\chi_T(\tau, \nu)| \quad (3A.3a)$$

$$^2|\chi_T(\tau + nT_r, \nu)| = {}^2|\chi_T(\tau, \nu)| \quad (3A.3b)$$

$$^1|\chi_T(\tau + 2nT_r, \nu)| = {}^1|\chi_T(\tau, \nu)| \quad (3A.3c)$$

Furthermore, the PAF magnitude symmetry with respect to the origin property holds only for the definition given in (3A.1). The other definitions of the PAF, given in (3.37) and (3A.2), are magnitude symmetric with respect to the origin only for $\nu = k/NT_r$ and $\tau = NT$, with k an integer and N the number of periods used.

As in the nonperiodic ambiguity function, there is a constraint on the volume under ambiguity function magnitude. For all three definitions of the periodic ambiguity function described here the volume within a strip of width T_r on the delay axis is equal to 1. For a larger filter length this volume is reduced according to the period-to-filter length ratio (N). The unlimited reduction of the volume as N increases is the result of the unlimited improvement of the Doppler resolution.

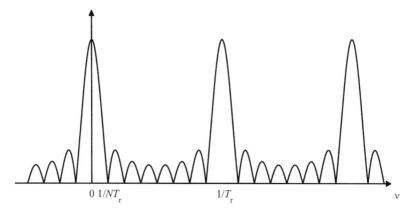

FIGURE 3.5 Function $|(\sin N\pi\nu T_r)/(N \sin \pi\nu T_r)|$ $(N=8)$.

It can be shown (Freedman and Levanon, 1994; Getz and Levanon, 1995) that a very simple and important relationship exists between equations (3.37) and (3.38):

$$|\chi_{NT}(\tau, \nu)| = |\chi_T(\tau, \nu)| \left| \frac{\sin N\pi\nu T_r}{N \sin \pi\nu T_r} \right| \quad (3.39)$$

Equation (3.39) suggests that it is sufficient to calculate the single-period PAF (3.38) and then multiply it by the function $|(\sin N\pi\nu T_r)/(N \sin \pi\nu T_r)|$ to get the N-period PAF. The multiplying function is a function of the Doppler shift only. Its version for $N = 8$ is plotted in Fig. 3.5.

Figure 3.5 demonstrates the main reason for using a coherent train of N pulses with a repetition interval T_r. Note that the Doppler resolution improves dramatically and becomes $1/NT_r$: namely, the inverse of the coherently processed time duration, and it is practically independent of the original pulse waveform. The penalty is recurrent lobes at Doppler intervals of $1/T_r$: namely, the inverse of the pulse repetition interval. Because the function plotted in Fig. 3.5 multiplies an ambiguity function, which is two-dimensional, it may help to point out that what multiplies the ambiguity function is an extension of Fig. 3.5 to all delays, as demonstrated in Fig. 3.6.

An interesting observation regarding the PAF states that for delays shorter than the pulse duration, $|\tau| \leq T$, the ambiguity function $|\chi(\tau, \nu)|$ is equal to the periodic ambiguity function $|\chi_{NT}(\tau, \nu)|$. This result makes it easier to calculate and plot the partial ambiguity function (for $|\tau| \leq T$) of a signal constructed from N identical pulses. The statement above does not hold for continuous-wave signals nor for pulse trains in which $T_r < 2T$. Coherent pulse trains were introduced here to serve as an example of the need and use of the periodic ambiguity function. The subject of coherent pulse trains will be expanded with much more detail in later chapters.

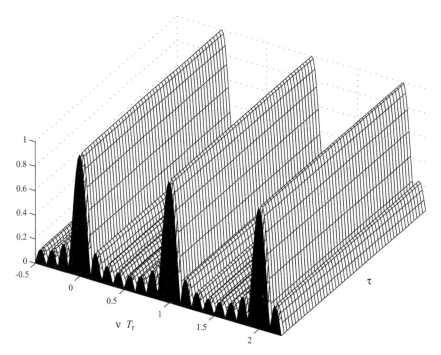

FIGURE 3.6 Extension of $|(\sin N\pi\nu T_r)/(N \sin \pi\nu T_r)|$ to all delays ($N = 8$).

3.7 DISCUSSION

Many types of signals are used for various radar applications and systems. Modern pulse radars generally use pulse compression waveforms (i.e., expanded pulses with larger time–bandwidth products). These kinds of waveforms are applied to obtain high pulse energy (with no increase in peak power) and large pulse bandwidth and, consequently, high range resolution without sacrificing maximum range, which is related to the pulse energy.

Unfortunately, no analytic method exists for calculating a signal given its ambiguity function (*inverse ambiguity transform*); thus the design of a radar signal with desirable characteristics of the ambiguity function is based primarily on the radar designer's prior knowledge of radar waveforms and his or her expertise in such designs. In the following chapters we acquire this knowledge. In Chapter 4 we develop the AF of several basic radar signals. The specific examples will contribute further understanding of the ambiguity function, its properties, and its significance to radar.

The ambiguity function of some of the radar signals to be discussed can be derived analytically. However, many signals are too complicated, and only numerical calculation of their AF is feasible. The most practical means for displaying the numerical result is a three-dimensional plot. The appendix that follows contains a MATLAB program for plotting ambiguity functions (Mozeson and Levanon, 2002).

APPENDIX 3A: MATLAB CODE FOR PLOTTING AMBIGUITY FUNCTIONS

A MATLAB code capable of plotting the ambiguity function $|\chi(\tau, \nu)|$ of many different radar signals is presented. The program makes use of MATLAB's sparse matrix operations, and avoids loops. The code makes it possible to input many different signals and provides control over many plot parameters. The program allows oversampling of the signal with much finer resolution than needed for calculating the delayed signal. This makes it possible to compute a diluted picture (fewer delay–Doppler grid points) of the ambiguity function with low computational effort while sampling the signal with a sufficiently large sampling rate.

Because of the symmetry of the AF with respect to the origin, the code plots only two of the four quadrants. This provides an opportunity to display the zero-Doppler cut of the AF, which is the magnitude of the autocorrelation function. The program also produces a second figure with subplots of three characteristics of the signal: amplitude, phase, and frequency.

Comment on the choice of r: Since the signal is described by a vector, with a well-defined length (number of elements, referred to as M), it is often necessary to increase the number of samples (repeats) during each of these elements (bits), in order to meet the Nyquist criterion. This is the function of r.

For a Costas signal with M elements, the signal bandwidth is approximately M/t_b. Therefore, the sampling interval should be

$$t_s < \frac{t_b}{2M}$$

Hence,

$$\frac{t_b}{t_s} = r > 2M$$

For a phase-coded signal, the main spectral lobe ends at $f = 1/t_b$. However, the spectral sidelobes extend much farther at a rate of approximately 6 dB/octave. A typical spectral skirt crosses the -30 dB level at $f = 10/t_b$. Hence, choosing $r = 2$ is the minimum setting, but using $r > 10$ is recommended.

Note: When a continuation ellipsis (...) appears within a single quote, such as a string, it indicates that the string and what follows it (up to a continuation ellipsis that is not within a string) must be completed in the line on which it was started.

The Code

```
% The matlab code plots the ambiguity function of a signal u_basic (row vector)
% The m-file returns a plot of quadrants 1 and 2 of the ambiguity function
%     of a signal
% The ambiguity function is defined as:
%
```

```
% a(t,f) = abs ( sum i( u(k)*u'(i+t)*exp(j*2*pi*f*i) ) )
%
% The user is prompted for the signal data:
% u_basic is a row complex vector representing amplitude and phase
% f_basic is a corresponding frequency coding sequence
%
% The duration of each element is tb (total duration of the signal is
%    tb*(m_basic-1))
%
% F is the maximal Doppler shift
% T is the maximal Delay
% K is the number of positive Doppler shifts (grid points)
% N is the number of delay shifts on each side (for a total of 2N+1 points)
% The code allows r samples within each bit
%
% Written by Eli Mozeson and Nadav Levanon, Dept. of EE-Systems, Tel Aviv
%    University

clear all

% prompts for signal data

u_basic=input(' Signal elements (row complex vector, each element last tb ...
    sec) = ? ');
m_basic=length(u_basic);

fcode=input(' Allow frequency coding (yes=1, no=0) = ? ');
if fcode==1
    f_basic=input(' Frequency coding in units of 1/tb (row vector of same ...
        length) = ? ');
end

F=input(' Maximal Doppler shift for ambiguity plot [in units of 1/Mtb] ...
    (e.g., 1)= ? ');
K=input(' Number of Doppler grid points for calculation (e.g., 100) = ? ');
df=F/K/m_basic;

T=input(' Maximal Delay for ambiguity plot [in units of Mtb] (e.g., 1)= ? ');

N=input(' Number of delay grid points on each side (e.g. 100) = ? ');

sr=input(' Over sampling ratio (>=1) (e.g. 10)= ? ');

r=ceil(sr*(N+1)/T/m_basic);

if r==1
    dt=1;
    m=m_basic;
    uamp=abs(u_basic);
    phas=uamp*0;
    phas=angle(u_basic);
    if fcode==1
        phas=phas+2*pi*cumsum(f_basic);
    end
    uexp=exp(j*phas);
    u=uamp.*uexp;
else                            % i.e., several samples within a bit
    dt=1/r;                     % interval between samples
    ud=diag(u_basic);
    ao=ones(r,m_basic);
```

```
        m=m_basic*r;
        u_basic=reshape(ao*ud,1,m);    % u_basic with each element repeated r times
        uamp=abs(u_basic);
        phas=angle(u_basic);
        u=u_basic;
        if fcode==1
            ff=diag(f_basic);
            phas=2*pi*dt*cumsum(reshape(ao*ff,1,m))+phas;
            uexp=exp(j*phas);
            u=uamp.*uexp;
        end
    end
end

t=[0:r*m_basic-1]/r;

tscale1=[0 0:r*m_basic-1 r*m_basic-1]/r;

dphas=[NaN diff(phas)]*r/2/pi;

% plot the signal parameters

figure(1), clf, hold off

subplot(3,1,1)
plot(tscale1,[0 abs(uamp) 0],'linewidth',1.5)
ylabel(' Amplitude ')
axis([-inf inf 0 1.2*max(abs(uamp))])

subplot(3,1,2)
plot(t, phas,'linewidth',1.5)
axis([-inf inf -inf inf])
ylabel(' Phase [rad] ')

subplot(3,1,3)
plot(t,dphas*ceil(max(t)),'linewidth',1.5)
axis([-inf inf -inf inf])
xlabel(' \itt / t_b ')
ylabel(' \itf * Mt_b ')

% calculate a delay vector with N+1 points that spans from zero delay to
%    ceil(T*t(m))
% notice that the delay vector does not have to be equally spaced but must
%    have all
% entries as integer multiples of dt
dtau=ceil(T*m)*dt/N;
tau=round([0:1:N]*dtau/dt)*dt;

% calculate K+1 equally spaced grid points of Doppler axis with df spacing
f=[0:1:K]*df;

% duplicate Doppler axis to show also negative Doppler's (0 Doppler is
%    calculated twice)
f=[-fliplr(f) f];

% calculate ambiguity function using sparse matrix manipulations (no loops)

% define a sparse matrix based on the signal samples u1 u2 u3 ... um
% with size m+ceil(T*m) by m (notice that u' is the conjugate transpose of u)
% where the top part is diagonal (u*) on the diagonal and the bottom part is a
%    zero matrix
```

```
%            [u1*  0   0  0 ...  0  ]
%            [ 0  u2*  0  0 ...  0  ]
%            [ 0   0  u3* 0 ...  0  ]  m rows
%            [ .              .  .  ]
%            [ .              .  .  ]
%            [ .   0   0  .  ... um*]
%            [ 0                  0 ]
%            [ .                . . ]    N rows
%            [ 0   0   0  0 ...  0  ]

mat1=spdiags(u',0,m+ceil(T*m),m);

% define a convolution sparse matrix based on the signal samples u1 u2 u3 ...
%   um
% where each row is a time(index) shifted versions of u.
% each row is shifted tau/dt places from the first row
% the minimal shift (first row) is zero
% the maximal shift (last row) is ceil(T*m) places
% the total number of rows is N+1
% number of columns is m+ceil(T*m)

% for example, when tau/dt=[0 2 3 5 6] and N=4
%
%            [u1 u2 u3 u4 ...          ... um  0  0  0  0  0]
%            [ 0  0 u1 u2 u3 u4 ...        ... um  0  0  0  0]
%            [ 0  0  0 u1 u2 u3 u4 ...        ... um  0  0  0]
%            [ 0  0  0  0  0 u1 u2 u3 u4 ...        ... um  0]
%            [ 0  0  0  0  0  0 u1 u2 u3 u4 ...        ... um]

% define a row vector with ceil(T*m)+m+ceil(T*m) places by padding u with zeros
% on both sides
u_padded=[zeros(1,ceil(T*m)),u,zeros(1,ceil(T*m))];

% define column indexing and row indexing vectors
cidx=[1:m+ceil(T*m)];
ridx=round(tau/dt)';

% define indexing matrix with Nused+1 rows and m+ceil(T*m) columns
% where each element is the index of the correct place in the padded version
%    of u
index = cidx(ones(N+1,1),:) + ridx(:,ones(1,m+ceil(T*m)));

% calculate matrix
mat2 = sparse(u_padded(index));

% calculate the ambiguity matrix for positive delays given by
%
%   [u1 u2 u3 u4 ...    ... um  0  0  0  0  0] [u1*  0   0  0 ...  0  ]
%   [ 0  0 u1 u2 u3 u4 ...   ... um  0  0  0  0] [ 0  u2*  0  0 ...  0  ]
%   [ 0  0  0 u1 u2 u3 u4 ...  ... um  0  0  0]*[ 0   0  u3* 0 ...  0  ]
%   [ 0  0  0  0  0 u1 u2 u3 u4 ...   ... um  0] [ .              .  . ]
%   [ 0  0  0  0  0  0 u1 u2 u3 u4 ...   ... um] [ .              .  . ]
%                                                 [ .   0   0  .  ... um*]
%                                                 [ 0                  0 ]
%                                                 [ .                . . ]
%                                                 [ 0   0   0  0 ...  0  ]
%
% where there are m columns and N+1 rows and each element gives an element
% of multiplication between u and a time shifted version of u*. each row gives
```

```
% a different time shift of u* and each column gives a different entry in u.
%

uu_pos=mat2*mat1;
clear mat2 mat1

% calculate exponent matrix for full calculation of ambiguity function.
%   The exponent
% matrix is 2*(K+1) rows by m columns where each row represents a possible
%   Doppler and
% each column stands for a different place in u.

e=exp(-j*2*pi*f'*t);

% calculate ambiguity function for positive delays by calculating the integral
%    for each
% possible delay and Doppler over all entries in u.
% a_pos has 2*(K+1) rows (Doppler) and N+1 columns (Delay)
a_pos=abs(e*uu_pos');

% normalize ambiguity function to have a maximal value of 1
a_pos=a_pos/max(max(a_pos));

% use the symmetry properties of the ambiguity function to transform the
%    negative Doppler
% positive delay part to negative delay, positive Doppler
a=[flipud(conj(a_pos(1:K+1,:))) fliplr(a_pos(K+2:2*K+2,:))];

% define new delay and Doppler vectors
delay=[-fliplr(tau) tau];
freq=f(K+2:2*K+2)*ceil(max(t));

% excludes the zero Delay that was taken twice
delay=[delay(1:N) delay(N+2:2*N)];
a=a(:,[1:N,N+2:2*N]);

% plot the ambiguity function and autocorrelation cut
[amf amt]=size(a);

% create an all blue color map
cm=zeros(64,3);
cm(:,3)=ones(64,1);

figure(2), clf, hold off
mesh(delay, [0 freq], [zeros(1,amt);a])

hold on
surface(delay, [0 0], [zeros(1,amt);a(1,:)])

colormap(cm)
view(-40,50)
axis([-inf inf -inf inf 0 1])
xlabel(' {\it\tau}/{\itt_b}','Fontsize',12);
ylabel(' {\it\nu}*{\itMt_b}','Fontsize',12);
zlabel(' |{\it\chi}({\it\tau},{\it\nu})| ','Fontsize',12);
hold off
```

PROBLEMS

3.1 Ambiguity function of a time-scaled signal

If the ambiguity function of $u(t)$ is $|\chi(\tau, v)|$, prove that the ambiguity function $|\chi_1(\tau, v)|$ of $u_1(t) = u(at)$ is given by

$$|\chi_1(\tau, v)| = \frac{1}{a}\left|\chi\left(a\tau, \frac{v}{a}\right)\right|$$

3.2 Calculation of the AF using a signal's Fourier transform

Prove that the ambiguity function can also be written in the form

$$|\chi(\tau, v)| = \left|\int_{-\infty}^{\infty} U^*(f)U(f-v)\exp(j2\pi f\tau)\,df\right|$$

where $U(f)$ is the Fourier transform of $u(t)$.

3.3 Doppler resolution connection to signal amplitude

Show that for a normalized signal ($E = 1$),

$$\frac{\partial^2 \chi(0, 0)}{\partial v^2} = -4\pi^2 \int_{-\infty}^{\infty} t^2|u(t)|^2\,dt$$

3.4 Prove equation (3.39).

3.5 Another form of the ambiguity function

In the ambiguity function (3.1), change the integration variable t to t_1 by using

$$t = t_1 - \frac{\tau}{2}$$

and obtain another form of the ambiguity function.

REFERENCES

Freedman, A., and N. Levanon, Properties of the periodic ambiguity function, *IEEE Transactions on Aerospace and Electronic Systems*, vol. 30, no. 3, May 1994, pp. 938–941.

Getz, B., and N. Levanon, Weight effects on the periodic ambiguity function, *IEEE Transactions on Aerospace and Electronic Systems*, vol. 31, no. 1, January 1995, pp. 182–193.

Mozeson, E., and N. Levanon, MATLAB code for plotting ambiguity functions, *IEEE Transactions on Aerospace and Electronic Systems*, vol. 38, no. 3, July 2002, pp. 1064–1068.

Papoulis, A., *Signal Analysis*, McGraw-Hill, New York, 1977, Chap. 8.

Rihaczek, A. W., *Principles of High Resolution Radar*, McGraw-Hill, New York, 1969.

4

BASIC RADAR SIGNALS

Three basic radar signals are studied in this chapter: a constant-frequency pulse, a linear-FM pulse, and a coherent train of identical constant-frequency pulses. The ambiguity function will be the main tool in this study. For these three signals we will be able to develop closed-form expressions of the ambiguity functions (AF), but their three-dimensional plots will be obtained numerically using the MATLAB code given in Appendix 3A.

4.1 CONSTANT-FREQUENCY PULSE

The complex envelope of a constant-frequency (or unmodulated) pulse appears in Fig. 4.1 and is given by

$$u(t) = \frac{1}{\sqrt{T}} \text{rect}\left(\frac{t}{T}\right) \qquad (4.1)$$

The ambiguity function is obtained by using (4.1) in

$$\chi(\tau, \nu) = \int_{-\infty}^{\infty} u(t) u^*(t + \tau) \exp(j2\pi\nu t)\, dt \qquad (4.2)$$

yielding

$$\chi(\tau, \nu) = \begin{cases} \dfrac{1}{T} \displaystyle\int_{-(T/2)+\tau}^{T/2} \exp(j2\pi\nu t)\, dt, & 0 \leq \tau \leq T \\ \dfrac{1}{T} \displaystyle\int_{-T/2}^{(T/2)+\tau} \exp(j2\pi\nu t)\, dt, & -T \leq \tau < 0 \\ 0, & \text{elsewhere} \end{cases} \qquad (4.3)$$

Radar Signals, By Nadav Levanon and Eli Mozeson
ISBN 0-471-47378-2 Copyright © 2004 John Wiley & Sons, Inc.

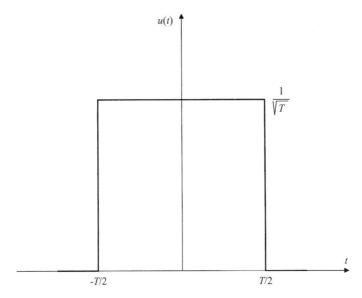

FIGURE 4.1 Complex envelope of a constant-frequency pulse.

Solving the integrals and taking absolute value yields

$$|\chi(\tau, \nu)| = \left| \left(1 - \frac{|\tau|}{T}\right) \frac{\sin\left[\pi T \nu (1 - |\tau|/T)\right]}{\pi T \nu (1 - |\tau|/T)} \right|, \quad |\tau| \leq T, \quad \text{zero elsewhere} \tag{4.4}$$

The cut along the delay axis is obtained by setting $\nu = 0$, yielding

$$|\chi(\tau, 0)| = 1 - \frac{|\tau|}{T}, \quad |\tau| \leq T, \quad \text{zero elsewhere} \tag{4.5}$$

The cut along the Doppler axis is obtained by setting $\tau = 0$, yielding

$$|\chi(0, \nu)| = \left| \frac{\sin \pi T \nu}{\pi T \nu} \right| \tag{4.6}$$

Note that the cut $|\chi(0, \nu)|$ extends from $-\infty$ to ∞.

The first two quadrants of the ambiguity function are plotted in Fig. 4.2. A contour plot of the AF, covering all four quadrants, appears in Fig. 4.3. Two contour levels are plotted; the solid line represents $|\chi(\tau, \nu)| = 0.707$ and the dotted contours represent $|\chi(\tau, \nu)| = 0.1$. Figure 4.2 clearly shows the triangular zero-Doppler cut of the ambiguity function, described in (4.5). The delay response reaches zero at the pulse width T. The zero-delay cut is less obvious from Fig. 4.2 and is plotted separately in Fig. 4.4. The first Doppler null is at the inverse of the pulse duration: namely, $|\chi(0, 1/T)| = 0$. We can therefore approximately state

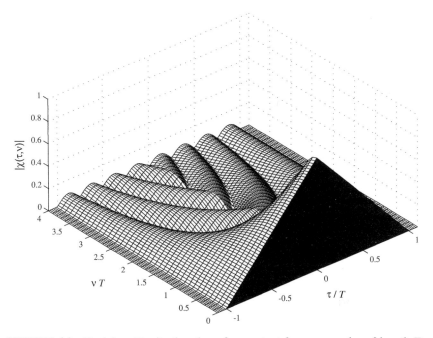

FIGURE 4.2 Partial ambiguity function of a constant-frequency pulse of length T.

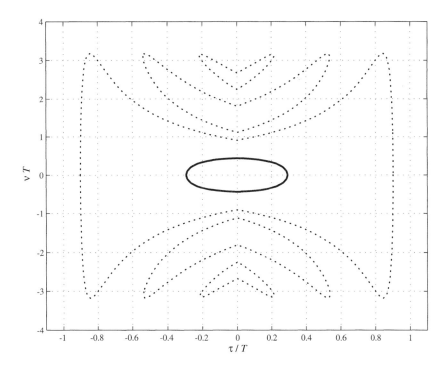

FIGURE 4.3 Contours 0.1 (dotted) and 0.707 (solid) of the AF of a pulse.

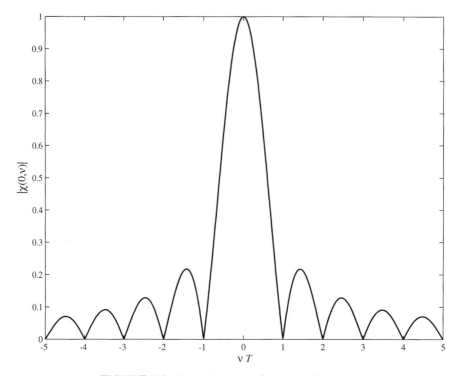

FIGURE 4.4 Zero-delay cut of the AF of a pulse.

that the delay resolution is T and the Doppler resolution is $1/T$. A numerical example will be useful here.

Consider $T = 5\,\mu s$ and $f_c = 5\,\text{GHz}$, which yields $\Delta R = C_p T/2 = 750\,\text{m}$. With a pulse compression ratio of 100, the range resolution will be reduced to an acceptable 7.5 m. With regard to the velocity resolution, we get

$$V_{\text{null}}(\text{m/s}) = v_{\text{null}} \frac{C_p}{2 f_c} = \frac{1}{T} \frac{C_p}{2 f_c} = \frac{3 \times 10^8}{(5 \times 10^{-6})(2 \times 5 \times 10^9)} = 6 \times 10^3\,\text{m/s} = 6\,\text{km/s}$$

As velocity resolution, 6 km/s is clearly unacceptable. This velocity encompasses all possible terrestrial velocities, reaching typical satellite velocities. The conclusion is that a single uncompressed pulse cannot usually provide sufficient range resolution or velocity (Doppler) resolution. Satisfactory range resolution will be reached using pulse compression and acceptable velocity resolution by using a coherent pulse train. Examples of both are discussed.

Resolution was practically defined by the first null, but beyond the first null the response may build up again (see Fig. 4.4) and produce sidelobes. The sidelobes of a return from a strong target can mask a small target. The first (and highest) Doppler sidelobe in Fig. 4.4 or equation (4.6) occurs at $vT = 1.43$, where $|\chi(0, 1.43/T)| = 0.2172$ or $-13.26\,\text{dB}$. Amplitude weighting techniques are usually used to reduce such high sidelobes. To avoid variable amplitude in the typical high-power (class C) radar transmitter, the amplitude weighting is usually introduced only in the receiver, which ceases to be matched.

One more function of interest is the spectrum of the complex envelope of the signal. The voltage spectral density is the Fourier transform of $u(t)$. Because in the constant-frequency pulse $u(t)$ is a real constant, the Fourier transforms of $u(t)$ and $|u(t)|^2$ exhibit the same result, and Fig. 4.4 also describes the magnitude of the voltage spectral density of our signal. Here again we see poor performance. The signal occupies its bandwidth inefficiently, with relatively high spectral sidelobes. We saw poor performances of a constant-frequency rectangular pulse in many aspects: poor range resolution, poor Doppler resolution and high Doppler sidelobes, and inefficient spectral use. Our next basic signal, a linear-FM pulse, improves on some of those weaknesses.

4.2 LINEAR FREQUENCY-MODULATED PULSE

Linear frequency modulation (LFM) is the first and probably still is the most popular pulse compression method. It was conceived during World War II, independently on both sides of the Atlantic, as can be deduced from German, British, and U.S. patents (Cook and Bernfeld, 1967; Cook and Seibert, 1988). The basic idea is to sweep the frequency band B linearly during the pulse duration T (Fig. 4.5).

The complex envelope of a linear-FM pulse is given by

$$u(t) = \frac{1}{\sqrt{T}} \operatorname{rect}\left(\frac{t}{T}\right) \exp(j\pi k t^2), \qquad k = \pm \frac{B}{T} \qquad (4.7)$$

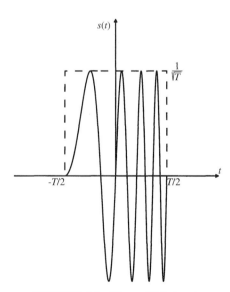

FIGURE 4.5 Linear-FM signal.

The instantaneous frequency $f(t)$ is obtained by differentiating the argument of the exponential,

$$f(t) = \frac{1}{2\pi} \frac{d(\pi k t^2)}{dt} = kt \tag{4.8}$$

The instantaneous frequency is indeed a linear function of time. The frequency slope k has the dimension s^{-2}.

The ambiguity function (AF) is obtained by applying property 4 to the AF of an unmodulated pulse. Replacing ν in (4.4) with $\nu - k\tau$ yields the AF of a linear-FM (LFM) pulse:

$$|\chi(\tau, \nu)| = \left| \left(1 - \frac{|\tau|}{T}\right) \frac{\sin\left[\pi T \left(\nu \mp B(\tau/T)\right)(1 - |\tau|/T)\right]}{\pi T \left(\nu \mp B(\tau/T)\right)(1 - |\tau|/T)} \right|,$$

$$|\tau| \leq T, \quad \text{zero elsewhere} \tag{4.9}$$

Figure 4.6 presents an example of the AF of an LFM pulse calculated using the MATLAB code in Appendix 3A. The phase and frequency of the complex envelope are shown in Fig. 4.7. The effective time–bandwidth product of the signal is $kT^2 = BT = 10$, where B is the total frequency deviation. Note that the total deviation of the normalized frequency plot is BT, and the total phase deviation is $BT\pi/4$.

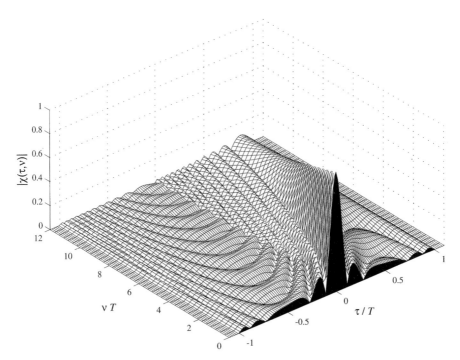

FIGURE 4.6 Partial ambiguity function of linear-FM pulse ($BT = 10$).

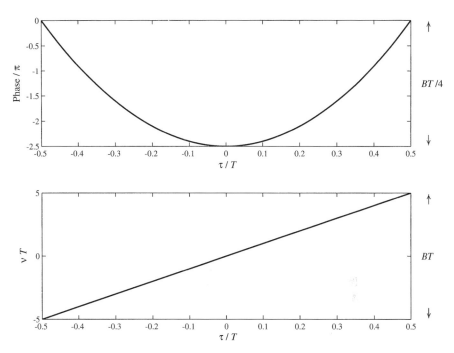

FIGURE 4.7 Phase and frequency characteristic of the LFM pulse used in Fig. 4.6.

The zero-Doppler cut of the AF is obtained by setting $\nu = 0$ in (4.9). Using $kT = B$ yields

$$|\chi(\tau, 0)| = \left|\left(1 - \frac{|\tau|}{T}\right) \frac{\sin\left[\pi B \tau (1 - |\tau|/T)\right]}{\pi B \tau (1 - |\tau|/T)}\right|, \quad |\tau| \leq T, \quad \text{zero elsewhere}$$

(4.10)

For a large time–bandwidth product ($kT^2 = BT \gg 4$) the first null of $|\chi(\tau, 0)|$ occurs at

$$\tau_{1'\text{st null}} \approx \frac{1}{|k|T} = \frac{1}{B} \quad (4.11)$$

The compression ratio will be defined as $T/\tau_{1'\text{st null}} \cong BT$, and it is approximately equal to the time–bandwidth product. Returning to Fig. 4.6, since $BT = 10$, the first null should occur at $\tau/T = 0.1$. A plot of $|\chi(\tau, 0)|$ for $BT = 10$ appears in Fig. 4.8.

The issue of spectral efficiency is demonstrated in Fig. 4.9. The horizontal scale is frequency normalized with respect to the pulse width. The vertical scale is the spectral density in decibels. To obtain similar range resolution, the unmodulated pulse width was one-tenth of the LFM pulse width. The absolute horizontal scales are therefore identical in both plots. Comparing the two plots clearly shows more efficient spectrum use in the LFM case. The spectral efficiency of

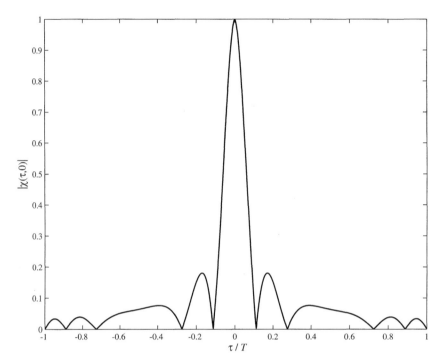

FIGURE 4.8 Zero-Doppler cut of the AF of LFM pulse with a time–bandwidth product of 10.

LFM improves as the time–bandwidth product increases, because the spectral density then approaches a rectangular shape. Note that the spectrums plotted in Fig. 4.9 are of the signals' complex envelopes and are therefore centered about zero. Due to symmetry, it suffices to plot positive frequencies only.

The improved delay resolution of LFM does come with a penalty, delay–Doppler coupling. It is expressed by the diagonal ridge seen in the three-dimensional plot of the AF (Fig. 4.6). A contour plot (Fig. 4.10) emphasizes the coupling problem. From (4.9) we find that for small Doppler shift ν, the delay location of the peak response is shifted from true delay by

$$\tau_{\text{shift}} = \frac{\nu}{k} \tag{4.12}$$

The physical interpretation is that when $k > 0$, a target with positive Doppler appears closer than its true range. In many applications the resulting range error is acceptable. The delay error of the shifted peak response is accompanied by a small decrease in the height of the peak, as evident in Fig. 4.6. It can be shown that near the origin, the peak height decreases according to

$$|\chi(\tau_{\text{peak}}, \nu)| = 1 - \left|\frac{\nu}{kT}\right| = 1 - \left|\frac{\nu}{B}\right| \tag{4.13}$$

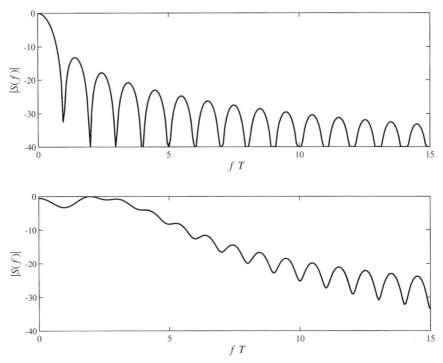

FIGURE 4.9 Spectral density (in dB) of unmodulated pulse (top) and LFM pulse, $BT = 10$ (bottom).

Since the typical Doppler shift ν is usually much smaller than the signal's frequency deviation B, the AF peak height decreases very moderately. This behavior is responsible for the Doppler tolerance property attributed to LFM. The importance of Doppler tolerance is discussed in Section 4.3.

4.2.1 Range Sidelobe Reduction

Adding linear frequency modulation has increased the bandwidth and thus improved the range resolution of the signal by a factor equal to the time–bandwidth product. However, relatively strong sidelobes remain in the autocorrelation function (ACF), as seen, for example, in Fig. 4.8. The ACF is related to the power spectral density of the signal through the Fourier transform. ACF sidelobes can be reduced by shaping the spectrum. The shaping should alter the spectrum from its nearly rectangular window shape to one of the well-known (Harris, 1978; Nuttall, 1981) weight windows (e.g., Hann, Hamming). Spectral reshaping of LFM can be done using two different basic approaches: amplitude weighting or frequency weighting. Amplitude weighting is discussed here; frequency weighting is covered in Section 5.2.

Spectral shaping through amplitude weighting makes use of the linear relation between instantaneous frequency and time along the pulse. At any given time, a

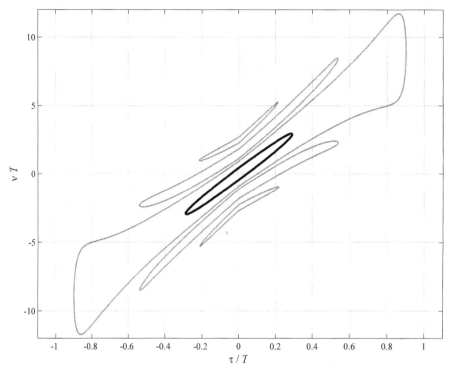

FIGURE 4.10 Contours 0.1 and 0.707 of the AF of an LFM pulse ($BT = 10$).

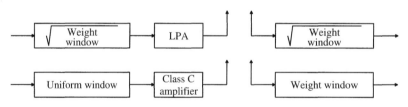

FIGURE 4.11 Implementing weight windows: matched weighting (top); mismatched weighting (bottom).

particular frequency is transmitted. If the amplitude of the signal at that instance is higher, the power spectral density of the corresponding frequency is also higher. Hence, all that is needed is to shape the amplitude of the pulse according to the desired weight window (Harris, 1978; Nuttall, 1981).

To maintain matched filtering, the weight should be split between the transmitter and the receiver. Hence, the amplitude should be shaped using the square root of the window. This requires a linear power amplifier in the transmitter, which is relatively inefficient (Fig. 4.11, top). The other alternative is to implement the entire weight window at the receiver, accepting a loss due to mismatched filtering (Fig. 4.11, bottom).

In addition to the mismatch loss, the performances of the two approaches are similar but *not identical*, as will be shown shortly. The contributions of matched weighting to sidelobe reduction are demonstrated in Figs. 4.12 and 4.13. The figures present a comparison between an unweighted LFM pulse and a square root of Hamming-weighted LFM pulse, both with $BT = 100$. Note how the weighted amplitude (top subplots) affects the ACF peak sidelobes (middle subplots), which were reduced from about -13 dB in the unweighted LFM pulse (Fig. 4.12) to approximately -40 dB in the amplitude-weighted LFM pulse (Fig. 4.13).

Also evident in the two figures (bottom subplots) is the effect of the amplitude weighting on the shape of the spectral density. The spectrums were determined numerically using FFT. Theoretical analysis is available for the Fourier transform of an unweighted LFM signal. According to Klauder et al. (1960) the Fourier transform $U(f)$ of the LFM complex envelope $u(t)$ defined in (4.7) is

$$U(f) = \frac{1}{\sqrt{T}} \int_{-T/2}^{T/2} \exp[j\pi(kt^2 - 2ft)]\,dt$$

$$= \frac{1}{\sqrt{2B}} \exp\left(-\frac{j\pi f^2}{k}\right) [Z(x_2) - Z(x_1)] \qquad (4.14)$$

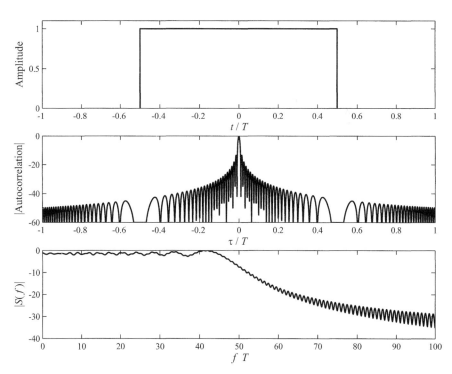

FIGURE 4.12 Amplitude, autocorrelation, and spectrum of unweighted LFM pulse, $BT = 100$.

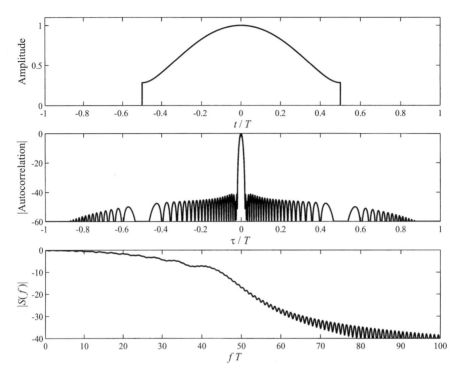

FIGURE 4.13 Amplitude, autocorrelation, and spectrum of a square-root of Hamming-weighted LFM pulse, $BT = 100$.

where $Z(x)$ is the complex Fresnel integral

$$Z(x) = \int_0^x \exp\left(\frac{j\pi\alpha^2}{2}\right) d\alpha = C(x) + jS(x)$$
$$= \int_0^x \cos\left(\frac{\pi\alpha^2}{2}\right) d\alpha + j \int_0^x \sin\left(\frac{\pi\alpha^2}{2}\right) d\alpha \qquad (4.15)$$

and the arguments in (4.14) are

$$x_2 = -2f\sqrt{\frac{T}{2B}} + \sqrt{\frac{TB}{2}}, \qquad x_1 = -2f\sqrt{\frac{T}{2B}} - \sqrt{\frac{TB}{2}} \qquad (4.16)$$

The partial ambiguity functions (zoom in delay and in Doppler) for an unweighted LFM pulse and a square root of Hamming-weighted LFM pulse, both with $BT = 20$, are plotted in Fig. 4.14. The penalty of mainlobe broadening is clearly evident. It is also important to note that the reduction in range sidelobes due to weighting holds well even at higher Doppler frequencies. This will not be the case when we study a different approach for spectral shaping: nonlinear FM.

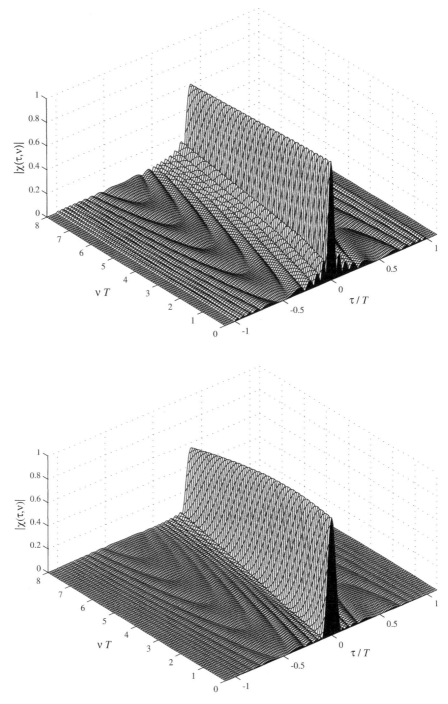

FIGURE 4.14 Partial ambiguity function of LFM (top) and Hamming-weighted LFM (bottom) pulse, both with $BT = 20$.

66   BASIC RADAR SIGNALS

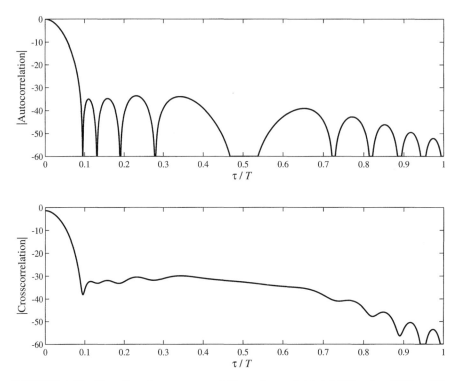

FIGURE 4.15 Zero-Doppler response of Hamming-weighted LFM pulse ($BT = 20$): matched weighting (top); weighting on receipt (bottom).

Finally, in Fig. 4.15, we demonstrate the difference between the autocorrelation of matched weighting and the cross-correlation of mismatched weighting. Both cases involve an LFM pulse with $BT = 20$ and Hamming weighting. In the top subplot both the transmitted and reference signals in the receiver include a square root of Hamming amplitude weighting. In the lower subplot, the signal transmitted has uniform amplitude, while the amplitude of the reference signal is weighted according to the (full) Hamming window. Both weighting approaches exhibit considerable sidelobe reduction compared to unweighted LFM. However, the correlation of the mismatched weighting is slightly worse than that of the matched weighting. The topic of weighting LFM on receive is expanded in Chapter 7.

4.2.2 Mismatch Loss

The mismatch loss caused by implementing the weight window only at the receiver depends on the specific weight window and the noise (assumed here to be additive white Gaussian noise). Assume that the weight window is implemented as an FIR filter with N real coefficients:

$$\text{weight window} = \{w_1, w_2, \ldots, w_N\} \qquad (4.17)$$

At the input to the weight filter, the signal samples are uniform with amplitude A, and the noise samples' root mean square (RMS) value is σ. In the filter the signal samples add up coherently (power of a sum), whereas the noise samples add up noncoherently (a sum of powers). The signal-to-noise ratio (SNR) is therefore

$$\text{SNR} = \frac{A^2 \left(\sum_{n=1}^{N} w_n \right)^2}{\sigma^2 \sum_{n=1}^{N} w_n^2} \quad (4.18)$$

The SNR loss will be calculated relative to a filter with uniform coefficients (matches the absent weight filter in the transmitter). Setting $w_n = 1$ in (4.18) yields

$$\text{SNR}_{\text{uni}} = \frac{A^2 N^2}{\sigma^2 N} = \frac{A^2}{\sigma^2} N \quad (4.19)$$

The SNR loss is defined as

$$\text{SNR}_{\text{loss}} = \frac{\text{SNR}}{\text{SNR}_{\text{uni}}} \quad (4.20)$$

Using (4.18) and (4.19) in (4.20) yields

$$\text{SNR}_{\text{loss}} = \frac{\left(\sum_{n=1}^{N} w_n \right)^2}{N \sum_{n=1}^{N} w_n^2} \quad (4.21)$$

In a Hamming weight filter, for example, whose coefficients are

$$w_n = a_0 - a_1 \cos \frac{2\pi(n-1)}{N},$$
$$n = 1, 2, \ldots, N, \quad a_0 = 0.53836, \quad a_1 = 0.46164 \quad (4.22)$$

the SNR loss is 1.36 dB.

4.3 COHERENT TRAIN OF IDENTICAL UNMODULATED PULSES

In the unmodulated and LFM pulses discussed so far, the volume of the ambiguity function (AF) remained concentrated around the origin. Furthermore, the Doppler resolution remained $1/T$, where T is the pulse duration. In the third basic signal, discussed in this section, most of the AF volume spreads away from the origin. A narrow peak of unity at the origin obviously remains. The title of this section emphasizes that the coherent pulse train is constructed from *identical* pulses. This is not always the case. In more sophisticated signals some diversity between the pulses is introduced, which gains some advantages. The title also points out the *unmodulated* (or constant-frequency, constant-amplitude) nature of the pulses. Again, there are more advanced signals in which the pulses in the train are modulated (e.g., LFM). In order not to drag out the long title of this signal, we will refer to it as a *coherent pulse train* (CPT), but imply both restrictions.

The coherency of the signal (same carrier frequency and known initial phase of each pulse) is accounted for by the fact that the signal (not the envelope) is described by

$$s(t) = \text{Re}[u_N(t) \exp(j2\pi f_c t)] \qquad (4.23)$$

The complex envelope of a train of N identical pulses is described by

$$u_N(t) = \frac{1}{\sqrt{N}} \sum_{n=1}^{N} u_1[t - (n-1)T_r] \qquad (4.24)$$

where T_r is the pulse repetition interval. The *unmodulated* nature of each pulse is implied by

$$u_1(t) = \frac{1}{\sqrt{T}} \text{rect}\left(\frac{t}{T}\right) \qquad (4.25)$$

where T is the pulse duration. A plot of $u_N(t)$ appears in Fig. 4.16.

The ambiguity function of a coherent pulse train has been developed in many textbooks. Using the results in (Levanon, 1988), we get for the practical case in which $T < T_r/2$ the expression

$$|\chi(\tau, \nu)| = \frac{1}{N} \sum_{p=-(N-1)}^{N-1} |\chi_T(\tau - pT_r, \nu)| \left| \frac{\sin[\pi \nu(N - |p|)T_r]}{\sin \pi \nu T_r} \right|,$$

$$|\tau| \leq NT_r, \quad \text{zero elsewhere} \qquad (4.26)$$

where $|\chi_T(\tau, \nu)|$ is the AF of an individual pulse. For our unmodulated pulse $|\chi_T(\tau, \nu)|$ was developed in Section 4.1 and is given by

$$|\chi_T(\tau, \nu)| = \left| \left(1 - \frac{|\tau|}{T}\right) \frac{\sin[\pi T \nu(1 - |\tau|/T)]}{\pi T \nu(1 - |\tau|/T)} \right|, \quad |\tau| \leq T, \quad \text{zero elsewhere} \qquad (4.27)$$

If we limit the delay to the mainlobe area, namely to $|\tau| \leq T$, then (4.26) reduces to

$$|\chi(\tau, \nu)| = |\chi_T(\tau, \nu)| \left| \frac{\sin N\pi\nu T_r}{N \sin \pi\nu T_r} \right|, \quad |\tau| \leq T \qquad (4.28)$$

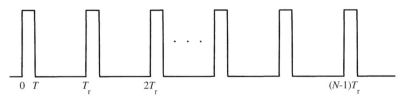

FIGURE 4.16 Envelope of a coherent pulse train.

This result resembles equation (3.39), which was obtained for the periodic ambiguity function. There it applied to the entire PAF, not only to the mainlobe.

The two AF cuts can be obtained from (4.26) and (4.27), yielding

$$|\chi(\tau, 0)| = \sum_{p=-(N-1)}^{N-1} \left(1 - \frac{|p|}{N}\right)\left(1 - \frac{|\tau - pT_r|}{T}\right), \qquad |\tau - pT_r| \leq T, \quad \text{zero elsewhere} \tag{4.29}$$

and

$$|\chi(0, \nu)| = \left| \frac{\sin \pi \nu T}{\pi \nu T} \frac{\sin N \pi \nu T_r}{N \sin \pi \nu T_r} \right| \tag{4.30}$$

The ambiguity function of a coherent train of six pulses with a duty cycle of 0.2 is plotted in Fig. 4.17 and the corresponding contour plot (positive and negative Doppler shifts) in Fig. 4.18 (with contours at 0.72 and 0.1). The ambiguity function of a coherent pulse train is often called a *bed of nails*. The nails refer to the mesh of recurrent lobes, at intervals of T_r in delay and $1/T_r$ in Doppler. Had we picked N much larger than 6 and a duty cycle much smaller than 0.2, Fig. 4.17 would have indeed looked like a fakir bed.

The zero-Doppler cut, given in (4.29) and displayed clearly in Fig. 4.17, is a set of triangles, all with a base of $2T$, but with linearly decreasing height. This

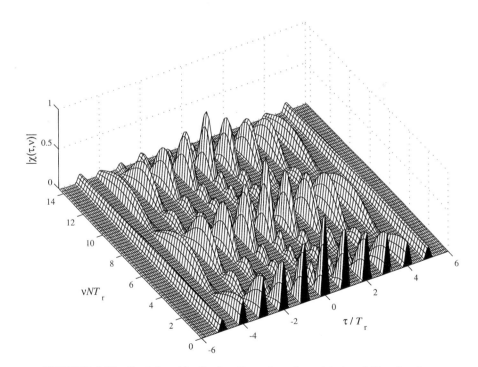

FIGURE 4.17 Partial ambiguity function of a coherent train of $N = 6$ pulses.

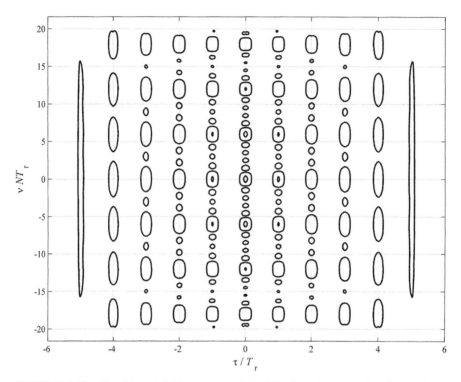

FIGURE 4.18 The 0.1 and 0.72 contours of the AF of a coherent train of six pulses.

can be explained easily. Around $\tau = 0$, all six ($= N$) received pulses overlap the six reference pulses. Around $\tau = T_r$ only five ($= N - 1$) pulses overlap, and so on. Because the pulses are identical, the correlation between individual pulses maintains the triangular shape. Only the heights of the triangles change to reflect the decreasing number of pulses involved in the cross-correlation between the received and reference signals.

The zero-delay cut given in (4.30) is a product of the zero-delay cut of a single pulse, given in (4.16), with the expression $|(\sin N\pi\nu T_r)/(N \sin \pi\nu T_r)|$. The cut appears in Fig. 4.19 (solid line). The dashed line is the AF zero-delay cut of a single pulse. Figure 4.19 demonstrates the large improvement in Doppler resolution, gained by coherent processing of N pulses. It also demonstrates the Doppler ambiguity caused by the recurrent Doppler peaks at $1/T_r$ intervals. Not marked in Fig. 4.19, but marked in Fig. 4.20, the first Doppler null is located at $1/NT_r$ (i.e., at the inverse of the total coherent signal duration).

As can be seen in Fig. 4.20, the coherent pulse train provides independent control of the delay and Doppler resolutions that was not possible in the single-pulse case. The delay resolution is controlled by the pulse duration T, while the Doppler resolution is controlled by the total signal length NT_r. On the other hand, the Doppler and delay ambiguities are tied; both are functions of the pulse repetition interval T_r. Note that their product, which is the area of the rectangle

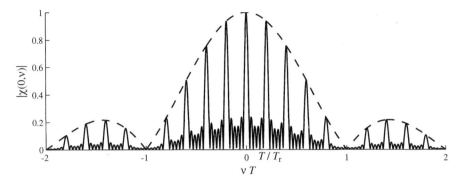

FIGURE 4.19 Zero-delay cut of the ambiguity function of six pulses.

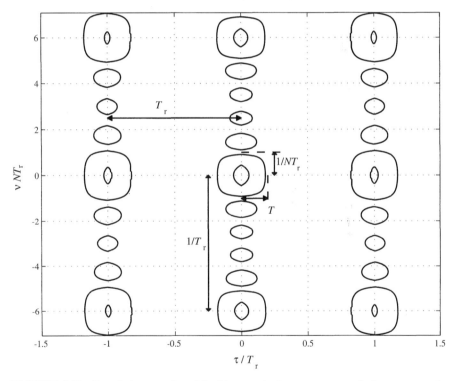

FIGURE 4.20 Resolutions and ambiguities marked on a zoom of the contours in Fig. 4.18.

connecting four recurrent lobes, is given by $T_r \cdot 1/T_r = 1$. This trade-off between Doppler (velocity) and delay (range) ambiguities is an inherent difficulty in radar. It is the cause for the design parameter referred to as low, medium, or high pulse repetition frequency (PRF).

PROBLEMS

4.1 Ambiguity function of a Gaussian
  (a) Calculate the ambiguity function (AF) of a signal with the envelope $u(t) = A \exp(-t^2/T)$.
  (b) What value of A will make the signal of unit energy?
  (c) Calculate the AF of a signal with the envelope $u(t) = A \exp(-t^2/T) \exp(j\pi k t^2)$.

4.2 Symmetry property of the ACF nulls of an LFM pulse
Show that the autocorrelation function (ACF) nulls of linear FM are located symmetrically around $\tau = T/2$.

4.3 Positive-definite ACF of an LFM pulse
In an LFM pulse, what is the largest time–bandwidth product BT in which $\chi(\tau, 0)$ does not cross zero [i.e., $|\chi(\tau, 0)|$ exhibits no null]?

4.4 LFM pulse AF ridge peak
In an LFM pulse, what is the value of $|\chi(\tau, \nu_{\text{ridge}})|$ (i.e., the peak value along the ridge) as a function of τ/T ?

4.5 Train of frequency-stepped LFM pulses
The signal described in Fig. P4.5 is a coherent train of N LFM pulses. Each pulse exhibits a total frequency deviation B. In addition, the center frequency of each pulse is frequency stepped by Δf relative to the preceding pulse. The pulse width is T and the pulse repetition interval is T_r.
  (a) Write the complex envelope of the signal (note that there is no phase-shift accumulation in the interval between pulses).

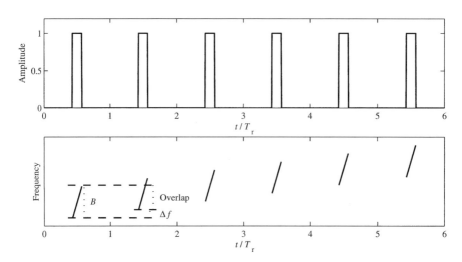

FIGURE P4.5 Amplitude and frequency of stepped-frequency train of LFM pulses.

(b) Show that the ambiguity function of this signal, within the delay range $|\tau| \leq T$, is given by

$$|\chi(\tau, \nu)| = \left|\left(1 - \frac{|\tau|}{T}\right) \operatorname{sinc}\left[T\left(\nu - B\frac{\tau}{T}\right)\left(1 - \frac{|\tau|}{T}\right)\right]\right|$$
$$\times \left|\frac{\sin\{N\pi[\nu - \Delta f(\tau/T_r)]T_r\}}{N\sin\{\pi[\nu - \Delta f(\tau/T_r)]T_r\}}\right|, \qquad |\tau| \leq T$$

where $\operatorname{sinc}(x) = \sin \pi x / \pi x$.

(c) Plot the ambiguity function for the case $T\Delta f = 3$, $TB = 18$, $N = 8$, $T_r/T = 9$. The boundaries of the plot should be $|\tau/T| \leq 1$, $0 \leq \nu N T_r \leq 10$.

(d) Repeat part (c) for the case $B = 0$.

(e) Repeat part (c) for the case $\Delta f = 0$.

REFERENCES

Cook, C. E., and M. Bernfeld, *Radar Signals: An Introduction to Theory and Application*, Academic Press, New York, 1967.

Cook, C. E., and W. M. Seibert, The early history of pulse compression radar, *IEEE Transactions on Aerospace and Electronic Systems*, vol. AES-24, no. 6, November 1988, pp. 825–837.

Harris, F. J., On the use of windows for harmonic analysis with the discrete Fourier transform, *Proceedings of the IEEE*, vol. 66, no. 1, January 1978, pp. 51–83.

Klauder, J. R., A. C. Price, S. Darlington, and W. J. Albersheim, The theory and design of chirp radars, *Bell System Technical Journal*, vol. 39, no. 4, July 1960, pp. 745–808.

Levanon, N., *Radar Principles*, Wiley, New York, 1988.

Nuttall, A. H., Some windows with very good sidelobe behavior, *IEEE Transactions on Acoustics, Speech, and Signal Processing*, vol. ASSP-29, no. 1, February 1981, pp. 84–91.

5

FREQUENCY-MODULATED PULSE

Linear frequency modulation (LFM), discussed in Chapter 4, is not the only frequency modulation law used in radar. In LFM the volume of the ambiguity function was concentrated in a slowly decaying diagonal ridge. This may be considered an advantage when Doppler resolution is not expected from a single pulse. LFM exhibited relatively high autocorrelation sidelobes, and some form of amplitude weighting was necessary in order to reshape the spectrum and thus reduce the autocorrelation sidelobes. In this chapter we describe two other frequency modulation schemes. The first is Costas coding, which results in a rather randomlike frequency evolution. The second is nonlinear FM, in which the spectrum is shaped not by amplitude weighting but by spending more time in the frequencies that need to be emphasized.

5.1 COSTAS FREQUENCY CODING

John P. Costas (1984) suggested a discrete frequency coding that is practically the opposite of the linear law used in LFM. The difference is demonstrated by the binary matrices in Fig. 5.1. The columns represent M contiguous time slices (each of duration t_b), and the rows represent M distinct frequencies, equally spaced (by Δf). In both signals we find one dot in each column and in each row. This means, respectively, that at any one of the M time slices, only one frequency is transmitted, and each frequency is used only once.

Radar Signals, By Nadav Levanon and Eli Mozeson
ISBN 0-471-47378-2 Copyright © 2004 John Wiley & Sons, Inc.

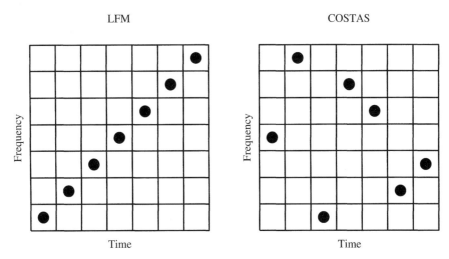

FIGURE 5.1 Binary matrix representation of quantized linear FM (left) and Costas coding (right).

The hopping orders described in Fig. 5.1 are only two out of $M!$ possible orders that meet the restriction of one dot per column and per row. The hopping order strongly affects the ambiguity function (AF) of the signal. The AF can be predicted roughly by overlaying a copy of the binary matrix on itself, and then shifting one relative to the other according to the desired delay (horizontal shifts) and Doppler (vertical shifts). When a given delay–Doppler shift results in a coincidence of N points, the ambiguity function is expected to yield a peak of approximately N/M at the corresponding delay–Doppler coordinate.

In the LFM case it is easily observed that only delay and Doppler shifts of equal number of units $[\tau = mt_b, \nu = m\Delta f, m = 0, \pm 1, \ldots, \pm(M-1)]$ will cause an overlap of dots, and the number of coinciding dots will be $N = M - |m|$. This hints at a diagonal ridge in the ambiguity function, along the line $\nu = \Delta f \tau / t_b$. What is unique for a Costas signal is that the number of coinciding dots cannot be larger than one for all but the zero-shift case, where all dots coincide ($N = M$). This property implies a narrow peak of the AF at the origin and low sidelobes elsewhere. If $\Delta f = 1/t_b$, the exact AF values at the grid points will be either 1 or 0, according to the corresponding number of coinciding dots. By *grid points* we refer to delay and Doppler shifts which are integer multiples of t_b and Δf, respectively. An example of an overlap in a Costas signal of length 7 is shown in Fig. 5.2.

5.1.1 Costas Signal Definition and Ambiguity Function

Construction algorithms for Costas signals were discussed by Golomb and Taylor (1984). However, a simple approach is to perform an exhaustive search of all

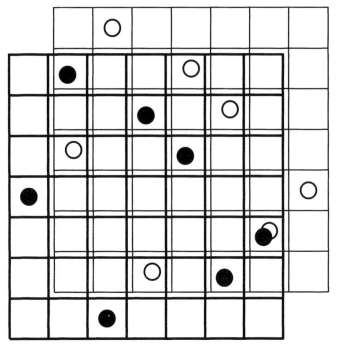

FIGURE 5.2 Example of one coincidence occurring at $\tau/t_b = 1, \nu/\Delta f = 1$.

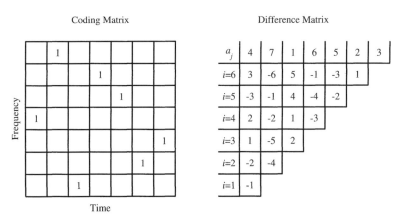

FIGURE 5.3 Coding matrix and difference matrix of a Costas signal. Coding sequence {4 7 1 6 5 2 3}.

possible signals of a given dimension M. Checking if a signal is Costas can be performed easily with the help of the *difference matrix*. An example is shown in Fig. 5.3. The *coding sequence*, the order of frequencies used, is a concise way to describe the *coding matrix*. With regard to the difference matrix, note that the top row and the leftmost column are headings and not part of the matrix. The element of the difference matrix in row i and column j is

$$D_{i,j} = a_{i+j} - a_j, \qquad i + j \leq M \tag{5.1}$$

where a_i is the ith element of the coding sequence. The remaining locations (where $i + j > M$) are left blank. Equation (5.1) says that the first row is formed by taking differences between adjacent elements in the coding sequence, the second row by taking differences between next-adjacent elements, and so on.

How the *sidelobe matrix* (Fig. 5.5) is derived from the difference matrix will be demonstrated by an example. Consider the first element in the first row of the difference matrix ($D_{1,1} = a_2 - a_1 = 7 - 4 = 3$). It says that at a positive normalized delay of 1 there is a coincidence if the normalized Doppler shift is 3. This result will prompt adding 1 to the value accumulated in the {delay = +1, Doppler = +3} location of the sidelobe matrix. For the signal to be Costas, there should not be accumulated values larger than 1. Another way to say the same thing: If all the elements in a row of the difference matrix are different from each other, the signal is Costas. Since the serial number of a row in the difference matrix represents the normalized delay (a column) in the sidelobe matrix, how do we fill the columns representing negative delays? To complete the left-hand side of the sidelobe matrix, we simply apply the AF rule of symmetry with respect to the origin. The number of 1's in the sidelobe matrix is $M(M-1)$. The number of 0's in the sidelobe matrix is $3M(M-1)$. To complete the resemblance between the sidelobe matrix and the grid point values of the ambiguity function, we add a value of M ($= 7$) at the origin. This is the only nonzero entry in the zero-delay column and the zero-Doppler row. This brings the total number of elements to $(2M-1)^2$. Finally, to normalize the peak value to 1, we divide all the entries by M.

The complex envelope of a Costas signal whose hopping sequence is

$$a = \{a_1, a_2, \ldots, a_M\} \tag{5.2}$$

can be described as

$$u(t) = \frac{1}{\sqrt{Nt_b}} \sum_{m=1}^{M} u_m[t - (m-1)t_b] \tag{5.3}$$

where

$$u_m(t) = \begin{cases} \exp(j2\pi f_m t), & 0 \leq t \leq t_b \\ 0 & \text{elsewhere} \end{cases} \tag{5.4}$$

and

$$f_m = \frac{a_m}{t_b} \tag{5.5}$$

Equation (5.5) is rather critical in a Costas signal design. It states that the frequency spacing is equal to the inverse of the time duration at each frequency.

This implies orthogonality and forces ambiguity function (AF) zeros at all grid points except those in which the sidelobe matrix had a "1" element. Since there were no such elements in the zero-Doppler row, we should expect nulls of the zero-Doppler cut of the AF at all integer multiples of chip duration t_b.

A closed-form expression of the ambiguity function was given by Costas (1984):

$$\chi(\tau, \nu) = \frac{1}{M} \sum_{m=1}^{M} \exp[j2\pi(m-1)\nu t_b] \left\{ \Phi_{mm}(\tau, \nu) + \sum_{\substack{n=1 \\ m \neq n}}^{M} \Phi_{mn}[\tau - (m-n)t_b, \nu] \right\}$$
(5.6)

where

$$\Phi_{mn}(\tau, \nu) = \left(1 - \frac{|\tau|}{t_b}\right) \frac{\sin \alpha}{\alpha} \exp(-j\beta - j2\pi f_n \tau), \quad |\tau| \leq t_b, \quad \text{zero elsewhere}$$
(5.7)

in which

$$\alpha = \pi(f_m - f_n - \nu)(t_b - |\tau|)$$ (5.8)

$$\beta = \pi(f_m - f_n - \nu)(t_b + \tau)$$ (5.9)

Despite having a theoretical expression of the AF, we will apply our numerical AF plotting program (Appendix 3A) to plot the AF of the Costas signal with the coding sequence used in Fig. 5.3 and look for similarities to the sidelobe matrix. The partial (positive Doppler only) AF is shown in Fig. 5.4. We first note that the zero-Doppler cut indeed displays zero values at multiples of t_b. Off the grid points the AF exhibits many local sidelobe peaks (even in areas located between grid points having AF value of zero). The sidelobe peaks around the origin are especially high. In general, the thumbtack nature of the AF is clearly evident.

To demonstrate agreement between the sidelobe matrix and the ambiguity function, Fig. 5.5 displays the sidelobe matrix next to a contour map of $|\chi(\tau, \nu)| = 0.125$. To help with orientation, consider the leftmost closed contour. It corresponds to the two 1's located at $\{\text{delay} = -6, \text{Doppler} = 1\}$ and $\{-5, 2\}$ in the sidelobe matrix. That relatively well-defined small sidelobe ridge can also be identified in the lower left corner of the three-dimensional plot in Fig. 5.4. A contour level of $\frac{1}{8}$ was chosen because it is below the grid point sidelobe peaks, whose level is exactly $1/N$ ($= \frac{1}{7}$). The agreement between the ambiguity function that results from coherent processing and the sidelobe matrix, which has nothing to do with phase coherence, indicates that noncoherent processing will work relatively well with Costas signals. Indeed, Costas's original application was sonar, where coherency is poorly preserved.

From the autocorrelation plot (Fig. 5.6) we can conclude that the first null is located at $\tau_{1'\text{st null}} \cong T/M^2 = t_b/M = t_b/7$. This implies that the pulse compression of a Costas signal is M^2. To reduce the sidelobe pedestal, M will have to be increased. Where will the volume beneath the AF squared go to? The answer is

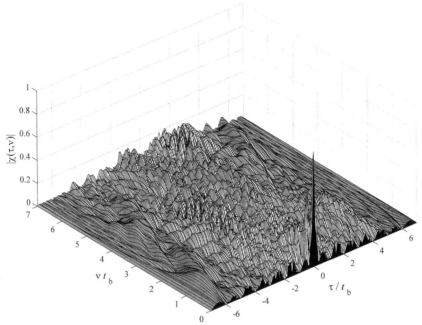

FIGURE 5.4 Partial ambiguity function of a Costas signal with code sequence {4 7 1 6 5 2 3}.

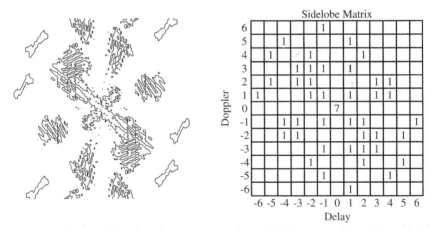

FIGURE 5.5 Ambiguity function contour at 0.125 (left) compared with the sidelobe matrix (right).

simple: It will spread farther in Doppler because the spread in Doppler extends as far as $\nu \approx M/t_b$. This value, M/t_b, is also approximately the spectral width of the signal, because $1/t_b$ is the spacing between two adjacent frequencies. In Fig. 5.6 we plotted the zero-Doppler cut of the ambiguity function. We did not bother to plot the zero-delay cut because, according to the AF properties, that cut

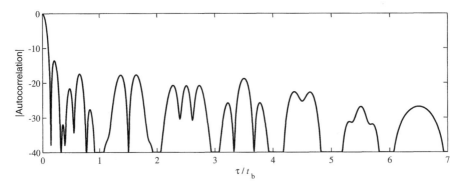

FIGURE 5.6 Autocorrelation function of a Costas signal with code sequence {4 7 1 6 5 2 3}.

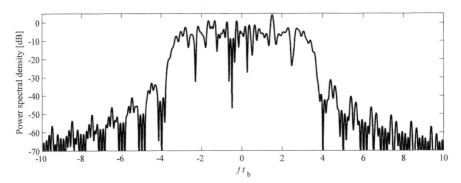

FIGURE 5.7 Power spectral density of the complex envelope of the Costas signal.

is a function only of the magnitude of the pulse. Adding frequency modulation (or hopping) does not affect the magnitude. Hence the zero-delay cut is identical to that of an unmodulated pulse of the same duration.

Figure 5.7 displays the power spectral density of the complex envelope. Note that the envelope spectrum extends, in normalized frequency (ft_b), from $-M/2$ to $+M/2$; hence, the total bandwidth is approximately M/t_b. Note also the relatively flat spectrum, which should be expected since the duration at each frequency is the same (t_b). Finally, note that increasing M by a factor k, while keeping the total pulse duration fixed, implies reducing the chip duration t_b, also by k. Hence the total bandwidth will increase by k^2.

5.1.2 On the Number of Costas Arrays and Their Construction

The total number of different $M \times M$ binary arrays with one dot per row and per column is $M!$. How many of them are Costas arrays is not easily determined. Early exhaustive searches revealed that the ratio decreases fast. For $M = 7$ about one out of 25 such arrays is Costas, while for $M = 12$ only one out of 61,000 is Costas. Despite the decreasing density, it was initially believed that the number of different Costas arrays of a given size M increases monotonically with M. Later exhaustive searches revealed that the trend reverses at $M = 17$. Silverman et al. (1988) predicted that the number will behave according to the solid line in Fig. 5.8. Above $M = 25$ it is believed that very few Costas arrays will be found for a given M. As a matter of fact, none are known for $M = 32$ and $M = 33$.

For large M, exhaustive search is not a practical way to find Costas arrays. Golomb and Taylor (1984) present several construction methods, of which we repeat the method labeled as *Welch 1*. This construction method is applicable for $M = p - 1$, where p can be every prime larger than 2. GF(p) is the Galois field of order p, and α is a primitive element in GF(p). (See Appendix 6A for the properties and definitions of Galois fields.) Numbering the columns of the array $j = 0, 1, 2, \ldots, p - 2$ and the rows $i = 1, 2, \ldots, p - 1$, we put a dot in position $\{i, j\}$ if and only if $i = \alpha^j$.

Note that with Welch 1 construction the coding sequence always begins with 1. Removing the first element and subtracting 1 from the remaining elements in the sequence creates a Costas array with size $M = p - 2$. This construction method

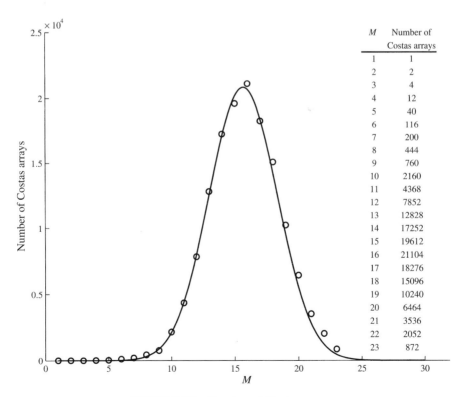

FIGURE 5.8 Number of Costas arrays.

is labeled *Welch 2*. When $\alpha = 2$, the Welch 1 construction begins with 1, 2. Removing both and subtracting 2 results in a Costas array with size $M = p - 3$. This construction method is labeled *Welch 3*.

We give an example of finding Costas arrays of size $M = 4$. First, since $M + 1$ is a prime integer higher than 2, we can use the Welch 1 method with $p = M + 1 = 5$. We write GF(5) as {0 1 2 3 4} and test the elements to be primitive. Using $\alpha = 2$, we get

$$j = 0; \quad i = 2^0 = 1$$
$$j = 1; \quad i = 2^1 = 2$$
$$j = 2; \quad i = 2^2 = 4$$
$$j = 3; \quad i = 2^3 = 8 \equiv 3 \quad (\text{mod } 5)$$

which gives the coding sequence {1 2 4 3}. Using $\alpha = 3$, we get

$$j = 0; \quad i = 3^0 = 1$$
$$j = 1; \quad i = 3^1 = 3$$
$$j = 2; \quad i = 3^2 = 9 \equiv 4 \quad (\text{mod } 5)$$
$$j = 3; \quad i = 3^3 = 27 \equiv 2 \quad (\text{mod } 5)$$

which give the coding sequence {1 3 4 2}. However, using $\alpha = 4$ gives $4 \times 4 = 1$ (mod 5); thus, 4 is not a primitive element of GF(5).

Using the Welch 1 method gives only two Costas arrays for $M = 4$. The Welch 2 method cannot be used here since $M + 2 = 6$ is not prime. Finally, we try to use Welch 3 ($M + 3 = 7$ is a prime). Testing that $\alpha = 2$ is primitive, we get that $2 \times 2 = 4$, $4 \times 2 = 1$ (mod 7). Thus $\alpha = 2$ is not primitive in GF(7) and Welch 3 cannot be used here.

Using the Welch 1 method we found only two of the 12 possible Costas arrays of size 4, as indicated in Fig. 5.8. The remaining arrays are obtained by flipping up–down, left–right, clockwise–anticlockwise rotation, or from other construction methods. All 12 arrays are shown in Table 5.1. Examples of Costas coding sequences resulting from Welch 1, 2, and 3 constructions, for $M \leq 18$, are given in Table 5.2. A Welch construction algorithm coded in MATLAB appears in Appendix 5A.

TABLE 5.1 All 12 Costas Arrays of Size 4

1000	0001	0010	0010	1000	0001
0010	1000	1000	0100	0001	0100
0001	0010	0100	0001	0100	1000
0100	0100	0001	1000	0010	0010
0100	0100	1000	0001	0100	0010
0010	0001	0100	0010	1000	0001
1000	0010	0001	1000	0010	0100
0001	1000	0010	0100	0001	1000

TABLE 5.2 Examples of Welch-Constructed Costas Coding Sequences for $M = 2$ to 18[a]

M																		
2	1	2																
3	1	3	2															
4	1	2	4	3														
5	2	1	5	3	4													
6	1	3	2	6	4	5												
8	2	6	3	8	7	5	1	4										
9	1	3	7	4	9	8	6	2	5									
10	1	2	4	8	5	10	9	7	3	6								
11	1	3	7	2	5	11	10	8	4	9	6							
12	1	2	4	8	3	6	12	11	9	5	10	7						
15	2	8	9	12	4	14	10	15	13	7	6	3	11	1	5			
16	1	3	9	10	13	5	15	11	16	14	8	7	4	12	2	6		
17	1	3	7	15	12	6	13	8	17	16	14	10	2	5	11	4	9	
18	1	2	4	8	16	13	7	14	9	18	17	15	11	3	6	12	5	10

[a] Welch constructions cannot be applied to $M = 7$, 13, and 14.

5.1.3 Longer Costas Signals

The properties of Costas signals become more prominent in longer signals. This section contains an example of Costas coding of length $M = 40$ (Table 5.3), implying a pulse compression of 1600. A plot of the ambiguity function exactly

TABLE 5.3 Costas Code of Length 40

i	j	i	j	i	j
1	1	15	33	28	19
2	12	16	27	29	23
3	21	17	37	30	30
4	6	18	34	31	32
5	31	19	39	32	15
6	3	20	17	33	16
7	36	21	40	34	28
8	22	22	29	35	8
9	18	23	20	36	14
10	11	24	35	37	4
11	9	25	10	38	7
12	26	26	38	39	2
13	25	27	5	40	24
14	13				

on the grid points $|\chi(mt_b, n/t_b)|$ yields a three-dimensional rendition of the sidelobe matrix (Fig. 5.9), while a more detailed zoom over $-2 \leq \tau/t_b \leq 2$ and $0 \leq \nu M t_b \leq 10$ shows sidelobes higher than $1/M$ in between grid points, especially near the origin (Fig. 5.10). The behavior on the delay axis is presented by the magnitude of the autocorrelation function (in dB).

The complete ACF appears in the top subplot of Fig. 5.11, while the middle subplot zooms on the first two chips (bits). Note that the ACF sidelobes are usually below -26 dB [$= 20 \log_{10}(2/M)$]. However, the near sidelobes ($t_b/M < \tau < t_b$) are higher, decaying from -13.7 dB in a manner typical of the ACF sidelobes of a signal with a rectangular spectrum. Indeed, the spectrum of our relatively long Costas signal is nearly rectangular, as shown in the bottom subplot of Fig. 5.11.

The last example of a relatively long Costas signal indicates how close its ambiguity function approaches the thumbtack shape. However, the Doppler resolution of the single Costas pulse remains $1/T$, where $T = Mt_b$ is the pulse duration. That Doppler resolution can be improved by using a coherent train of Costas pulses. Figure 5.12 displays a partial AF of a train of four Costas 40 pulses, with a duty cycle of $T/T_r = \frac{1}{4}$. The delay axis extends only as far as $\tau = 0.6 t_b = T/65$, and the Doppler axis extends slightly beyond $2/T_r$. The first recurrent Doppler lobe is already attenuated because of the relatively large duty cycle. If the radar is expected to work at higher (ambiguous) Doppler shifts, the

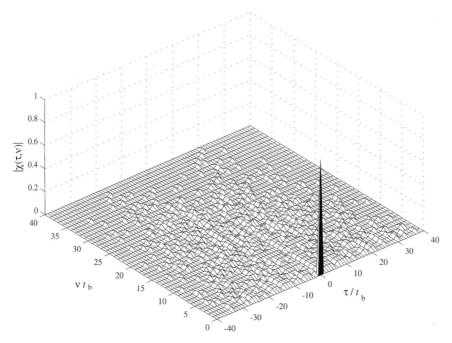

FIGURE 5.9 Ambiguity function of a Costas signal (length $M = 40$) at all relevant grid points.

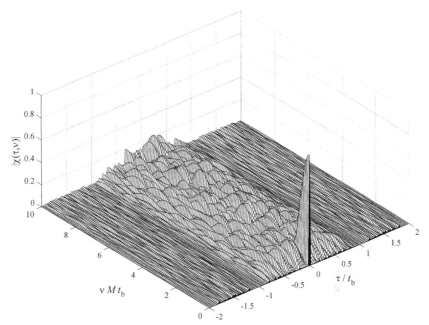

FIGURE 5.10 Ambiguity function of a Costas signal (length $M = 40$) zoom near the origin.

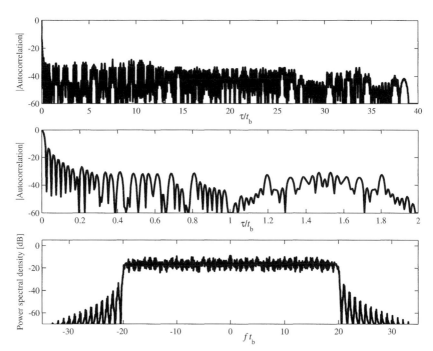

FIGURE 5.11 ACF (top and middle) and the spectrum (bottom) of a Costas signal (length 40).

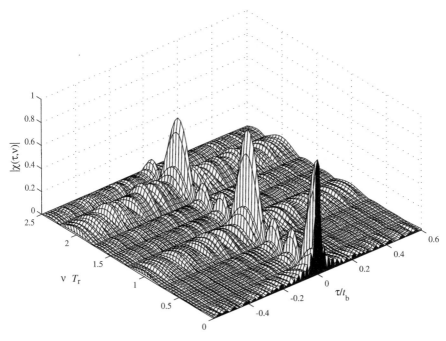

FIGURE 5.12 Partial AF of a coherent train of four Costas pulses (length 40), $T/T_r = \frac{1}{4}$.

single-pulse AF should be wider in Doppler (Doppler tolerant). In that situation, the preferred shape of the AF of the individual pulse is a ridge that decays slowly with Doppler. LFM pulses, discussed in Chapter 4, and NLFM pulses, discussed next, provide such a ridge.

5.2 NONLINEAR FREQUENCY MODULATION

Pulse compression achieved with unweighted linear FM (see Section 4.2) suffers from relatively high autocorrelation (ACF) sidelobes. Since the ACF is the inverse Fourier transform of the power spectrum, the ACF sidelobes could be reduced by shaping the power spectrum according to one of the many sidelobe suppression weight windows (Hann, Hamming, etc.). One method of shaping the spectrum was described in Section 4.2.1. The shaping was performed by changing the pulse amplitude along the time axis. Since in LFM instantaneous frequency is linearly related to time, that was equivalent to changing the amplitude along the frequency axis. Indeed, the resulting spectral shaping was very close to the desired spectrum, yielding the expected ACF sidelobe pattern.

Shaping the spectrum by amplitude weighting an LFM pulse has a serious drawback. In a matched transmitter–receiver pair, it results in variable amplitude of the pulse transmitted. Variable amplitude requires linear power amplifiers, which are less efficient than saturated power amplifiers. This problem can be alleviated by performing amplitude weighting only at the receiver. The resulting mismatch causes SNR loss.

In linear FM the transmitter spends equal time at each frequency, hence the nearly uniform spectrum. Another method for shaping the spectrum is to deviate from the constant rate of frequency change and to spend more time at frequencies that need to be enhanced. This approach was termed *nonlinear FM* (NLFM). Early works on NLFM (Fowle, 1964; Cook and Bernfeld, 1967) suggest finding the nonlinear frequency law through use of the stationary-phase concept. Let the complex envelope of the NLFM signal be given by

$$u(t) = g(t) \exp[j\phi(t)] \qquad (5.10)$$

and its Fourier transform by

$$U(f) = |U(f)| \exp[j\Phi(f)] = \int_{-\infty}^{\infty} g(t) \exp[j(-2\pi ft + \phi(t))] \, dt \qquad (5.11)$$

At time t_k the instantaneous frequency is determined by the time derivative of the phase at t_k:

$$f_k = \frac{1}{2\pi} \phi'(t_k) \qquad (5.12)$$

The *stationary-phase concept* says that the energy spectral density at that frequency is relatively large if the rate of change of the frequency at that time is relatively small. This intuitively appealing inverse relationship between spectral density and frequency rate of change is approximated by the simple expression

$$|U(f_k)|^2 \approx 2\pi \frac{g^2(t_k)}{|\phi''(t_k)|} \qquad (5.13)$$

In addition to being related to the inverse of the frequency rate of change $|\phi''(t_k)|$, the spectral density is clearly also dependent on the amplitude of the signal at that time $g(t_k)$. In LFM, $|\phi''(t_k)|$ is a constant and the spectrum is shaped through $g(t_k)$. In NLFM we wish to keep $g(t_k)$ a constant and shape the spectrum through $|\phi''(t_k)|$.

Let us try to approximate $U(f)$ by a function $V(f)$ specified independent of the amplitude function $g(t)$. The stationary-phase principle says, resembling an inverted (5.13), that

$$\Phi''(f) = 2\pi \frac{V^2(f)}{g^2(t)} \qquad (5.14)$$

If $V(f)$ is defined over the frequency interval $-B/2 \le f \le B/2$, the first derivative $\Phi'(f)$ is obtained by integrating the second derivative,

$$\Phi'(f) = \int_{-B/2}^{f} \Phi''(x) \, dx \qquad (5.15)$$

The group time delay function $T(f)$ is then given by

$$T(f) = -\frac{1}{2\pi}\Phi'(f) \tag{5.16}$$

The instantaneous frequency as a function of time is the inverse of $T(f)$: namely,

$$f(t) = T^{-1}(f) \tag{5.17}$$

A typical qualitative relationship among the three functions $V(f), T(f)$, and $f(t)$ is presented in Fig. 5.13.

Finally, the phase function of the designed signal is obtained from the frequency function

$$\phi(t) = 2\pi \int_0^t f(x)\,dx \tag{5.18}$$

$\phi(t)$ and the predetermined amplitude $g(t)$ are used in (5.10) to generate the complex envelope of the NLFM signal, which according to the stationary-phase principle should have a Fourier spectrum $U(f)$ closely resembling the design target $V(f)$.

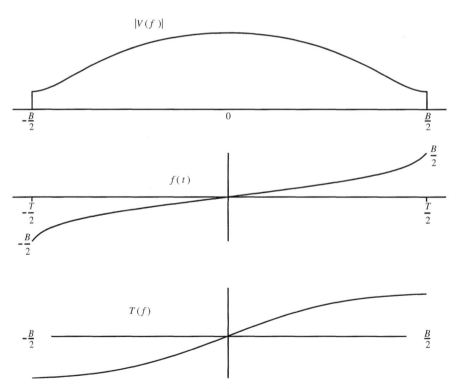

FIGURE 5.13 Transforming a desired spectrum to a NLFM.

NONLINEAR FREQUENCY MODULATION

We will use the approach outlined above to generate a NLFM signal with uniform amplitude and a spectrum following a square root of raised cosine with power n:

$$V(f) = \left[k + (1-k) \cos^n \left(\pi \frac{f}{B} \right) \right]^{1/2} \qquad (5.19)$$

Choosing $k = 0.015$ and $n = 4$, $V(f)$ is the solid line in Fig. 5.14. Following the procedure outlined in (5.14) to (5.18), the instantaneous frequency and phase obtained for a pulse with a time–bandwidth product of 100 appear in Fig. 5.15. Numerical calculation of the resulting spectrum $U(f)$, plotted as a dashed line in Fig. 5.14, reveals rather poor matching to the desired spectrum $V(f)$. Somewhat better agreement will be obtained if we allow the designed spectrum to extend beyond $(-B/2 \leq f \leq B/2)$. Increasing the spectrum boundaries by a factor of 1.3 results in the spectrum shown in Fig. 5.16.

The major discrepancy between the designed and achieved spectrums is the ripple at the middle frequencies. It turns out that the ripple cannot be removed as long as the envelope $g(t)$ remains a constant. The discrepancy causes higher autocorrelation sidelobes than the ideal raised cosine spectrum should yield. Figure 5.17 presents the autocorrelation function (log scale, positive delays),

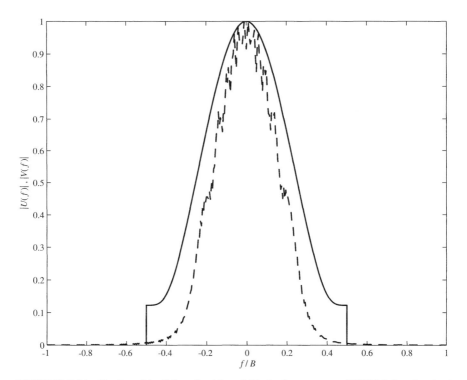

FIGURE 5.14 Designed (solid) and achieved (dashed) spectrum of NLFM signal: raised cosine $n = 4, k = 0.015$, B-factor $= 1$.

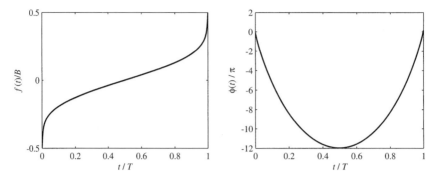

FIGURE 5.15 Instantaneous frequency (left) and phase (right) of the NLFM signal generated: raised cosine $n = 4$, $k = 0.015$, B-factor $= 1$.

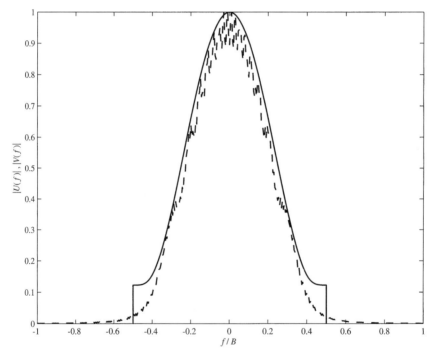

FIGURE 5.16 Designed (solid) and achieved (dashed) spectrum of NLFM signal, allowing spectrum spreading of 30% (B-factor $= 1.3$).

resulting from the signal whose spectrum appears in Fig. 5.16 (dashed) and whose time–bandwidth product is $BT = 130$. With an ideal raised cosine spectrum, the ACF sidelobes would have been below $-60\,\text{dB}$.

Figure 5.18 presents a partial ambiguity function of the NLFM signal whose spectrum appears in Fig. 5.16. The delay axis extends over $|\tau| < T/5$ and the Doppler axis over $0 \leq \nu \leq 7/T$. Comparing the ambiguity function (AF) of

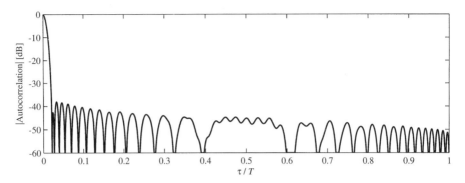

FIGURE 5.17 ACF of the NLFM signal whose spectrum appears in Fig. 5.16.

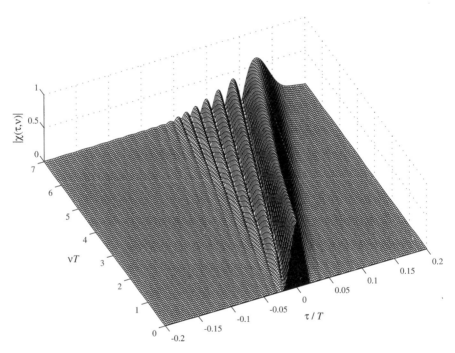

FIGURE 5.18 Partial AF (zoom in delay and Doppler) of the NLFM signal whose spectrum appears in Fig. 5.16.

NLFM (Fig. 5.18) to that of LFM (Fig. 4.15) reveals significant delay sidelobes at higher Doppler cuts. This is another drawback of NLFM. This disintegration of the AF ridge can be explained with the help of Fig. 5.19. On its left side two identical NLFM frequency histories are superimposed with delay and Doppler shift matching the slope at the central frequency. It demonstrates that the alignment holds over a limited duration of the signals. At a different combination

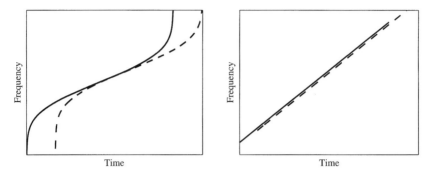

FIGURE 5.19 Alignment of frequency characteristics of two NLFM signals (left) and two LFM signals (right).

of delay and Doppler shift (with different ratio), other parts of the frequency characteristics will match, giving rise to other small AF ridges off the main ridge. In LFM (right-hand side of Fig. 5.19) we seek an alignment of two straight parallel lines, and that would happen only for one specific ratio of delay to Doppler shift—hence the single ridge.

We saw that a procedure based on applying the stationary-phase principle to a desired good spectral shape did not yield exactly that shape, resulting in poorer ACF. This problem was not unique to the raised cosine window. Other windows (e.g., Chebyshev, Kaiser) yielded similar results. This led to an empirical search for other nonlinear frequency characteristics. One example, suggested by Price (1979), combines LFM and NLFM according to

$$f(t) = \frac{t}{T}\left(B_\text{L} + B_\text{C}\frac{1}{\sqrt{1-4t^2/T^2}}\right), \qquad -\frac{T}{2} \leq t \leq \frac{T}{2} \qquad (5.20)$$

where B_L is the total frequency sweep of the LFM part (left term) and B_C is the total frequency sweep that would have been caused by an LFM having the slope of the second term at $t = 0$. To study this signal more easily using our numerical ambiguity function code, we adapt (5.20) to a piecewise NLFM signal where the frequency of the mth chip is f_m, the chip duration is t_b, and there are M chips, implying a pulse duration of $T = Mt_b$. With this notation the mth normalized frequency, taken as the frequency of the continuous NLFM at the center of the bit, is

$$f_m t_b = \frac{2m+1-M}{2M^2}\left[B_\text{L}T + B_\text{C}T\frac{1}{\sqrt{1-(2m+1-M)^2/M^2}}\right] \qquad (5.21)$$

where $m = 0, 1, \ldots, M - 1$.

Note that $B_\text{L}T$ would have been the time–bandwidth product of the pulse if only the linear term existed. Similarly, $B_\text{C}T$ is the time–bandwidth product of the NLFM component. Recall that in LFM the compression ratio equals the time–bandwidth product. The frequency and phase characteristics of a stepped NLFM signal using $M = 50$, $B_L T = 20$, $B_C T = 40$ are shown in Fig 5.20. Note that the signal frequency span obtained from the vertical extent of the frequency characteristic in Fig. 5.20 (left) is approximately 210 and is a function of M. When M is increased, the frequency extent is higher. The continuous NLFM case is obtained with infinitely high M and an infinite frequency span.

The resulting ambiguity function is shown in Fig. 5.21. At higher Doppler shifts, the AF exhibits the same disintegration of the ridge into multiple ridges,

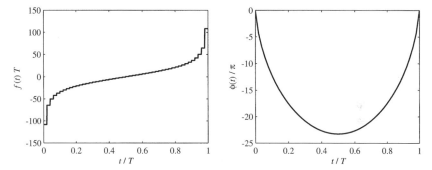

FIGURE 5.20 Phase and frequency characteristic of a combined stepped LFM and NLFM characteristic ($M = 50$ steps, $B_L T = 20$, $B_C T = 40$).

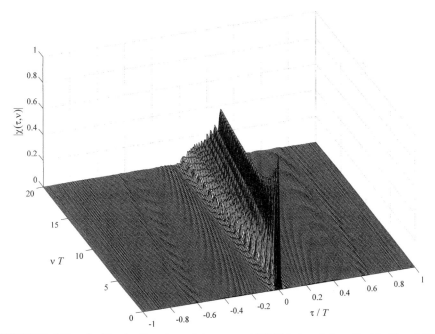

FIGURE 5.21 Ambiguity function of the stepped NLFM signal described in Fig. 5.20.

as also shown in Fig. 5.18. The magnitude of the autocorrelation and the spectrum (one-sided) of the signal are shown in Fig. 5.22. Note from Fig. 5.22 (center) that the first ACF null is at approximately $T/39$, indicating an effective compression ratio of 39, a value between $B_L T$ and $B_C T$. Note also the signal bandwidth extends far more than what could be expected from the compression ratio ($39/T$). This is probably due to the NLFM frequency span, which can also be seen on the left-hand side of Fig. 5.20.

The phase and frequency characteristics of a *continuous* NLFM signal having the same frequency law as the stepped NLFM of Fig. 5.20 is shown in Fig. 5.23. The resulting ACF and spectrum are shown in Fig. 5.24. Note that the first ACF null is at approximately $T/34$, indicating an effective compression ratio of 34, which is still between $B_L T$ and $B_C T$ but lower than the one of the stepped NLFM signal. Note also the much lower sidelobes, having an almost flat pedestal of less than -65 dB (excluding the first sidelobe at -56 dB).

None of the NLFM laws is able to reach outstanding sidelobe reduction, mostly because of the constant-amplitude requirement [$g(t) =$ constant]. A useful rule (Collins and Atkins, 1999) says that the peak range sidelobe level cannot be suppressed below a limit of $-20 \log_{10}(BT) - 3$ dB. Indeed, note that the ACF plotted in Fig. 5.17 was obtained with a $BT = 130$ signal. The expected peak sidelobe level is therefore $-20 \log_{10}(130) - 3 = -39.3$ dB, which agrees with the

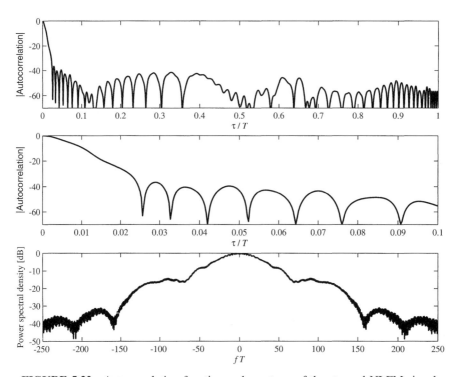

FIGURE 5.22 Autocorrelation function and spectrum of the stepped NLFM signal.

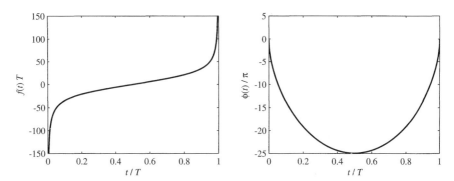

FIGURE 5.23 Phase and frequency characteristic of a combined LFM and NLFM characteristic ($B_L T = 20$, $B_C T = 40$).

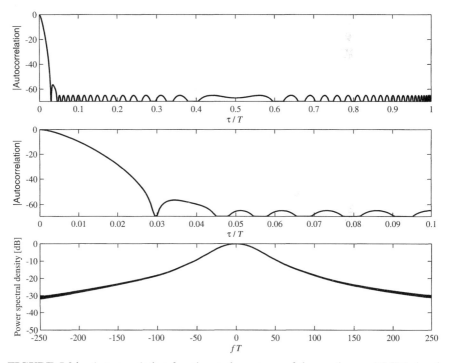

FIGURE 5.24 Autocorrelation function and spectrum of the continuous NLFM signal.

peak sidelobe observed on the ACF plot. Similarly, the ACF in Fig. 5.22 (top) was obtained with a $BT = 210$ signal (see the left-hand side of Fig. 5.20). The expected sidelobe level cannot be lower than $-20\log_{10}(210) - 3 = -43.4$ dB, and indeed, the ACF sidelobe is slightly higher than -40 dB. The ACF sidelobe level of the continuous NLFM signal (-56 dB) corresponds to an effective time–bandwidth product of more than 440.

This limited sidelobe suppression led to a compromise in which both the pulse amplitude $g(t)$ and frequency rate are used to achieve the desired spectral shape and (through the Fourier relation) the desired autocorrelation function (ACF). This approach was termed *hybrid FM*. Johnston and Fairhead (1986) and Collins and Atkins (1999) suggest hybrid systems that also attempt to control the range sidelobes at higher Dopplers.

APPENDIX 5A: MATLAB CODE FOR WELCH CONSTRUCTION OF COSTAS ARRAYS

```
function positions=costas(N)
% calculate the positions/frequencies of a Costas array with size NxN
% for a given N there are N! possible frequency-time arrays (sequences)
% there can be more then one Costas sequence for any value of N.
% the sequences are calculated following Welch construction
%
% type 1 - when N=p-1 and p is an odd prime (N=2,4,6,10,12,16,...)
%
% If j=0,1,2,..,p-2 and i=1,2,3,...,p-1 are the row and column indexes
% then a dot is marked in the Costas array iff i=alfa^j where alfa is
% a primitive root of GF(p)
%
% Type 2 - obtained from type 1 by deleting the first row and first
% column from the Type 1 array. (N=p-2=1,3,5,9,11,15,...)
%
% Type 3 - when alfa=2 (2 is a primitive element of GF(p)). Deleting the two
% first rows and columns produce a Costas array. In this case
%     N=p-3=2,4,8,10,14,...
index=0;
if (isprime(N+1))*(rem(N+1,2)==1),
    % calculate Costas array using Welch construction type 1
    p=N+1;
    % check primitive elements is GF(p)
    isprimitive=[0 ones(1,p-2)];
    e_pow_n=ones(p-1,p-1);
    for e=2:p-1,
       for idx=[2:p-1],
          e_pow_n(e,idx)=mod(e*e_pow_n(e,idx-1),p);

          isprimitive(e)=isprimitive(e)*(e_pow_n(e,idx) =1);
       end
    end
    % loop over all possible primitive elements of GF(p)
    for pr=find(isprimitive==1),
       index=index+1;
       positions(index,1:N)=e_pow_n(pr,1:p-1);
    end
elseif (isprime(N+2))*(rem(N+2,2)==1),
    % calculate Costas array using Welch construction type 2
    p=N+2;
    % check primitive elements is GF(p)
    isprimitive=[0 ones(1,p-2)];
    e_pow_n=ones(p-1,p-1);
    for e=2:p-1,
       for idx=[2:p-1],
          e_pow_n(e,idx)=mod(e*e_pow_n(e,idx-1),p);
```

```
        isprimitive(e)=isprimitive(e)*(e_pow_n(e,idx) =1);
      end
  end
  % loop over all possible primitive elements of GF(p)
  for pr=find(isprimitive==1),
      index=index+1;
      positions(index,1:N)=e_pow_n(pr,2:p-1)-1;
  end
elseif (isprime(N+3))*(rem(N+3,2)==1),
  % calculate Costas array using Welch construction type 3
  p=N+3;
  % check 2 is primitive
  isprimitive=[0 zeros(1,p-2)];
  e_pow_n=ones(p-1,p-1);
  for e=2,
    for idx=[2:p-1],
        e_pow_n(e,idx)=mod(e*e_pow_n(e,idx-1),p);
        isprimitive(e)=isprimitive(e)*(e_pow_n(e,idx) =1);
    end
  end
  for pr=find(isprimitive==1),
      index=index+1;
      positions(index,1:N)=e_pow_n(pr,3:p-1)-2;
  end
end
```

PROBLEMS

5.1 Costas sidelobe matrix

(a) Draw the sidelobe matrix for the Costas code $\{5\ 3\ 2\ 7\ 1\ 8\ 4\ 6\ 9\}$.

(b) Find five additional Costas codes of length $M = 9$ using the one in part (a).

5.2 Number of 1's and 0's in a Costas matrix

Consider a Costas code of order M.

(a) Show that the number of 1's in the sidelobe matrix is $M(M-1)$.

(b) Show that the number of 0's in the sidelobe matrix is $3M(M-1)$.

(c) Show that the total volume in a sidelobe matrix of the Costas code is M^2.

5.3 Sidelobe structure of a Costas matrix

(a) Show that the sidelobe matrix of a Costas signal of order M has $M - |j|$ sidelobes at the jth delay column $j = \pm 1, \pm 2, \ldots, \pm(M-1)$. Demonstrate this with two examples.

(b) Show that the sidelobe matrix of a Costas signal of order M has $M - |j|$ sidelobes at the jth Doppler row $j = \pm 1, \pm 2, \ldots, \pm(M-1)$. Demonstrate this with two examples.

5.4 Welch algorithm

(a) Using the Welch algorithm, find two different Costas sequences of length $M = 6$ (based on two different primitive elements).

(b) Plot the sidelobe matrices and ambiguity functions of the two Costas signals.

5.5 Frequency-coded pulse

(a) Create the sidelobe matrix and plot the ambiguity function (zoom on $|\tau/t_b| < 3$ and $0 < \nu M t_b < 6$) of the frequency-coded signal, of length $M = 40$, given below (not Costas).

1 4 7 10 13 48 45 42 39 36 11 14 17 20 23 38 35 32 29 26 18
15 12 9 6 31 34 37 40 43 21 24 27 30 33 28 25 22 19 16

(b) Comment on the sidelobe distribution. Note that the frequency spacing between neighboring frequency subcarriers is $1/t_b$.

5.6 Frequency-coded pulse

(a) Create the sidelobe matrix and plot the ambiguity function (zoom on $|\tau/t_b| < 3$ and $0 < \nu M t_b < 6$) of the frequency-coded signal of length 10 given by { 1 6 4 3 9 2 8 7 5 10 } (Maric and Titlebaum, 1992).

(b) Comment on the sidelobe distribution. Note that the frequency spacing between neighboring frequency subcarriers is $1/t_b$.

5.7 AF of a Manchester frequency-keyed pulse

Consider a frequency-keyed pulse where the second half of the pulse is frequency shifted relative to the first half. The complex envelope $u(t)$ of the frequency-keyed pulse is given by

$$u(t) = \frac{1}{\sqrt{2T}} \begin{cases} 1, & -T \leq t < 0 \\ \exp(j2\pi \Delta f t), & 0 \leq t \leq T \\ 0, & \text{elsewhere} \end{cases}$$

(a) Calculate the ambiguity function $|\chi(\tau, \nu)|$ of $u(t)$.

(b) For the case $\Delta f = 1/T$, draw the following three cuts: $|\chi(\tau, 0)|$, $|\chi(0, \nu)|$, and $|\chi(T, \nu)|$.

5.8 Costas X-band waveforms

For a Costas signal with the parameters $f_c = 10\,\text{GHz}$, $t_b = 0.4\,\mu\text{s}$, and $M = 40$, find:

(a) The signal bandwidth.
(b) The signal duration.
(c) The range resolution (first null) in meters.
(d) The velocity resolution (first null) in m/s.

5.9 Train of Costas pulses
Eight pulses from Problem 5.8 are processed coherently. The duty cycle is $T/T_r = \frac{1}{10}$.
(a) What is the new velocity resolution (first null) in m/s?
(b) What is the peak level of the first Doppler recurrent lobe $|\chi(0, 1/T_r)|$?

5.10 NLFM pulse
Design a nonlinear FM pulse based on the frequency characteristic in equation (5.21) using $B_L T = 55$, $B_C T = 18$, and $M = 200$. Plot the autocorrelation function (log scale) and the ambiguity function (linear scale).

5.11 V-shaped FM pulse
Plot the ambiguity function of an FM pulse with a V-shaped frequency characteristic and a total time–bandwidth product of 100.

REFERENCES

Collins, T., and P. Atkins, Non-linear frequency modulation chirp for active sonar, *IEE Proceedings: Radar, Sonar and Navigation*, vol. 146, no. 6, December 1999, pp. 312–316.

Cook, C. E., and M. Bernfeld, *Radar Signals: An Introduction to Theory and Application*, Academic Press, New York, 1967.

Costas, J. P., A study of a class of detection waveforms having nearly ideal range–Doppler ambiguity properties, *Proceedings of the IEEE*, vol. 72, no. 8, August 1984, pp. 996–1009.

Fowle, E. N., The design of FM pulse-compression signals, *IEEE Transactions on Information Theory*, vol. IT-10, no. 1, January 1964, pp. 61–67.

Golomb, S. W., and H. Taylor, Constructions and properties of Costas arrays, *Proceedings of the IEEE*, vol. 72, no. 9, September 1984, pp. 1143–1163.

Johnston, J. A., and A. C. Fairhead, Waveform design and Doppler sensitivity analysis for nonlinear FM chirp pulses," *IEE Proceedings*, vol. 133, pt. F, no. 2, April 1986, pp. 163–175.

Maric, S. V., and E. L. Titlebaum, A class of frequency hop codes with nearly ideal characteristics for use in multiple-access spread-spectrum communications and radar and sonar systems, *IEEE Transactions on Communications*, vol. 40, no. 9, September 1992, pp. 1442–1447.

Price, R., Chebyshev low pulse compression sidelobes via a nonlinear FM, National Radio Science Meeting of URSI, Seattle, WA, June 18, 1979.

Silverman, J., V. E. Vickers, and J. M. Mooney, On the number of Costas arrays as a function of array size, *Proceedings of the IEEE*, vol. 76, no. 7, July 1988, pp. 851–853.

6

PHASE-CODED PULSE

One of the early methods for pulse compression is by phase coding. We start from a pulse of duration T. The pulse is divided into M bits of identical duration $t_b = T/M$, and each bit is assigned (or *coded*) with a different phase value. The complex envelope of the phase-coded pulse is given by

$$u(t) = \frac{1}{\sqrt{T}} \sum_{m=1}^{M} u_m \text{rect}\left[\frac{t - (m-1)t_b}{t_b}\right] \quad (6.1)$$

where $u_m = \exp(j\phi_m)$ and the set of M phases $\{\phi_1, \phi_2, \ldots, \phi_M\}$ is the *phase code* associated with $u(t)$.

Finding optimal sets of M phases (or codes) for different radar application has kept radar engineers busy from the early days of radar. The number of possibilities of generating phase codes of length M is unlimited. The criteria for selecting a specific code are the resolution properties of the resulting waveform (shape of the ambiguity function), frequency spectrum, and the ease with which the system can be implemented. Sometimes the design is even more complicated by using different phase codes for the transmitted pulse and the reference pulse used at the receiver (possibly even with different lengths). This can improve resolution at the expense of a suboptimal signal-to-noise ratio.

The problem of finding a code that leads to a predetermined range–Doppler resolution is very complicated. A more manageable problem is finding a code with a good correlation function rather than an ambiguity function. The correlation function of a phase-coded pulse is a continuous function of the delay τ. In general,

---

Radar Signals, By Nadav Levanon and Eli Mozeson
ISBN 0-471-47378-2   Copyright © 2004 John Wiley & Sons, Inc.

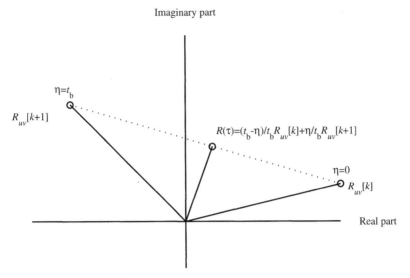

FIGURE 6.1 Interpolation in the complex plane.

when examining the properties of the correlation function, one should look at the correlation function for all $-T < \tau < T$. However, as we show in Box 6A, it is sufficient to calculate the correlation function at integer multiples of the bit duration. The values in between are then obtained by connecting the values at $\tau = nt_b$ using straight lines in the complex plane. It is easy to see using Fig. 6.1 that the interpolation in the complex plane results with concave sections of the correlation function magnitude (i.e., the autocorrelation peaks are obtained at an integer multiples of the bit duration). Thus, minimizing the peak value, or the area (integral) under the *continuous* correlation function $|R(\tau)|$, is simplified to finding the peak value and sum of the *discrete* correlation function $|R_k|$.

BOX 6A: Aperiodic Correlation Function of a Phase-Coded Pulse

Consider a transmitted pulse $u(t)$ with M_u phase elements defined by u_m ($1 \leq m \leq M_u$) and a reference pulse $v(t)$ with M_v elements defined by v_n ($1 \leq n \leq M_v$). The cross-correlation function of the two phase-coded pulses is defined by

$$R_{uv}(\tau) = \int_{-\infty}^{\infty} u(t) v^*(t + \tau) \, d\tau$$

$$= \int_{-\infty}^{\infty} \frac{1}{\sqrt{M_u t_b}} \sum_{m=1}^{M_u} u_m \text{rect}\left[\frac{t - (m-1) t_b}{t_b}\right] \frac{1}{\sqrt{M_v t_b}}$$

$$\sum_{n=1}^{M_v} v_n^* \text{ rect}\left[\frac{t+\tau-(n-1)t_b}{t_b}\right] d\tau \qquad (6A.1)$$

$$= \frac{1}{t_b\sqrt{M_u M_v}} \sum_{m=1}^{M_u} \sum_{n=1}^{M_v} u_m v_n^*$$

$$\int_{-\infty}^{\infty} \text{rect}\left[\frac{t-(m-1)t_b}{t_b}\right] \text{rect}\left[\frac{t+\tau-(n-1)t_b}{t_b}\right] d\tau$$

Since

$$\int_{-\infty}^{\infty} \text{rect}\left[\frac{t-(m-1)t_b}{t_b}\right] \text{rect}\left[\frac{t+\tau-(n-1)t_b}{t_b}\right] d\tau$$

$$= \begin{cases} t_b - |\tau-(n-m)t_b|, & |\tau-(n-m)t_b| < t_b \\ 0, & |\tau-(n-m)t_b| \geq t_b \end{cases}$$

the integral in (6A.1) represents a triangle having a peak of t_b and centered at $(n-m)t_b$ with a base width of $2t_b$. If we write the delay as $\tau = kt_b + \eta$, where $0 \leq \eta < t_b$ and k is an integer, the integral will be nonzero only for $-t_b < kt_b + \eta - (n-m)t_b < t_b$. Since $0 \leq \eta < t_b$, we can solve for n as a function of m and k and find that when $\eta > 0$, the integral is nonzero only for $n_1 = k+m$ and $n_2 = k+m+1$. When $\eta = 0$, the integral is nonzero only for $n = k+m$. Using n_1 and n_2 in (6A.1), we can write the correlation function as

$$R_{uv}(kt_b + \eta) = \frac{1}{t_b\sqrt{M_u M_v}} \left[(t_b - \eta) \sum_{\max(1,1-k)}^{\min(M_u, M_v-k)} u_m v_{m+k}^* \right.$$

$$\left. + \eta \sum_{\max(1,-k)}^{\min(M_u, M_v-k-1)} u_m v_{m+k+1}^* \right] \qquad (6A.2)$$

Note that the single sums in (6A.2) are now the discrete aperiodic cross-correlations between u_m and v_m evaluated at $\tau = k$ (first sum) and $\tau = k+1$ (second sum). Thus,

$$R_{uv}(\tau) = R_{uv}(kt_b + \eta) = \frac{1}{t_b\sqrt{M_u M_v}}[(t_b - \eta)R_{uv}[k] + \eta R_{uv}[k+1]] \qquad (6A.3)$$

where $R_{uv}[k]$ is the discrete aperiodic cross-correlation function of sequences u and v evaluated at $\tau = k$. Note that the interpolation is done in the complex plane (see Fig. 6.1).

PHASE-CODED PULSE

The discrete cross-correlation of two sequences can be calculated using a procedure demonstrated below. To simplify the example, sequences with binary (± 1) values are used; the same procedure can be implemented on complex-valued sequences. Consider the phase-coded pulse defined by the sequence $\{u_m\} = \{-1\ 1\ -1\ -1\ 1\ 1\ -1\ 1\ -1\}$ and the reference phase-coded pulse $\{v_m\} = \{-1\ 1\ -1\ -1\ 1\ 1\ -1\ -1\ -1\}$ with the same number of elements $M_u = M_v = 9$. To calculate the cross-correlation we multiply the conjugate of the reference signal by the elements of the transmitted sequence and add the shifted sequences as demonstrated in Table 6.1.

Figure 6.2 shows the full (continuous) cross-correlation function of $u(t)$ and $v(t)$ obtained by connecting the discrete values of R_{uv} using straight lines (bottom). The figure also shows the autocorrelation function where no phase coding is used (dashed) and the autocorrelation function of $u(t)$ (top). Note how phase coding lowers the mainlobe width. Note also that the autocorrelation of $u(t)$ is symmetric, while the cross-correlation of $u(t)$ and $v(t)$ is not symmetric.

Different phase codes that yield identical aperiodic autocorrelation function magnitude are called *equivalent*. Using the four properties of the cross-correlation function described in Box 6B, it is easy to see that the following operations on a phase code u_m give equivalent phase codes:

1. A reversal transformation: $\hat{u}_m = u_{M-m}$.
2. A conjugate transformation: $\hat{u}_m = u_m^*$.
3. A constant multiplication transformation: $\hat{u}_m = \eta u_m$, where $|\eta| = 1$.
4. A progressive multiplication transformation: $\hat{u}_m = \rho^m u_m$, where $|\rho| = 1$.

It is also easy to show that every polyphase code is equivalent to one that begins with $\phi_1 = 0$, $\phi_2 = 0$, ϕ_3, where $0 \leq \phi_3 \leq \pi$. This form of the polyphase code is known as the *normalized form*.

TABLE 6.1 Construction Table of the Discrete Cross-Correlation

$\{v_m^*\}$									−1	1	−1	−1	1	1	−1	−1	−1
$\{u_m\}$																	
−1									1	−1	1	1	−1	−1	1	1	1
1								−1	1	−1	−1	1	1	−1	−1	−1	
−1							1	−1	1	1	−1	−1	1	1	1		
−1						1	−1	1	1	−1	−1	1	1	1			
1					−1	1	−1	−1	1	1	−1	−1	−1				
1				−1	1	−1	−1	1	1	−1	−1	−1					
−1			1	−1	1	1	−1	−1	1	1	1						
1		−1	1	−1	−1	1	1	−1	−1	−1							
−1	1	−1	1	1	−1	−1	1	1	1								
$\{R_{uv}\}$	1	−2	3	−2	−1	2	−1	−2	7	−2	−3	0	1	0	1	0	1

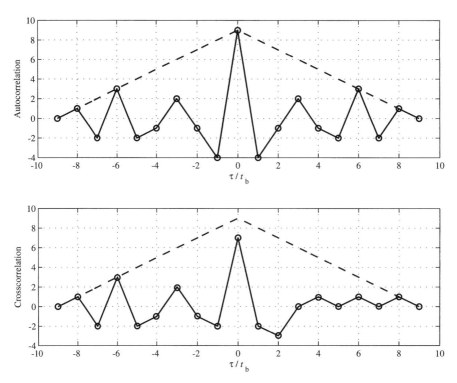

FIGURE 6.2 Autocorrelation function (top) and cross-correlation (bottom) of a phase-coded pulse.

BOX 6B: Properties of the Cross-Correlation Function of a Phase Code

Property I: *Reversal transformation*: The cross-correlation of reversed-order sequences is the reversed-order cross-correlation of the original sequences:

$$R_{\hat{u}\hat{v}}[k] = \sum_m \hat{u}_m \hat{v}^*_{m+k} = \sum_m u_{M-m} v^*_{M-m-k} = \sum_{m'} u_{m'} v^*_{m'-k}$$

$$= R_{uv}[-k] = R^*_{uv}[k]$$

Note that due to symmetry property of the autocorrelation the autocorrelation of the reversed-order sequence equals the conjugate of the original sequence autocorrelation.

Property II: *Conjugation transformation.* The cross-correlation of conjugated sequences is the conjugated cross-correlation of the original sequences:

$$R_{\hat{u}\hat{v}}[k] = \sum_m \hat{u}_m \hat{v}^*_{m+k} = \sum_m u^*_m v_{m+k} = R^*_{uv}[k]$$

Property III: *Constant multiplication transformation.* The cross-correlation of sequences multiplied by a constant is the multiplied cross-correlation of the original sequences:

$$R_{\hat{u}\hat{v}}[k] = \sum_m \hat{u}_m \hat{v}^*_{m+k} = \sum_m \eta_u u_m \eta_v^* v^*_{m+k} = \eta_u \eta_v^* \sum_m u_m v^*_{m+k} = \eta_u \eta_v^* R_{uv}[k]$$

Property IV: *Progressive multiplication transformation.* The cross-correlation of progressively multiplied sequences is the progressively multiplied cross-correlation of the original sequences:

$$R_{\hat{u}\hat{v}}[k] = \sum_m \hat{u}_m \hat{v}^*_{m+k} = \sum_m \rho_u^m u_m \left(\rho_v^{m+k}\right)^* v^*_{m+k} = \left(\rho_v^k\right)^* \sum_m \left(\rho_u \rho_v^*\right)^m u_m v^*_{m+k}$$

$$= \left(\rho^k\right)^* \sum_m u_m v^*_{m+k} = \left(\rho^k\right)^* R_{uv}[k]$$

where we used $\rho_u = \rho_v = \rho$.

In some applications the resolution properties are determined by *periodic* correlation of the phase-coded pulse. In addition to the previous transformations, the following transformations preserve the periodic autocorrelation function (the aperiodic correlation function is not necessarily preserved):

5. Cyclically shifting the sequence: $\hat{u}_m = u_{m+a \bmod M}$.
6. Decimating the sequence by d, which is relatively prime to the sequence length M results with decimating the periodic autocorrelation function: $\hat{u}_m = u_{md \bmod M}$. Note that some authors refer to the codes obtained by decimation as *permutation codes*.

Although it is not possible to design phase-coded pulses with zero aperiodic correlation sidelobes, the periodic correlation of a phase-coded signal can be zero for all nonzero shifts. Phase codes having zero periodic autocorrelation sidelobes are called *perfect* codes. One method for designing low aperiodic autocorrelation codes is to start with a perfect code. Transformations (5) and (6) are used to find optimal aperiodic autocorrelation function codes while keeping the code perfect. Note that for perfect codes the aperiodic correlation function has sidelobes symmetric around $\tau = Mt_b/2$ [i.e., $R_{uv}(\tau) = -R_{uv}(M - \tau)$ for $t_b \leq |\tau| \leq Mt_b$].

6.1 BARKER CODES

Probably the most famous family of phase codes is named *Barker*, after its inventor (Barker, 1953). Originally, the Barker codes were designed as the sets

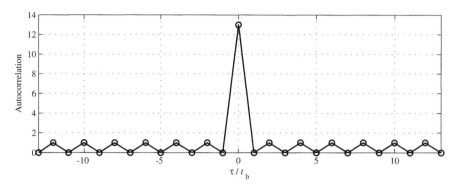

FIGURE 6.3 Autocorrelation function of phase-coded pulse using the 13-element Barker code.

TABLE 6.2 All Known Binary Barker Codes

Code Length	Code
2	11 or 10
3	110
4	1110 or 1101
5	11101
7	1110010
11	11100010010
13	1111100110101

of M binary phases yielding a peak-to-peak sidelobe ratio of M. For example, the autocorrelation function of the $M = 13$-element Barker code is shown in Fig. 6.3. All known binary sequences yielding a peak-to-peak sidelobe ratio of M were reported by Barker (1953) and Turyn (1963) and are given in Table 6.2. Although it was only proved that no binary Barker codes exist for $13 < M < 1,898,884$ and that no binary Barker codes exist for all odd $M > 13$ (Turyn, 1963; Eliahou and Kervaire, 1992), it is common belief that no Barker codes exist for all $M > 13$. In Barker and other phase-coded signals, the instantaneous phase switching causes extended spectral sidelobes. The effect of a phase-switching slope on bandwidth and autocorrelation function is discussed in Section 6.8. A Barker signal is used as an example.

6.1.1 Minimum Peak Sidelobe Codes

Binary codes that yield minimum peak sidelobes but do not meet the Barker condition (i.e., the peak-to-peak sidelobe ratio is less than M) are often called *minimum peak sidelobe* (MPS) codes. Finding MPS codes involves exhaustive computer search. Results up to $M \leq 69$ were reported by Lindner (1975), Cohen et al. (1989, 1990) and Coxson et al. (2001). Codes with a peak sidelobe of 2 were reported for $M \leq 28$. The MPS codes reported for $28 < M \leq 48$ and $M = 51$ have a sidelobe level of 3, and the MPS codes of length $M = 50$ and $52 \leq M \leq 69$ have a sidelobe level of 4. Table 6.3 gives a single MPS code for each length up to 69. The table also gives the peak sidelobe level. For $M \leq 48$ the listed codes are those that have, from all those with minimum peak sidelobe, the minimum integrated sidelobe.

It seems that for any peak sidelobe level there is a limit of the maximal value of M for which a binary sequence with that sidelobe level exists (i.e., $M = 13$ for peak sidelobe level of 1, $M = 28$ for peak sidelobe level of 2, $M = 51$ for peak sidelobe level of 3, etc.). The main advantage of the binary Barker and minimum peak sidelobe codes is in their relatively low complexity (no multiplications are needed at the receiver). The main drawback is that such codes are only known for limited values of M, and finding additional codes with the *minimum* peak sidelobe for any given length M involves an exhaustive search with a size growing exponentially with M. Some methods for finding potentially good suboptimal peak sidelobe codes were reported (e.g., Schroeder, 1970), and some good codes (not necessarily optimal) were reported for higher values of M [e.g. $M = 51, 69$, and 88 by Kerdock et al. (1986), and for $M = 53, 59, 61, 67, 71, 73, 79, 83, 89, 97, 101, 103, 107, 109$, and 113 by Boehmer (1967)].

6.1.2 Nested Codes

One early method of generating low peak sidelobe sequences for large values of M is by nesting codes of shorter length. We demonstrate the concept of nested codes by forming a 39-element code using the three-element Barker code ($\{v_m\} = \{1\ 1\ -1\}$, $M_v = 3$) and the 13-element Barker code ($\{u_m\} = \{1\ 1\ 1\ 1\ 1\ -1\ -1\ 1\ 1\ -1\ 1\ -1\ 1\}$, $M_u = 13$). There are two forms of generating elements of the nested code. The first is given by $u \otimes v$ and the second is given by $v \otimes u$, where $\otimes$ is the Kronecker product. When the code is formed by $v \otimes u$, u is called the *inner code* and v is referred to as the *outer code*. The nested codes resulting from the two forms defined above are given by

$$v \otimes u = \{\boxed{1\ 1\ 1\ 1\ -1\ -1\ 1\ 1\ -1\ 1\ -1\ 1}\ \boxed{1\ 1\ 1\ 1\ -1\ -1\ 1\ 1\ -1\ 1\ -1\ 1}$$
$$\boxed{-1\ -1\ -1\ -1\ -1\ 1\ 1\ -1\ -1\ 1\ -1\ 1\ -1}\}$$

and

$$u \otimes v = \{\boxed{1\ 1\ -1}\ \boxed{1\ 1\ -1}\ \boxed{1\ 1\ -1}\ \boxed{1\ 1\ -1}\ \boxed{1\ 1\ -1}\ \boxed{-1\ -1\ 1}\ \boxed{-1\ -1\ 1}\ \boxed{1\ 1\ -1}\ \boxed{1\ 1\ -1}$$
$$\boxed{-1\ -1\ 1}\ \boxed{1\ 1\ -1}\ \boxed{-1\ -1\ 1}\ \boxed{1\ 1\ -1}\}$$

TABLE 6.3 Minimum Peak Sidelobe Codes[a]

M	PSL	Sample Code
6	2	110100
8	2	10010111
9	2	011010111
10	2	0101100111
12	2	100101110111
14	2	01010010000011
15	2	001100000101011
16	2	0110100001110111
17	2	00111011101001011
18	2	011001000011110101
19	2	1011011101110001111
20	2	01010001100000011011
21	2	101101011101110000011
22	3	0011001101101010111111
23	3	01110001111110101001001
24	3	011001001010111111100011
25	2	1001001010100000011100111
26	3	10001110000000101011011001
27	3	010010110111011101110000111
28	2	1000111100010001000100101101
29	3	10110010010101000000011100111
30	3	100011000101010010010000001111
31	3	0101010010010011000110000001111
32	3	00000001111001011010101011001100
33	3	011001100101010100101100001111111
34	3	1100110011111111100001101001010101
35	3	00000000111100101101010101100110011
36	3	001100110001010010100000100000111110
37	3	0010101110100001001110110111110011110
38	3	00000000111100001101001010101001100110
39	3	001001100110101000010111110111100111100
40	3	0010001000100011110111000011101001011010
41	3	00011100011101010010100100000001101100100
42	3	000100010001000111101110000111010010110100
43	3	0000000010110110010101011001100111000011100
44	3	00001111111011001110110010110010101011010111
45	3	000101010111100001100110001101101101111110110
46	3	0000111100000011001111011110110110010101010110
47	3	00001101001101001111110100000101000110011000011000
48	3	000101010110101101100001111001100100111111110011
49	4	0000100101010101111101100011110011110010001101111
50	4	00001001011000011000111010101111000010011001101111
51	3	000111000111111100010001100100010010101001001001011
52	4	0000100101000101101011100000111100110010010001101111
53	4	00001001100101010101001111111100011010010011000110011
54	4	000010011001101010001010000000100101100111100011011111
55	4	0000100110000100110101010100001111000110010010001101111
56	4	00001001100110111010101010010110100011110111100011011111

TABLE 6.3 (*continued*)

M	PSL	Sample Code
57	4	000010010011010001010100011101101011000100011110001101111
58	4	0000100011110011100101010001011100100111101101011001101111
59	4	00001001001110100111000000100101000101000011101110001101111
60	4	000010101011000110111110000110010010111001100100101111
61	4	0000001011011010001001100010011000111100111101010001101010000
62	4	00000000101101011001100110001101001100101100000111010001010000
63	4	000010011001111010110100010010001110001011001010111110001101111
64	4	0100000010010000101000101110100111100110001100100011011111000010
65	4	00000001011011100000010110000110110011011110011100101010001010000
66	4	000000011010011011010001010100011100111001111100010010101101000010
67	4	0100000010100000110110010011010101100011110100100001001110011000010
68	4	00000000100111100100100111100011011001100010101010001110101001010000
69	4	000100110111111011011000010011010100000111010000100011000111000010101

a Barker codes are excluded.

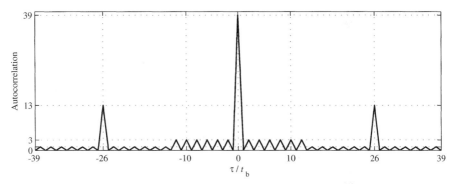

FIGURE 6.4 Autocorrelation function of the nested Barker 13 and Barker 3 code.

Figure 6.4 shows the autocorrelation magnitude of the nested code formed by $v \otimes u$. The resulting 39-element code autocorrelation function has two high peak sidelobes with a level of 13, and the remainder of the sidelobes have a peak level of 3. Note that excluding the two sidelobes at $\pm 26t_b$, the autocorrelation function has a peak sidelobe level identical to that of the 39-element minimal peak sidelobe code.

6.1.3 Polyphase Barker Codes

Allowing any phase values (nonbinary) can lead to lower sidelobes. However, the outermost sidelobe is always 1 (for any polyphase or binary code). The *polyphase* sequence with minimal peak-to-sidelobe ratio excluding the outermost sidelobe are called *generalized Barker sequence* or *polyphase Barker*. Systematic methods to construct polyphase Barker sequences were not yet found. Searches for generalized Barker sequences with no restriction on the values of the sequence phases have been carried out using numerical optimization techniques. Examples of such sequences were found for all $M \leq 45$ (Bomer and Antweiler, 1989; Friese and Zottmann, 1994; Friese, 1996). Table 6.4 gives the normalized form (first two phases are zeroed) of such polyphase Barker sequences for $M \leq 45$.

TABLE 6.4 Normalized Form of Some Polyphase Barker Codes with Large Alphabet[a]

M	PSL	Phase Values (deg)
4	0.50	104.52, 313.47
5	0.77	73.04, 225.31, 90.62
6	1.00	60.00, 180.00, 0.00, 240.00
7	0.53	106.48, 93.06, 316.72, 60.61, 270.86
8	0.66	72.33, 28.48, 294.09, 151.07, 250.77, 62.87
9	0.11	53.57, 42.22, 270.79, 215.59, 41.51, 161.92, 335.47
10	0.83	56.97, 127.04, 137.24, 12.74, 6.67, 224.63, 19.27, 233.50
11	0.89	34.17, 259.06, 266.63, 327.97, 158.47, 13.78, 22.74, 221.64, 94.65
12	0.91	104.89, 163.15, 171.04, 344.57, 241.31, 185.77, 282.58, 147.97, 209.41, 79.19
13	0.72	115.84, 114.84, 248.44, 213.38, 123.12, 154.90, 140.20, 12.75, 149.65, 303.48, 121.65
14	0.97	66.96, 133.73, 202.45, 100.74, 37.89, 236.27, 167.69, 86.72, 169.45, 34.20, 143.95, 14.33
15	0.80	17.81, 5.51, 5.37, 142.33, 211.98, 297.96, 123.75, 91.46, 1.09, 205.83, 314.02, 156.28, 23.66
16	0.93	26.46, 38.51, 97.32, 49.41, 305.85, 286.47, 197.00, 65.76, 241.32, 137.61, 319.19, 47.96, 178.58, 303.06
17	0.73	5.34, 18.49, 278.38, 307.59, 67.22, 148.91, 207.37, 70.46, 300.98, 282.64, 137.11, 6.31, 120.23, 327.59, 185.74
18	0.87	62.21, 45.41, 315.79, 282.94, 23.75, 37.27, 205.01, 186.58, 83.91, 155.50, 317.90, 337.25, 204.02, 11.17, 171.35, 31.96
19	0.96	54.66, 27.60, 91.21, 80.61, 235.94, 11.00, 333.09, 100.64, 241.78, 319.89, 162.37, 309.54, 162.20, 139.49, 33.30, 341.49, 218.99
20	0.98	99.16, 125.86, 232.99, 251.37, 133.93, 144.09, 354.74, 304.39, 192.21, 302.68, 219.51, 161.35, 283.77, 145.40, 250.28, 106.25, 228.47, 107.05
21	0.97	15.27, 30.14, 161.48, 203.92, 220.08, 190.88, 61.27, 126.36, 221.20, 340.41, 168.52, 153.07, 26.83, 255.87, 120.93, 209.07, 54.58, 239.74, 105.94
22	0.97	23.70, 53.81, 80.91, 74.08, 349.91, 264.11, 313.80, 245.26, 146.83, 73.98, 284.18, 159.95, 334.41, 77.11, 315.33, 145.77, 245.13, 343.88, 84.12, 206.36
23	0.91	7.39, 275.98, 286.43, 253.92, 256.71, 351.73, 58.40, 60.24, 226.34, 353.15, 100.57, 168.65, 41.05, 208.60, 347.90, 219.32, 126.05, 349.85, 315.45, 182.27, 56.51
24	0.99	5.05, 316.44, 257.26, 216.68, 202.65, 319.31, 311.49, 357.35, 297.32, 111.75, 36.71, 281.47, 137.65, 10.87, 116.56, 260.04, 135.27, 269.02, 29.14, 143.46, 209.62, 335.07

TABLE 6.4 (*continued*)

M	PSL	Phase Values (deg)
25	0.93	81.86, 64.96, 316.26, 273.05, 326.28, 339.78, 62.69, 18.77, 270.52, 198.01, 98.77, 126.61, 206.47, 350.69, 105.91, 270.80, 295.42, 162.27, 334.2400, 155.49, 339.81, 147.69, 4.40
26	0.88	51.32, 117.02, 138.11, 265.30, 266.83, 175.22, 117.20, 259.95, 199.68, 135.78, 153.84, 178.61, 75.37, 340.56, 186.87, 306.44, 193.87, 91.91, 189.55, 16.52, 109.34, 249.58, 37.87, 198.94
27	0.98	10.51, 21.75, 28.59, 324.46, 308.15, 280.33, 117.83, 98.81, 111.75, 284.00, 199.23, 313.20, 115.64, 326.00, 184.04, 52.60, 7.95, 193., 96.11, 239.93, 334.30, 101.85, 227.48, 330.80, 91.72
28	0.95	46.92, 84.28, 166.29, 145.70, 199.79, 105.11, 116.57, 58.71, 109.70, 325.89, 24.31, 189.90, 21.40, 196.18, 58.82, 326.50, 129.18, 258.98, 306.73, 123.51, 111.19, 312.71, 298.50, 173.83, 97.89, 327.81
29	0.87	6.99, 318.38, 240.11, 264.97, 239.45, 160.78, 302.00, 327.98, 19.20, 320.43, 85.58, 109.43, 224.94, 7.08, 32.32, 185.08, 168.89, 91.05, 326.61, 228.81, 146.73, 331.28, 93.10, 265.24, 95.64, 254.51, 61.3100
30	1.00	33.11, 34.66, 33.75, 11.97, 300.21, 281.63, 26.64, 54.40, 155.79, 212.11, 231.81, 134.65, 76.33, 318.00, 276.10, 67.92, 299.31, 184.97, 73.01, 154.21, 7.02, 263.07, 94.57, 243.28, 359.63, 150.24, 306.92, 72.0800
31	0.94	28.39, 117.68, 165.05, 236.45, 308.63, 304.92, 236.37, 216.25, 327.24, 279.34, 211.11, 246.98, 191.79, 95.21, 16.74, 272.73, 52.50, 330.82, 223.69, 303.38, 146.86, 21.37, 245.23, 28.92, 145.10, 296.68, 61.96, 190.35, 7.32
32	0.99	13.43, 16.18, 90.15, 109.65, 94.43, 59.91, 332.27, 306.11, 288.10, 280.54, 83.80, 162.77, 247.23, 333.62, 169.95, 72.93, 62.09, 219.53, 296.08, 107.98, 34.71, 270.22, 177.13, 16.92, 176.19, 285.32, 79.75, 289.11, 130.24, 326.30
33	0.97	143.07, 153.71, 339.20, 332.87, 180.92, 134.16, 19.42, 109.38, 166.70, 217.34, 226.71, 228.70, 319.75, 239.69, 185.92, 227.47, 143.26, 115.36, 76.97, 38.11, 187.58, 329.27, 228.90, 110.75, 304.75, 119.00, 275.49, 352.86, 190.92, 359.70, 167.67
34	0.96	11.60, 2.29, 308.93, 247.04, 202.65, 186.51, 234.77, 296.22, 303.84, 351.93, 49.89, 232.02, 253.31, 62.16, 340.45, 6.75, 133.71, 256.94, 76.56, 98.87, 323.47, 230.03, 65.62, 125.68, 248.19, 68.60, 297.53, 137.95, 284.29, 138.69, 17.64, 229.39
35	1.00	93.27, 65.44, 166.46, 132.46, 344.16, 279.49, 337.71, 301.42, 197.69, 56.27, 36.90, 9.33, 325.94, 334.44, 24.54, 157.98, 291.27, 301.26, 148.55, 113.09, 141.50, 296.81, 128.97, 125.66, 341.59, 130.11, 244.84, 74.10, 321.76, 157.86, 301.00, 107.79, 254.69
36	0.93	81.76, 117.41, 228.30, 227.39, 58.27, 59.76, 153.37, 108.01, 19.41, 233.18, 211.51, 260.82, 235.13, 195.44, 219.49, 114.51, 10.64, 224.56, 176.47, 119.93, 124.89, 74.97, 263.65, 112.81, 254.57, 106.85, 318.27, 98.35, 264.66, 28.37, 121.69, 244.37, 57.87, 183.84

(*continued overleaf*)

TABLE 6.4 (*continued*)

M	PSL	Phase Values (deg)
37	0.91	66.51, 90.83, 123.03, 235.61, 235.65, 325.40, 14.62, 279.47, 224.10, 332.97, 5.58, 280.26, 202.90, 334.58, 51.37, 330.14, 260.07, 293.54, 159.60, 150.16, 83.19, 356.98, 124.60, 25.28, 314.51, 129.47, 362.27, 242.00, 50.51, 181.69, 315.88, 111.51, 189.30, 363.49, 148.70
38	0.93	39.45, 95.89, 111.27, 227.27, 253.77, 283.15, 336.93, 233.52, 209.97, 277.20, 332.82, 222.51, 179.24, 290.76, 55.61, 218.04, 37.83, 286.84, 5.30, 165.25, 337.35, 16.57, 244.77, 288.43, 167.97, 127.73, 27.45, 27.16, 227.00, 42.48, 302.45, 185.14, 344.71, 204.83, 108.78, 344.40
39	0.95	39.36, 90.73, 83.15, 186.46, 246.21, 230.04, 301.71, 217.97, 269.86, 173.34, 233.32, 192.78, 308.41, 100.50, 196.44, 160.18, 327.80, 351.55, 180.86, 191.47, 358.49, 310.55, 218.60, 202.53, 77.73, 182.71, 1.76, 41.73, 260.78, 280.77, 129.13, 276.37, 132.96, 27.83, 252.61, 107.29, 331.78
40	0.94	28.66, 24.61, 46.23, 58.97, 95.28, 76.31, 337.40, 310.91, 258.76, 219.91, 49.64, 349.15, 197.79, 163.95, 216.11, 88.58, 44.59, 234.28, 23.36, 356.16, 201.31, 173.80, 252.90, 323.46, 104.99, 101.31, 318.13, 154.56, 259.66, 340.54, 85.72, 206.06, 66.94, 271.36, 80.67, 244.81, 34.78, 195.49
41	0.96	53.61, 63.59, 29.40, 3.64, 3.67, 24.31, 97.10, 114.90, 222.94, 238.18, 334.92, 304.48, 206.26, 143.67, 85.23, 80.54, 2.52, 260.82, 159.64, 309.06, 305.94, 123.34, 150.84, 297.45, 91.30, 82.12, 238.93, 349.39, 208.29, 70.77, 313.76, 167.10, 317.19, 92.70, 218.75, 23.80, 189.12, 357.01, 162.08
42	0.98	9.25, 39.87, 37.38, 136.87, 176.61, 191.79, 228.64, 85.22, 82.51, 113.02, 137.51, 57.10, 333.65, 70.13, 160.69, 364.86, 182.75, 360.18, 87.82, 232.68, 46.84, 358.94, 299.83, 212.66, 181.03, 31.53, 56.10, 278.54, 254.91, 30.65, 172.09, 256.54, 83.35, 26.66, 213.41, 372.71, 223.50, 100.92, 311.27, 161.96
43	0.97	47.01, 73.36, 91.12, 60.44, 51.29, 68.66, 91.88, 190.03, 246.45, 348.16, 17.39, 76.59, 328.08, 249.54, 161.54, 89.18, 261.26, 345.80, 359.22, 212.45, 176.39, 15.34, 172.90, 210.41, 75.23, 80.12, 1.44, 291.79, 72.16, 254.21, 310.20, 160.38, 282.71, 102.22, 284.57, 149.04, 61.46, 249.48, 45.27, 215.80, 37.36
44	0.97	30.29, 9.38, 42.55, 57.42, 313.42, 307.10, 351.41, 283.15, 48.75, 7.98, 190.94, 118.23, 120.89, 164.72, 313.19, 253.16, 133.43, 355.46, 46.04, 280.15, 247.72, 331.73, 137.00, 67.75, 157.72, 82.76, 298.75, 316.97, 191.27, 95.02, 233.35, 248.21, 50.51,, 59.16, 221.70, 353.62, 92.75, 284.48, 137.54, 257.20, 38.80, 165.29
45	0.95	28.88, 5.41, 305.70, 287.05, 307.70, 255.71, 229.28, 297.61, 354.33, 42.59, 42.95, 64.81, 109.31, 221.97, 258.33, 31.78, 141.23, 152.19, 353.50, 291.18, 174.43, 319.14, 244.62, 121.78, 53.24, 350.51, 243.66, 248.42, 104.62, 41.61, 229.91, 378.89, 215.27, 68.77, 296.56, 129.20, 275.14, 51.12, 175.68, 320.13, 105.83, 262.13, 79.18

[a] The first two phase elements in each code are 0 and are excluded

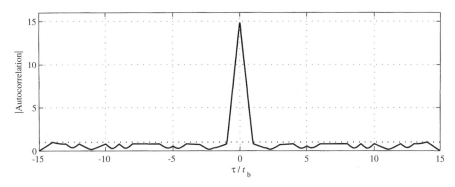

FIGURE 6.5 Autocorrelation function of the 15-element polyphase Barker-coded pulse.

In many applications, the phases are restricted to values that are the kth roots of unity (e.g., $k = 2$ gives the original Barker codes or $k = 6$ for sextic Barker codes). If such restrictions are made on the values of the generalized Barker sequence, there seems to be a limit on the maximal length M, depending on k, for such sequences (e.g., $M \leq 13$ for $k = 2$). Ein-dor et al. (2002) used statistical methods to show that for high values of M, a polyphase Barker sequence exists for any M as long as $k = M$.

Any unrestricted sequence can be approximated by a sequence of the kth roots of unity in such a way that the generalized Barker sequence condition is maintained for a sufficiently large value of k. Ternary ($k = 3$) Barker codes were shown to exist for $2 \leq M \leq 5$, $M = 7$ and 9. Quaternary ($k = 4$) Barker codes (Welti, 1960) were shown to exist for $2 \leq M \leq 5$, $M = 7$, 11, 13, and 15. Sextic ($k = 6$) Barker codes were shown to exist for all $2 \leq M \leq 13$ (Golomb and Scholtz, 1965). Sixty-phase generalized Barker codes were shown to exist for all $2 \leq M \leq 19$ (Zhang and Golomb, 1989) and for higher values of M up to $M = 37$ (Brenner, 1998). Finally, polyphase Barker codes with small alphabets (up to 120 phases) were shown to exist up to $M = 45$ (Brenner, 1998).

Figure 6.5 shows the autocorrelation function magnitude for the 15-element polyphase Barker sequence. Note that the peak sidelobe ($= 1$) is obtained at $\tau = \pm 14 t_b$. The magnitudes of all other sidelobes are less than 1. Note also that the autocorrelation function in the polyphase case is complex valued.

6.2 CHIRPLIKE PHASE CODES

One of the main drawbacks of the codes discussed in Section 6.1 is their Doppler tolerance. The polyphase Barker codes were derived using numerical methods that optimized only the correlation function sidelobes. Once the target return is Doppler shifted, the expected sidelobes (and interference level) are much higher than those predicted from observing only the zero-Doppler cut of the ambiguity function (the autocorrelation magnitude). For example, the ambiguity function of

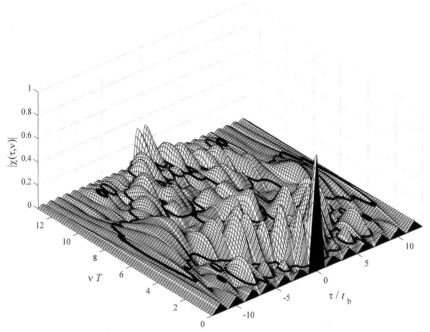

FIGURE 6.6 Partial ambiguity function of the 13-element Barker code showing some of the area where the range-Doppler sidelobes exceed $\frac{1}{13}$.

a Barker code with length 13 is shown in Fig. 6.6. A contour line marks the $\frac{1}{13}$ level (except for zero Doppler). Because of the random nature of the polyphase codes, the ambiguity function approaches a thumbtack shape.

Recall that the frequency-modulated signals described in Chapter 5 were shown to yield an ambiguity function exhibiting a ridge passing through the origin. Similar results are obtained for phase-coded pulses that have regularities similar to the frequency-modulated pulses of Chapter 5. In the following sections we describe some phase codes derived from the phase history of frequency-modulated pulses. The Frank code (Frank and Zadoff, 1962, 1963; Frank, 1963a,b) is derived from the phase history of a linearly frequency stepped pulse. The main drawback of the Frank code is that it only applies for codes of perfect square length ($M = L^2$). Frank and Zadoff first reported chirplike codes for any length M in a U.S. patent filed in 1957 (Frank and Zadoff, 1963). The patent specifically excludes claiming codes of perfect square length, which were described in a Frank patent filed earlier that year. Although Frank and Zadoff reported the chirplike code of square length in 1962 (now well known as Frank code), its application to any length was not reported at that time. At approximately the same time, similar codes of perfect square length were reported by Heimiller (1961, 1962), who constrained the sequence length to the square of a prime.

Twenty years later, apparently motivated by the work of Schroeder (1970) showing the favorable aperiodic properties of codes with quadratic phase dependence, chirplike sequences were reinvented by Chu in 1972 (Chu 1972, Frank 1973). Originally, the Frank and Zadoff–Chu codes were identified for their ideal *periodic* autocorrelation function. Later, the chirplike codes were also identified for their aperiodic properties (i.e., aperiodic correlation sidelobe level, bandwidth precompression limitations, etc.). Lewis and Kretschmer (1981a,b, 1982) and Kretschmer and Lewis (1983) reported specific versions of chirplike codes with good aperiodic properties. The specific versions were named P1, P2, P3, and P4. The P1 and P2 sequences are permutations of the Frank code and are applicable only for $M = L^2$. Several authors (e.g., Antweiler and Bomer, 1990; Zhang and Golomb, 1993; Wai Ho Mow and Li, 1997) studied the different variants of the Frank and Zadoff–Chu code. It was shown that the minimum aperiodic autocorrelation peak sidelobe and minimum integrated sidelobe codes are the code versions originally defined by Frank and Chu.

Other chirplike codes of length L^2 with good aperiodic properties (but not ideal periodic autocorrelation) were also reported by Rapajic (1998). Lewis and Kretschmer also report a palindromic variant of the P4 code (applicable to any length M) with better aperiodic properties (although not perfect).

6.2.1 Frank Code

There are several ways to describe the elements of the Frank code. For mathematical convenience, we define the elements of the *original* Frank code s_m ($1 \leq m \leq M$) of a square length $M = L^2$ as $s_{(n-1)L+k} = \exp(j\phi_{n,k})$, for $1 \leq n \leq L$ and $1 \leq k \leq L$, where $\phi_{n,k} = 2\pi(n-1)(k-1)/L$. Frank originally expressed the values of the code using the elements of an $L \times L$ discrete Fourier transform matrix given explicitly by

$$\begin{bmatrix} 0 & 0 & 0 & \cdots & 0 \\ 0 & 1 & 2 & \cdots & L-1 \\ 0 & 2 & 4 & \cdots & 2(L-1) \\ \vdots & \vdots & \vdots & \ddots & \vdots \\ 0 & L-1 & 2(L-1) & \cdots & (L-1)^2 \end{bmatrix}$$

The construction method is demonstrated for $M = 16$ ($L = 4$). To calculate the phase values of the 16-element Frank code, we first write the 4×4 Frank matrix given by

$$\begin{bmatrix} 0 & 0 & 0 & 0 \\ 0 & 1 & 2 & 3 \\ 0 & 2 & 4 & 6 \\ 0 & 3 & 6 & 9 \end{bmatrix}$$

The 16-element Frank code is formed by concatenating the rows of the Frank matrix and multiplying by $2\pi/L = 2\pi/4 = \pi/2$, resulting in the 16-element

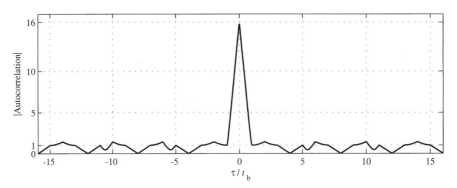

FIGURE 6.7 Autocorrelation function of a pulse coded using the 16-element Frank code.

phase code given by

$$\begin{bmatrix} 0 & 0 & 0 & 0 & 0 & \dfrac{\pi}{2} & \pi & \dfrac{3\pi}{2} & 0 & \pi & 2\pi & 3\pi & 0 & \dfrac{3\pi}{2} & 3\pi & \dfrac{9\pi}{2} \end{bmatrix}$$

Taking the phase value modulo 2π gives

$$\begin{bmatrix} 0 & 0 & 0 & 0 & 0 & \dfrac{\pi}{2} & \pi & \dfrac{3\pi}{2} & 0 & \pi & 0 & \pi & 0 & \dfrac{3\pi}{2} & \pi & \dfrac{\pi}{2} \end{bmatrix}$$

The aperiodic autocorrelation function of the 16-element Frank code is shown in Fig. 6.7. Note that the correlation function is zero for displacements of multiples of L since the rows of the Frank matrix are orthogonal. Note also that the autocorrelation has a magnitude of unity for a displacement of one more or one less than multiples of L. The original Frank code has two important properties: (1) the code is perfect (see Box 6C); (2) the aperiodic autocorrelation exhibits relatively low sidelobes (as can also be observed from Fig. 6.7).

Frank conjectured (1963a,b) that using cyclically shifted or decimated versions will not yield lower aperiodic autocorrelation sidelobes than the ones obtained using the original Frank code. This was also verified using a computer search (Antweiler and Bomer 1990, Popovic et al. 1991). Wai Ho Moe and Li (1997) use simple trigonometric identities to express the peak-to-peak sidelobe ratio of the Frank code as $[\sin(\pi/L)]^{-1}$ for L even and $[2\sin(\pi/2L)]^{-1}$ for L odd. Figure 6.8 shows the peak sidelobe level of the Frank code normalized by L. Note that it is given by two smooth monotonically decreasing curves, depending on whether L is even or odd, converging toward their common asymptote $1/\pi$ from above. Note also that the autocorrelation function peak for L even is higher than for L odd.

The connection between the Frank code and a frequency stepped pulse is demonstrated in Fig. 6.9, showing the phase history of a 16-element Frank code. Note that the code is made from four equal segments of linear phase dependence

BOX 6C: Perfectness of the Frank Code

The Frank code of length $M = L^2$ is written as $s_{(n-1)L+k} = \exp[j2\pi(n-1)(k-1)/L]$, for $1 \leq n \leq L$ and $1 \leq k \leq L$, where n is the row index and k is the column index of the Frank matrix. Since adding integer multiples of L to either k or n has no effect on the phase value (modulo 2π), the periodic autocorrelation of the Frank code for a delay corresponding to a shift of $0 \leq m < L$ rows and $0 \leq l < L$ columns is given by

$$\tilde{R}[Ll+m] = \sum_{n=1}^{L} \sum_{k=1}^{L} \exp[j2\pi(n-1)(k-1)/L]$$
$$\times \exp[-j2\pi(n+m-1)(k+l-1)/L]$$

which can be rearranged to give

$$\tilde{R}[Ll+m] = \exp[j2\pi(m-ml+l)/L] \sum_{n=1}^{L} \exp(-j2\pi nl/L)$$
$$\times \sum_{k=1}^{L} \exp(-j2\pi mk/L)$$

Using the closed-form expressions for the sum of a geometric progression, the sum over n can be expressed by

$$\sum_{n=1}^{L} \exp(-j2\pi nl/L) = \exp(-j2\pi l/L) \frac{1 - \exp(-j2\pi l)}{1 - \exp(-j2\pi l/L)}$$

The sum equals zero for all $0 < l < L$ and equals L for $l = 0$. In the same way we can show that the second sum is zero for all $0 < k < L$ and is equal to L for $k = 0$. Thus, the autocorrelation will take nonzero values only when both $l = 0$ and $k = 0$. In this case the autocorrelation value is $L^2 = M$.

(constant frequency). The segments' phase slope changes linearly from segment to segment implying linear frequency stepping between segments. If the frequency of the first section is zero, the frequency in the Lth section is $1/(Lt_b) = L/T$. Note that an artificial phase (integer multiple of 2π) was added to some of the sections for the plot. Note also that the first phase in each section is obtained by a linear prediction based on the phase values of the previous section. Finally, note that the resolution of phase values used for the Frank code is $2\pi/L$ compared to the resolution of π for the binary codes presented in Section 6.1.

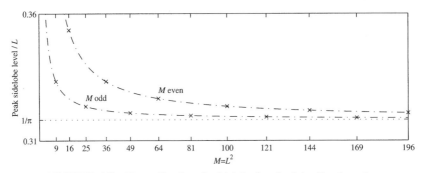

FIGURE 6.8 Normalized peak sidelobe level of the Frank code.

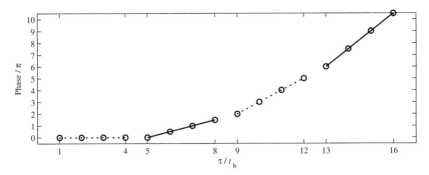

FIGURE 6.9 Phase history of a 16-element Frank code.

The partial ambiguity function of the 16-element Frank code is shown in Fig. 6.10. Note the diagonal ridge passing through the origin and extending to the first quadrant of the range–Doppler plane. Note also a secondary attenuated ridge duplicated $1/T$ in Doppler. Using decimated versions of the original Frank code yields a different slope of the diagonal ridge.

6.2.2 P1, P2, and Px Codes

The P1, P2, and Px codes are all modified versions of the Frank code, with the dc frequency term in the middle of the pulse instead of at the beginning. The Px code was introduced by Rapajic and Kennedy (1998) and was shown to yield the same aperiodic peak sidelobe as the Frank code but having lower integrated sidelobe level. The elements of the Px code are defined mathematically in a similar way to the Frank code as $s_{(n-1)L+k} = \exp(j\phi_{n,k})$, for $1 \leq n \leq L$ and $1 \leq k \leq L$, where in this case the $\phi_{n,k}$ are given by

$$\text{Px}: \quad \phi_{n,k} = \begin{cases} \dfrac{2\pi}{L}[(L+1)/2 - k][(L+1)/2 - n], & L \text{ even} \\ \dfrac{2\pi}{L}[L/2 - k][(L+1)/2 - n], & L \text{ odd} \end{cases} \quad (6.2)$$

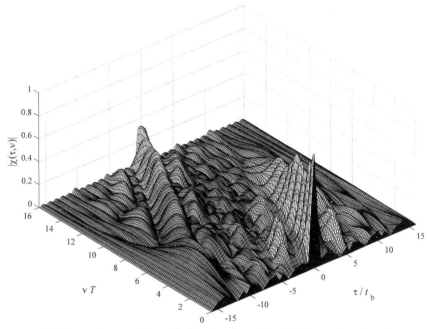

FIGURE 6.10 Partial ambiguity function of the 16-element Frank code.

In a similar way to calculating the Frank matrix, we can calculate an analogous Px matrix. For $M = 16$ ($L = 4$) the resulting matrix is

$$\frac{1}{4}\begin{bmatrix} 9 & 3 & -3 & -9 \\ 3 & 1 & -1 & -3 \\ -3 & -1 & 1 & 3 \\ -9 & -3 & 3 & 9 \end{bmatrix}$$

The 16-element Px code is formed by concatenating the rows of the Px matrix and multiplying by $2\pi/L = 2\pi/4 = \pi/2$, resulting in the 16-element phase code given by

$$\left[\frac{9\pi}{8} \quad \frac{3\pi}{8} \quad -\frac{3\pi}{8} \quad -\frac{9\pi}{8} \quad \frac{3\pi}{8} \quad \frac{\pi}{8} \quad -\frac{\pi}{8} \quad -\frac{3\pi}{8} \right.$$
$$\left. -\frac{3\pi}{8} \quad -\frac{\pi}{8} \quad \frac{\pi}{8} \quad \frac{3\pi}{8} \quad -\frac{9\pi}{8} \quad -\frac{3\pi}{8} \quad \frac{3\pi}{8} \quad \frac{9\pi}{8} \right]$$

Taking the phase value modulo 2π gives

$$\left[\frac{9\pi}{8} \quad \frac{3\pi}{8} \quad \frac{13\pi}{8} \quad \frac{7\pi}{8} \quad \frac{3\pi}{8} \quad \frac{\pi}{8} \quad \frac{15\pi}{8} \quad \frac{13\pi}{8} \quad \frac{13\pi}{8} \quad \frac{15\pi}{8} \quad \frac{\pi}{8} \right.$$
$$\left. \frac{3\pi}{8} \quad \frac{7\pi}{8} \quad \frac{13\pi}{8} \quad \frac{3\pi}{8} \quad \frac{9\pi}{8} \right]$$

120   PHASE-CODED PULSE

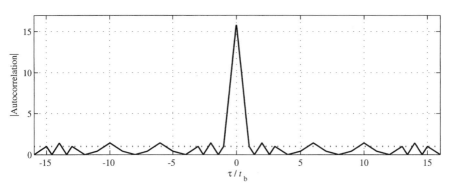

FIGURE 6.11 Autocorrelation function of a 16-element Px coded pulse.

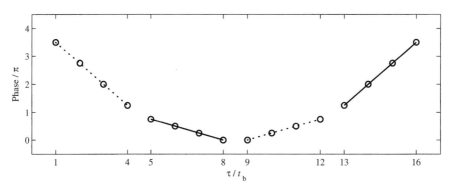

FIGURE 6.12 Phase history of a 16-element Px code.

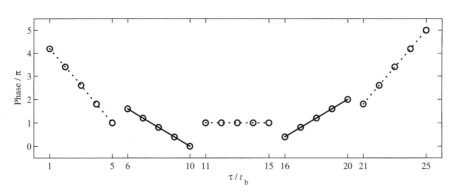

FIGURE 6.13 Phase history of a 25-element Px code.

The autocorrelation function and phase history of the 16-element Px code are shown in Figs. 6.11 and 6.12, respectively. The phase history of the 25-element Px code is shown in Fig. 6.13. Note that the sidelobes of the 16-element Px correlation function have a concave shape rather than the convex shape of the 16-element Frank code (compare Fig. 6.11 to Fig. 6.7). Although the peak sidelobe of the Px and Frank codes is the same, the integrated sidelobe of the Px code is superior to that of the Frank code (and to that of other codes discussed in later sections). Note that whereas the Frank code is a perfect code (having an ideal periodic correlation function), the Px code is not perfect.

The P1 and P2 codes are due to Lewis and Kretschmer (1981a,b). The two codes are derived based on the Frank code and are very similar to the Px code. Both codes, as the Frank and Px codes, are applicable only for perfect square length ($M = L^2$). The P2 code is valid only for L even and is defined exactly as the Px code for even L. Note that the P2 code (and Px for even M) is palindromic in that it has conjugate symmetry across each frequency (matrix row) and even symmetry about the center of the code (see Fig. 6.12).

The P1 code elements $s_m (1 \leq m \leq M)$ are defined as $s_{(n-1)L+k} = \exp(j\phi_{n,k})$, where $1 \leq n \leq L$, $1 \leq k \leq L$, and

$$\text{P1}: \quad \phi_{n,k} = \frac{2\pi}{L}[(L+1)/2 - n][(n-1)L + (k-1)] \qquad (6.3)$$

Figures 6.14 and 6.15 show the 16-element ($L = 4$) and 25-element ($L = 5$) P1 code phase history calculated using (6.3). Integer multiples of 2π were added

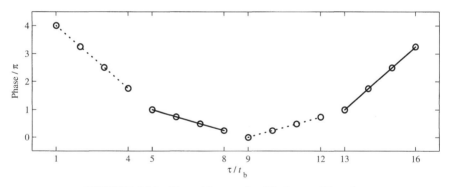

FIGURE 6.14 Phase history of a 16-element P1 code.

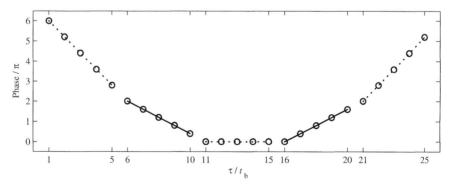

FIGURE 6.15 Phase history of a 25-element P1 code.

such that the phase difference between slices is less than π. Note that the P1 code has the same phase increments within each phase group as the P2/Px code, except that the starting phases are different. Note that while the Px and P2 codes are not perfect, the P1 code is perfect (as is the Frank code). The ambiguity function of the P1 code for odd L is identical to that of the Frank code. For L even the ambiguity function of the P2/Px code is very similar to that of the P1 code and to that of the Frank code.

Bandwidth limitations are found in all radar systems. Such limitations are usually the result of attempts to maximize signal to thermal noise and hardware limitations. The result of bandlimiting is to average the elements constituting the code waveform. For the P1 and P2/Px codes, the phase difference between adjacent code elements is low in the center of the code and higher closer to the ends of the code. Thus, bandlimiting results in amplitude intensifying the central part of the code relative to the code ends. As we already demonstrated in Chapter 4 and will also see in Chapter 7, this results in an increase to mainlobe width and lower sidelobes. The phase increments of the Frank code (taken modulo 2π) are higher in the central part of the code and lower close to the code ends. Thus, bandlimiting results in amplitude intensifying the ends of the code relative to the code center and yields a decrease in mainlobe width and higher sidelobes.

6.2.3 Zadoff–Chu Code

While the Frank, P1, P2, and Px codes are only applicable for perfect square lengths ($M = L^2$), the Zadoff code (Zadoff, 1963) is applicable for any length and is given by $s_m = \exp(j\phi_m)$, where

$$\text{Zadoff}: \quad \phi_m = \frac{2\pi}{M}(m-1)\left(r\frac{M-1-m}{2} - q\right) \quad (6.4)$$

$1 \leq m \leq M, 0 \leq q < M$ is any integer, and r is any integer relatively prime to M.

For any given length M, different variants of the Zadoff code are obtained by changing r or q and adding a constant phase shift to all elements. It can be shown that the permutations that preserve the ideal periodic autocorrelation function property (cyclic shift, decimation, conjugation, etc.) are equivalent to changing q and r. An important permutation of the Zadoff code was presented by Chu (1972) and is given by $s_m = \exp(j\phi_m)$, where

$$\text{Chu}: \quad \phi_m = \begin{cases} \dfrac{2\pi}{M}r'\dfrac{(m-1)^2}{2}, & M \text{ even} \\ \dfrac{2\pi}{M}r'\dfrac{(m-1)m}{2}, & M \text{ odd} \end{cases} \quad (6.5)$$

$1 \leq m \leq M$, and r' is any integer relatively prime to M. The Chu code can be obtained from the Zadoff code by setting $r = -r'$ and $q = -(M-2)r'/2$ for

M even or by setting $r = -r'$ and $q = -(M - 1)r'/2$ for M odd. We calculate the Zadoff–Chu phase code of length $M = 16$ and minimal chirp slope by using $r' = 1$ in (6.5) or $r = -1$ and $q = 9$ in (6.4). Figure 6.16 shows the phase values as a function of m. The autocorrelation function of the minimal chirp slope 16-element Zadoff–Chu phase-coded pulse is shown in Fig. 6.17.

Taking the phase values modulo 2π gives

$$\left[0 \quad \frac{3\pi}{16} \quad \frac{\pi}{2} \quad \frac{15\pi}{16} \quad \frac{3\pi}{2} \quad \frac{3\pi}{16} \quad \pi \quad \frac{31\pi}{16} \quad \pi \quad \frac{3\pi}{16} \quad \frac{3\pi}{2} \quad \frac{15\pi}{16} \right.$$
$$\left. \frac{\pi}{2} \quad \frac{3\pi}{16} \quad 0 \quad \frac{31\pi}{16} \right]$$

Note that taking the phase value modulo 2π shows close to palindromic phase behavior. Note also that many authors refer to the minimal chirp slope Zadoff–Chu code simply as Chu code. In the Chu code the phase is always zero for $m = 1$ and quadratic increasing or decreasing for higher m (as for the Frank code), while for the Zadoff code, choosing $q = 0$ gives a phase shape closer to that of the Px/P2 or P1 codes having a minimum or maximum phase in the center of the code.

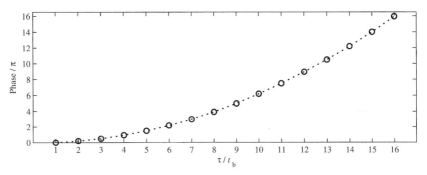

FIGURE 6.16 Phase history of a 16-element Zadoff–Chu code.

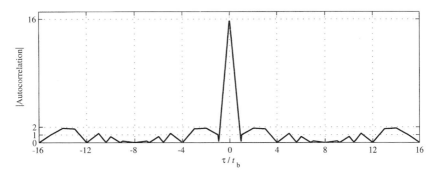

FIGURE 6.17 Autocorrelation function of a minimal chirp slope 16-element Zadoff–Chu coded pulse.

Like the Frank and P1 codes, the Zadoff–Chu code is perfect (see Box 6D). Antweiler and Bomer (1990) used an exhaustive computer search to show that considering all possible decimations (equivalent to changing the parameters of the Zadoff code), the minimal chirp slope Chu code obtained by setting $|r| = 1$ in (6.5) has the lowest aperiodic autocorrelation peak sidelobe. Note that all cyclic versions of the Zadoff–Chu code have the same aperiodic correlation magnitude (see Box 6E).

BOX 6D: Perfectness of the Zadoff–Chu Code

We assume that $s_m = \exp(j\phi_m)$, where $1 \le m \le M$ and where ϕ_m is given by (6.4). The condition for the code to be perfect is equivalent to showing that the $M - 1$ nonredundant equations

$$\tilde{R}[k] = \sum_{m=1}^{M} \exp[j(\phi_m - \phi_{m+k \bmod M})] = 0 \qquad (6D.1)$$

are satisfied for all $1 \le k \le M - 1$. Using (6.4) and simplifying, it is easy to show that

$$\phi_{m+M} - \phi_m = -\frac{2\pi}{M} qM + \frac{2\pi}{M} rmM \qquad (6D.2)$$

Thus, the set of equations can be rewritten as

$$\tilde{R}[k] = \sum_{m=1}^{M} \exp[j(\phi_m - \phi_{m+k})] = 0 \qquad (6D.3)$$

Using (6.4) again and simplifying gives

$$\phi_m - \phi_{m+k} = -\frac{2\pi}{M} qk + \frac{2\pi}{M} r \frac{k(-M + 2m + k)}{2}$$

$$= \frac{2\pi}{M}[-rmk + g(k)] \qquad (6D.4)$$

where $g(k)$ is some function of k. Substituting (6D.4) in (6D.3) and using the fact that r is relatively prime to M gives

$$\tilde{R}[k] = \exp\left[j\frac{2\pi}{M} g(k)\right] \sum_{m=1}^{M} \exp\left(-j\frac{2\pi}{M} rmk\right) = 0 \qquad (6D.5)$$

for all $1 \le k \le M - 1$.

BOX 6E: Rotational Invariance of the Zadoff–Chu Code Aperiodic ACF Magnitude

First we write the phase of the Zadoff–Chu code cyclically shifted by a positions in term of the phase values of the original Zadoff–Chu code:

$$\phi'_m = \phi_{(m+a)\bmod M} = \phi_m + \phi_{(m+a)\bmod M} - \phi_m \tag{6E.1}$$

Using (6D.2) and (6D.4), we can write

$$\phi'_m = \phi_m - \frac{2\pi}{M}[-rma + g(a)] \tag{6E.2}$$

Now, writing the aperiodic autocorrelation function (ACF) for positive delay k gives

$$R_a[k] = \sum_{m=1}^{M-k} \exp[j(\phi'_m - \phi'_{m+k})]$$

$$= \sum_{m=1}^{M-k} \exp\left[j\left(\phi_m - \frac{2\pi}{M}[-rma + g(a)] - \phi_{m+k}\right.\right.$$

$$\left.\left. + \frac{2\pi}{M}[-r(m+k)a + g(a)]\right)\right]$$

$$= \sum_{m=1}^{M-k} \exp\left[j\left(\phi_m - \phi_{m+k} - \frac{2\pi}{M}rka\right)\right] \tag{6E.3}$$

$$= \exp\left(-j\frac{2\pi}{M}rka\right) \sum_{m=1}^{M-k} \exp[j(\phi_m - \phi_{m+k})]$$

$$= \exp\left(-j\frac{2\pi}{M}rka\right) R[k]$$

Thus, $|R_a[k]| = |R[k]|$ for positive delay k. Since the ACF is conjugate symmetric around zero delay, $|R_a[k]| = |R[k]|$ for all positive and negative delays.

The autocorrelation sidelobe pattern of the Zadoff–Chu code shown in Fig. 6.17 is typical of all Zadoff–Chu codes with $|r| = 1$ (higher sidelobes close to the mainlobe and close to $\pm T$). When using different values of r (different chirp slopes), the area of higher aperiodic autocorrelation sidelobes "moves" on the delay axis to different locations. The position of the higher sidelobes can intuitively be predicted by examining the ambiguity function of the Zadoff–Chu phase-coded pulse. As can be seen from the example in Fig. 6.18, the higher sidelobes are obtained in the area where the secondary diagonal ridges, centered on $\pm n/T$ in Doppler ($n > 0$), cross the zero-Doppler axis.

Wai Ho Moe and Li (1997) show that the peak sidelobe of the aperiodic autocorrelation of the Chu code is given by $\left|\sin\left(\pi[\sqrt{\alpha M/\pi}]^2/M\right)/\sin\left(\pi[\sqrt{\alpha M/\pi}]/M\right)\right|$, where $\alpha = 1.16556118520721\ldots$ is the first positive root of the equation $\tan(\alpha) = 2\alpha$ and $[\cdot]$ stands for rounding toward the nearest integer. The expression is valid for all $M \geq 2$ except for $M = 33$, where $\sqrt{\alpha 33/\pi} \approx 3.499045$ and the sidelobe level is given by $|\sin(\pi 4^2/M)/\sin(\pi 4/M)|$. Figure 6.19 shows the peak sidelobe level of the Zadoff–Chu code as a function of the sequence length normalized to the square root of the sequence length. Note that the sidelobe level is higher than that for the Frank code (compare to Fig. 6.8). Asymptotically, the peak sidelobe converges to $\sin\alpha/\sqrt{\pi\alpha} \cong 0.480261$.

6.2.4 P3, P4, and Golomb Polyphase Codes

Analogously to the connection between P1 and P2/Px codes to the original Frank code, the P3, P4, and Golomb polyphase codes are specific cyclically shifted and decimated versions of the Zadoff–Chu code. Like the P1 and P2 codes, the P3 and P4 codes, are due to Lewis and Kretschmer (1982). Unlike the P1 and P2

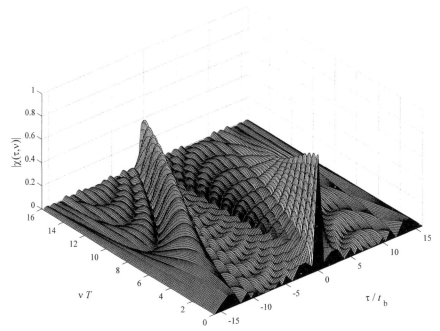

FIGURE 6.18 Partial ambiguity function of a 16-element Zadoff–Chu-coded pulse; $r = 1$ (above) and $r = 3$ (next page).

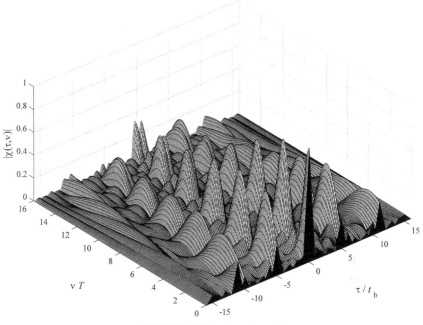

FIGURE 6.18 (*continued*)

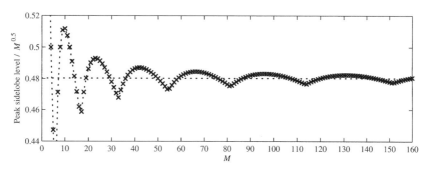

FIGURE 6.19 Normalized peak sidelobe level of the Chu code.

codes applicable only for square length (such as the Frank code), the P3 and P4 codes are, like the Zadoff–Chu code, applicable for any length M. The P3 and P4 codes are defined for any length M by

$$\text{P3}: \quad \phi_m = \frac{2\pi}{M} \frac{(m-1)^2}{2} \tag{6.6}$$

$$\text{P4}: \quad \phi_m = \frac{2\pi}{M}(m-1)\left(\frac{m-1-M}{2}\right) \tag{6.7}$$

where $1 \leq m \leq M$. The P3 code is identical to the Chu code for even M with $r' = 1$ and is perfect only for even values of M. The P4 code can be obtained by setting $r = -1$ and $q = 1$ in the expression for the Zadoff phase code. Like the Zadoff code, the P4 code has an ideal periodic correlation for both even and odd M.

A slightly different version of the Zadoff–Chu/P4 codes was defined by Zhang and Golomb (1993) and is given by

$$\text{Golomb polyphase code:} \quad \phi_m = \frac{2\pi}{M} r''' \frac{(m-1)m}{2} \quad (6.8)$$

where $1 \leq m \leq M$, and r''' is any integer relatively prime to M. Note that the Golomb polyphase sequences are identical to the Chu code for odd M. Zhang and Golomb showed that the polyphase Golomb code is perfect for all M.

When the target Doppler is unknown, using highly Doppler-tolerant FM waveforms allows for simplified receiver hardware with negligible degradation in performance. Kretschmer and Lewis (1983) showed that the P3 and P4 codes are much more Doppler tolerant than the Frank or P1 and P2/Px codes. This result is very intuitive and can be extended to the other Zadoff–Chu variants, based on sampling the phase history of a continuous frequency-modulated waveform other than stepped frequency-modulated waveforms.

The P4 code differs from the P3 code by having the largest code element to code element phase changes at the ends of the code instead of the middle, as in the P3 code. In this way, the P4 code differs from the P3 code in the same way that the P1 code differs from the Frank code, implying that the bandwidth tolerance of the P4 is better than the bandwidth tolerance of the P3 code. A palindromic P4 version having optimized bandwidth tolerance was defined by Kretschmer and Lewis (1983) and is given by

$$\text{Palindromic P4:} \quad \phi_m = \frac{\pi}{M}\left(m - \frac{1}{2}\right)^2 - \pi\left(m - \frac{1}{2}\right) \quad (6.9)$$

where $1 \leq m \leq M$. Note that the palindromic version of P4 (as the Px and P2 codes) is not a perfect code.

6.2.5 Phase Codes Based on a Nonlinear FM Pulse

The phase codes discussed in previous sections were shown to be strongly connected to the phase history of a linear frequency-modulated or frequency-stepped pulse. Other phase codes can be derived by sampling the phase history of a *nonlinear* frequency-modulated (or frequency-stepped) pulse. Selection of the phase (or frequency) history of the continuous pulse determines the properties (sidelobe level, mainlobe width, Doppler tolerance, bandwidth limitations, etc.) of the resulting phase code. As shown in Chapter 5, using nonlinear instead of linear FM can yield lower autocorrelation sidelobes but a higher mainlobe width.

CHIRPLIKE PHASE CODES

One example of a phase code derived from the phase history of a frequency-modulated pulse is the P(n, k) code (Felhauer, 1994). The P(n, k) code is based on step approximation of the phase function of a nonlinear FM chirp signal with the energy density function given by

$$W(f) = \begin{cases} k + (1+k)\cos^n\left(\dfrac{\pi f}{B}\right), & |f| \leq \dfrac{B}{2} \\ 0, & |f| > \dfrac{B}{2} \end{cases} \quad (6.10)$$

where k and n are free parameters and B is the bandwidth considered for the expanded pulse. Generation of the phase code first involves designing the continuous pulse yielding the desired energy density function with parameters n and k. The calculation is based on the principle of stationary phase (see Chapter 5). Next, the phase history of the continuous pulse is sampled, resulting in the phase code desired.

Figure 6.20 illustrates generation of the P(n, k) code. The upper parts of Fig. 6.20 show a typical weighting function $W(f)$ and the corresponding instantaneous frequency function $f(t)$ calculated using the principle of stationary phase.

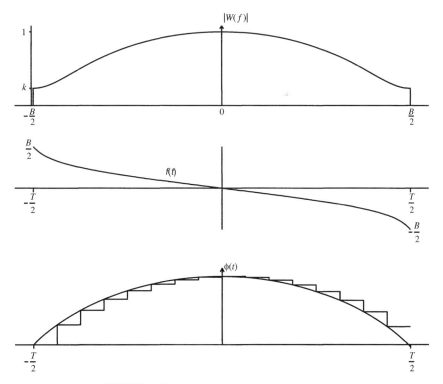

FIGURE 6.20 Generation of the P(n, k) code.

The lower part of Fig. 6.20 illustrates generation of the $P(n, k)$ code by step approximation of the phase function calculated from the instantaneous frequency $f(t)$. For $k = 1$, $W(f)$ is a constant rectangular spectrum, and the algorithm described here leads to the linear chirp signal with a corresponding phase function identical to the phase function of the P4 code. For arbitrary values of n and k, the corresponding phase code elements can only be calculated using numerical methods.

The $P(n, k)$ code of length N can be optimized for minimal peak sidelobe by controlling the free parameters n and k. For example, when $N = 100$, the optimal values of k are 0.05 and 0.015 for $n = 2$ and 4, yielding a peak sidelobe level of 34.8 and 37.1 dB, respectively. The autocorrelation function of the $P(2,0.05)$ and $P(4,0.015)$ codes are shown in Fig. 6.21 together with the P4 ACF. Note the peak sidelobe of the $P(4,0.015)$ is only 2.9 dB lower than the theoretical optimum for uniform phase codes and more than 10 dB better than the peak sidelobe of the P4 code. The price for this significant peak sidelobe improvement is a loss in range resolution in comparison with the P4 code, as seen by the widening of the ACF mainlobe [The mainlobe width to the chip length ratio is 1.8 for $P(2,0.05)$, 2.8 for $P(4,0.015)$ and 1 for P4.]

Table 6.5 compares the peak sidelobe and mainlobe width of the $P(n, k_{opt})$ codes for $1 \leq n \leq 5$, the Frank/P1/Px codes, and the P3/P4/Zadoff–Chu codes

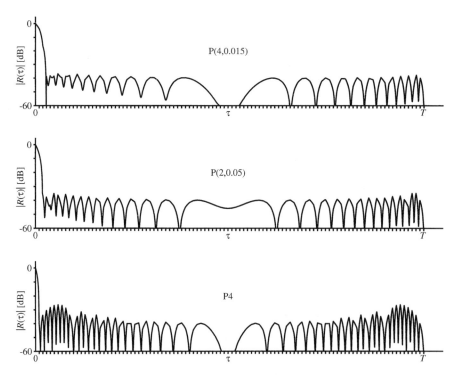

FIGURE 6.21 Autocorrelation function of the $P(4,k_{opt})$, $P(2,k_{opt})$, and P4 codes.

TABLE 6.5 Optimal Value k_{opt}, Peak Sidelobe Ratio, and Mainlobe Width for Some P(n,k) Codes

Phase Code	$N = 16$			$N = 64$			$N = 100$		
	k_{opt}	Peak Sidelobe	Mainlobe Width $\times t_b$	k_{opt}	Peak Sidelobe	Mainlobe Width $\times t_b$	k_{opt}	Peak Sidelobe	Mainlobe Width $\times t_b$
P(1, k_{opt})	0.0055	22.7	1.6	0.006	30.7	1.5	0.01	33	1.5
P(2, k_{opt})	0.12	23.5	1.7	0.039	31.9	1.9	0.05	34.8	1.8
P(3, k_{opt})	0.085	24.1	2.1	0.015	33	2.4	0.03	35.9	2.3
P(4, k_{opt})	0.11	24.1	2.1	0.025	34.2	2.7	0.015	37.1	2.8
P(5, k_{opt})	0.045	24.1	2.8	0.024	35.1	3.1	0.015	38.1	3.2
Frank/P1/Px	—	21.9	1	—	28	1	—	29.9	1
P3/P4/Chu	—	18.7	1	—	24.3	1	—	26.2	1

for $N = 16$, 64, and 100. Note that for suboptimal values of k the mainlobe width loss decreases with increasing k until for $k = 1$ there is no increase in mainlobe width. Note also that as n increases, k_{opt} is lower and the mainlobe width loss is higher. Finally, note that for lower values of N, the peak/peak sidelobe ratio approaches N (as for polyphase Barker codes). For example, Felhauer (1994) showed that for P(4,k_{opt}) the peak-sidelobe level is 1 for all $N \leq 25$.

The partial ambiguity function of a 16-element P(4,0.11) is shown in Fig. 6.22. Note that the Doppler tolerance is high (as expected). Felhauer (1994) showed

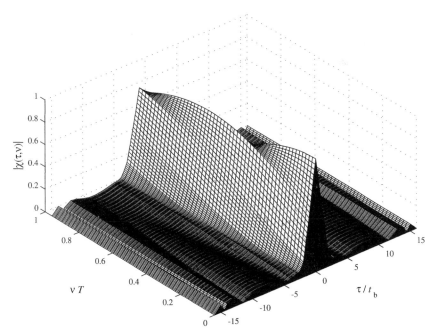

FIGURE 6.22 Partial ambiguity function of the 16-element P(4,k_{opt}) code.

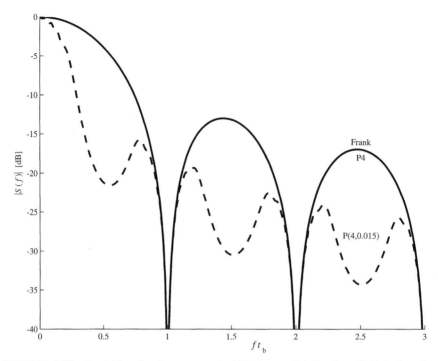

FIGURE 6.23 Partial baseband spectrum of a 100-element P4, Frank and P(4,k_{opt}) codes.

that P(n, k) is less affected by bandwidth limitation than even the P4 code. This is because the phase increments from code element to code element are smaller, compared to other phase codes, over most of the code length N. The result is further clarified by inspecting the Fourier spectra of the phase-coded pulses. Because of the rectangular chip of duration t_b, all phase-coded pulses have $\sin(\pi f t_b)/\pi f t_b$ baseband spectrum envelopes. The choice of the modulating phase code determines the fine structure in the baseband spectrum.

Figure 6.23 shows the partial baseband spectrum of a 100-element P4 coded pulse, 100-element Frank code baseband spectrum, and the 100-element P(4,0.015)-coded pulse spectrum. The figure illustrates that the P(4,0.015) has a much smaller effective bandwidth, which on the one hand, is the reason for the higher mainlobe width, and on the other hand, is the explanation for the improved tolerance to bandwidth limitations. Finally, note that other phase codes can be derived using a different energy density function than the one described in (6.10), yielding different ACF and frequency spectrum properties.

6.3 ASYMPTOTICALLY PERFECT CODES

The Frank and Zadoff–Chu codes discussed in previous sections are examples of perfect codes having zero periodic autocorrelation function for all nonzero shifts. Perfect codes are useful, for example, in radar applications employing CW waveforms. Asymptotically perfect codes have the property that the peak-to-peak PACF sidelobe goes to zero as the code length M increases. The main drawback of Frank and Zadoff–Chu codes is that they are polyphase. The asymptotically perfect codes described in this section are binary.

A prime example of such sequences are maximum-length linear feedback sequences, or *m-sequences*, having a two-valued PACF (i.e., the PACF magnitude is 1 for all nonzero shifts and M at integer multiples of the code period). All m-sequences are limited to length $M = 2^n - 1$ and are derived mathematically based on irreducible polynomials over $GF(2^n)$ (see Appendix 6A). Figure 6.24 shows the shift register that generates an m-sequence of period $2^3 - 1 = 7$. The PACF magnitude of the seven-element m-sequence is shown in Table 6.6.

The peak sidelobe of the aperiodic autocorrelation of m-sequence codes is usually suboptimal (e.g., for $M = 63$ we can find an m-sequence with a peak sidelobe of 6, whereas the MPS code has a peak sidelobe of 4). However, since the PACF sidelobe of the m-sequence is low, it is usually the case that the aperiodic autocorrelation sidelobe level is not high. Note that for different cyclic shifts of the m-sequence we get different ACF sidelobes. Thus, for each irreducible polynomial, it is necessary to check M different cyclic shifts of the periodic code to find the optimal aperiodic code. Recall that for large values of M this

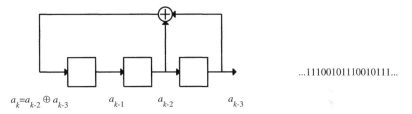

FIGURE 6.24 Generating the m-sequence of period $2^3 - 1 = 7$ using a shift register.

TABLE 6.6 Calculation of the PACF for the Seven-Element m-Sequence of Fig. 6.24

$\{s_m^*\}$							1	1	1	−1	−1	1	−1
$\{s_m\}$													
1	1	1	−1	−1	1	−1	1	1	1	−1	−1	1	−1
1	1	−1	−1	1	−1	1	1	1	−1	−1	1	−1	1
1	−1	−1	1	−1	1	1	1	−1	−1	1	−1	1	1
−1	1	−1	1	−1	−1	1	1	−1	1	−1	−1	1	−1	-1
−1	−1	1	−1	−1	1	1	−1	1	−1	−1	1	−1	1
1	−1	1	1	1	−1	−1	1	−1	1	1	1	−1	−1
−1	−1	−1	−1	1	1	−1	−1	−1	−1	1	1	1	−1
$\{\tilde{R}\}$	−1	−1	−1	−1	−1	−1	7	−1	−1	−1	−1	−1	−1

is significantly less than searching over 2^M possible binary codes (e.g., when $M = 63$ there are six irreducible polynomials).

Two-valued asymptotically perfect sequences correspond to combinatorial objects known as *cyclic Hadamard difference sets* (CDSs). Known constructions of CDS are the m-sequence, Legendre sequences, Hall's sextic residue sequence, twin prime sequence, GMW sequence, and the Maschietti construction (Maschietti, 1998). A cyclic $\{M, l, \lambda\}$ difference set is a set D of l elements, distinct modulo M, such that the congruence $d_i - d_j \equiv r \pmod{M}$ has exactly λ solution pairs (d_i, d_j) with d_i and d_j in D for each $1 \leq r \leq M - 1$. A simple example of cyclic difference set is given by $D = \{1\ \ 2\ \ 4\}$, which is a cyclic $\{7, 3, 1\}$ difference set. It is easy to verify that the periodic sequence with period M obtained by setting the sequence element to 1 where the element index is in D (mod M) and to -1 where it is not in D and has a two-valued periodic autocorrelation function given by

$$R[k] = \sum_{m=1}^{M} s_m s_{m+k}^* = \begin{cases} M, & k = 0 \pmod{M} \\ M - 4(l - \lambda), & k \neq 0 \pmod{M} \end{cases} \quad (6.11)$$

For example, $D = \{1\ \ 2\ \ 4\}$ implies $\{1\ \ 1\ \ -1\ \ 1\ \ -1\ \ -1\ \ -1\}$, which is the m-sequence generated by the shift register in Fig. 6.24.

6.4 GOLOMB'S CODES WITH IDEAL PERIODIC CORRELATION

A method for generating two-valued (biphase) perfect codes using cyclic difference sets was described by Golomb (1992). In general, the out-of-phase value $M - 4(l - \lambda)$ in (6.11) will not be zero for $M > 4$. However, if we replace the -1 elements in $\{s_m\}$ by a suitable complex number β, it is possible to obtain zero PACF sidelobes.

First we consider the case where $|\beta| = 1$ (i.e., we replace the $180°$ phase shifts between the two types of elements in the asymptotically perfect sequence described in Section 6.3 with a different phase). In this case it is possible to show (see Box 6F) that if we set $\beta = \exp(j\phi)$, where

$$\cos\phi = -\frac{M - 2l + 2\lambda}{2(l - \lambda)} \quad (6.12)$$

the resulting biphase code is perfect. Note that (6.12) does not have a solution ϕ for all CDS parameters. However, for the family of Hadamard cyclic difference sets satisfying $M = 4n - 1$, $l = 2n - 1$, and $\lambda = n - 1$, we get that

$$\cos\phi = -\frac{M - 1}{M + 1} \quad (6.13)$$

which always has a solution. Table 6.7 gives some Hadamard cyclic difference sets and the corresponding phase value used according to (6.13) to construct a Golomb biphase perfect code.

TABLE 6.7 Parameters of Some Nonequivalent Golomb Biphase Perfect Codes

M	Phase Shift	λ	D
2	90	0	{1}
3	120	1	{1 2}
4	180	2	{1 2 4}
7	138.6	2	{1 2 3 5}
11	146.4	3	{1 2 3 5 6 8}
15	151.0	4	{1 2 3 4 6 8 9 12} or {1 4 5 7 9 10 11 12}
19	154.2	5	{1 2 3 4 6 8 13 14 16 17}
23	156.4	6	{1 2 3 4 5 7 9 10 13 14 17 19}
31	159.6	7	{1 2 3 4 6 8 12 15 16 17 23 24 27 29 30}
35	160.8	8	{1 2 4 5 8 10 12 13 14 15 17 18 22 28 29 30 34}
43	162.7	10	{1 2 3 4 5 8 11 12 16 19 20 21 22 27 32 33 35 37 39 41 42}
63	165.6	15	{1 2 3 4 5 7 8 9 10 13 14 15 17 19 20 25 27 28 29 33 34 36 37 39 42 46 49 50 53 55 57} or {1 2 3 4 5 6 7 9 10 11 13 17 18 19 21 24 25 28 30 33 34 35 37 41 44 46 47 49 54 55 59}

BOX 6F: Deriving the Perfect Golomb Biphase Code

Assume a two-valued sequence $\{s_m\}$ with period M obtained by setting the sequence element to 1 where the element index is in the $\{M, l, \lambda\}$ cyclic difference set D (mod M) and to $\beta = \exp(j\phi)$ where it is not in D. Then when calculating the PACF for each shift $1 \leq k \leq M - 1$, we have the following four situations:

$$
\begin{array}{ll}
1 \times 1 & \lambda \text{ times} \\
1 \times \beta^* & l - \lambda \text{ times} \\
\beta \times 1 & l - \lambda \text{ times} \\
\beta \times \beta^* & M - 2l + \lambda \text{ times}
\end{array}
$$

We than have

$$\tilde{R}[k] = \sum_{m=1}^{M} s_m s_{m+k}^*$$

$$= \begin{cases} l + (M-l)|\beta|^2 = M, & k = 0 \\ \lambda + (M - 2l + \lambda)|\beta|^2 + (l - \lambda)(\beta^* + \beta) \\ \quad = M - 2l + 2\lambda + 2(l - \lambda)\cos\phi, & 1 \leq k \leq M - 1 \end{cases} \quad (6F.1)$$

Requiring that the code is perfect immediately yields (6.12).

An alternative method also described by Golomb employs both phase and amplitude coding. The idea is to use the original binary sequence in the transmitting end. The receiver reference signal amplitudes are altered from the original value of -1 to $-b$ such that the periodic cross-correlation is ideal. In this case it is possible to show (see Box 6G) that if

$$b = -\frac{2\lambda - l}{M - 3l + 2\lambda} \tag{6.14}$$

BOX 6G: Deriving the Golomb Two-Valued Code with Ideal Periodic Cross-Correlation

Assume a two-valued sequence $\{s_m\}$ with period M obtained by setting the sequence element to 1 where the element index is in the $\{M, l, \lambda\}$ cyclic difference set D (mod M) and to -1 where it is not in D. Next, assume a two-valued code $\{q_m\}$ formed by replacing the -1 elements in $\{s_m\}$ with $-b$, where $b > 0$ and real. Then, when calculating the periodic cross-correlation for each shift $1 \leq k \leq M - 1$, we have the following four situations:

$$\begin{array}{ll} 1 \times 1 & \lambda \text{ times} \\ 1 \times (-b) & l - \lambda \text{ times} \\ -1 \times 1 & l - \lambda \text{ times} \\ -1 \times (-b) & M - 2l + \lambda \text{ times} \end{array}$$

We than have

$$\tilde{R}_{sq}[k] = \sum_{m=1}^{M} s_m q_{m+k}^*$$

$$= \begin{cases} l + (M - l)b < M, & k = 0 \\ \lambda + (M - 2l + \lambda)b - (l - \lambda)(b + 1), & k \neq 0 \ (\text{mod } M) \end{cases} \tag{6G.1}$$

To have zero periodic cross-correlation sidelobes, we need to have

$$\lambda + (M - 2l + \lambda)b - (l - \lambda)(b + 1)$$
$$= 2\lambda - l + (M - 3l + 2\lambda)b = 0 \tag{6G.2}$$

which immediately gives (6.14).

---

the condition on zero periodic cross-correlation sidelobes is met. Note that using a reference pulse different from the transmitted pulse results in a mismatch loss given by $[l + (M - l)b]/M$. Note that for the Hadamard CDS, shown previously to yield good biphase sequences, we get $b = \infty$. Golomb showed that projective

TABLE 6.8 Parameters of Some CDS Yielding Real-Valued Codes with Ideal Cross-Correlation

M	b	λ	D
13	3/2	6	{1 2 3 4 5 8 9 11 13}
	2/3	1	{1 2 4 10}
21	3/8	1	{1 2 5 15 17}
31	4/15	1	{1 2 4 9 13 19}
57	6/35	1	{1 2 4 14 33 37 44 53}

planes CDS yield finite values of b. Table 6.8 gives some finite projective planes cyclic difference sets and the corresponding value of b used according to (6.14) to construct a Golomb real-valued code with ideal cross-correlation. Finally, note that due to the random nature of Golomb codes, the Doppler tolerance is similar to that of the Barker codes and is much lower than that of Frank/Zadoff–Chu or $P(n, k)$ codes.

6.5 IPATOV CODE

The Golomb codes discussed at the end of Section 6.4 resulted in a mismatch power loss. Ipatov (1992) considered the problem of finding the code pairs with minimal peak response loss. The search is constrained only for those code pairs where the code used at the transmitting end is binary, and the resulting binary code is referred to as an optimal code of length M. We start by observing that for a code to have an ideal periodic correlation function, the DFT of the code elements must be constant. When there is a mismatch between the transmitting and receiving ends, the two codes must have complementary DFTs such that the multiplication of the two transforms yields a constant value.

Using the DFT property for finding optimal codes is exemplified for $M = 3$. There are $2^3 = 8$ possible binary codes of length $M = 3$. For each possible code we calculate the DFT. For example, for the three-element Barker code {1 1 −1}, we get the following DFT values:

$$\{1 \quad 1 - j1.732 \quad 1 + j1.732\}$$

The reciprocal DFT values are given by

$$\{1 \quad 0.25 + j0.433 \quad 0.25 - j0.433\}$$

and the reference code yielding perfect periodic cross-correlation is obtained by taking the inverse DFT and is given by {0.5 0 0.5}. Normalizing the reference signal to get the same peak response yields the normalized reference code {1.5 0 1.5}. While the periodic autocorrelation of the original code is

TABLE 6.9 Globally Optimal Sequences

M	Sequence	Loss (dB)	M	Sequence	Loss (dB)
3	100	1.76	17	11001101011000000	0.85
4	1000	0.00	18	101011000011000000	1.09
5	10000	0.46	19	1100010111100100000	0.46
6	100000	1.19	20	10010001011100010000	0.46
7	1000000	1.88	21	111010001001000010000	0.45
8	11010000	1.25	22	1100010010101110100000	0.86
9	100100000	2.21	23	10001101001000111000000	0.46
10	1110100000	2.24	24	110000100101011101100000	0.25
11	11100100000	1.11	25	1011110010010100011000000	0.49
12	110010100000	0.51	26	10000101111001000101100000	0.51
13	1100101000000	0.17	27	100111101001010001001100000	0.40
14	11001010000000	0.85	28	1011001001110001101010000000	0.53
15	101001110100000	0.62	29	10011110000110101001000100000	0.50
16	1100110101000000	1.00	30	100101001111100001000100110000	0.51

{3 1 1}, the periodic cross-correlation of {1 1 −1} and {1.5 0 1.5} is {3 0 0}. The power loss involved with using {1.5 0 1.5} instead of {1 1 −1} as a reference is $(1.5^2 + 1.5^2)/3 = 1.5$ (1.76 dB).

Table 6.9 gives optimal codes for higher lengths. Note that it is not always possible to find a reference code yielding zero periodic cross-correlation (consider, for example, the case when one of the DFT values of the transmitted code is zero), but for any length there exists at least one binary code with all nonzero DFT elements. Note that the machine time needed to conduct the exhaustive search used to derive the codes of Table 6.9 becomes considerably large with M above 30.

Note also that the periodic autocorrelation function of the optimal codes given in Table 6.9 has an irregular form, the result of which is an extended size of alphabet (phase and amplitude values) of the reference code used at the receiving end (e.g., for $M = 30$ there are 30 different values in the reference code). To decrease the size of the reference code alphabet, Ipatov suggested using binary sequences with only two- or three-valued periodic autocorrelation function values. Ipatov showed that using specific families of codes with two- or three-valued periodic correlation yields a two- or three-valued spectrum (and reference alphabet) and extremely low mismatch power loss.

Ipatov code generation is based on constructing an m-sequence over GF(q) and replacing its elements by the binary symbols ± 1 according to some a priori selected law based on a corresponding cyclic difference set. Table 6.10 gives some Ipatov code parameters with the resulting mismatch loss. Consider, for example, the family for which $q = 5$, $v = 4$, and $r = 3$. Let $n = 2$. A corresponding binary sequence with length $M = (q^n - 1)v/(q - 1) = 24$ is generated on the basis of m-sequence with length $q^n - 1 = 24$. A primitive polynomial of degree

TABLE 6.10 Ipatov Parameters

q, v, r	Difference set $\{v, v-r, \lambda\}$	n	$M = (q^n - 1)v/(q-1)$	Mismatch loss (dB)
3, 1, 1	{1, 0, 0}	2	4	0
		3	13	0.17
		4	40	0.37
		5	121	0.46
		6	364	0.49
		7	1093	0.51
4, 1, 1	{1, 0, 0}	2	5	0.46
		3	21	1.00
5, 4, 3	{4, 1, 0}	2	24	0.28
		3	124	0.41
		4	624	0.45
7, 3, 2	{3, 1, 0}	2	24	0.29
		3	171	0.35
		4	1200	0.39
8, 7, 3	{7, 4, 2}	2	63	0.30
		3	511	0.44
13, 4, 3	{4, 1, 0}	2	56	0.69
		3	732	0.80
16, 15, 7	{15, 8, 4}	2	255	0.05
23, 11, 6	{11, 5, 2}	2	264	0.15
29, 7, 4	{7, 3, 1}	2	210	0.90
31, 15, 8	{15, 7, 3}	2	480	0.12
32, 31, 15	{31, 16, 8}	2	1023	0.01
47, 23, 12	{23, 11, 5}	2	1104	0.08
53, 13, 9	{13, 4, 1}	2	702	0.73
61, 15, 8	{15, 7, 3}	2	930	0.52

$n = 2$ over GF(5) is, for example, $\pi(x) = x^2 + 4x + 2$. The m-sequence $\{d_i\}$ is generated by the recursion $d_i = d_{i-1} + 3d_{i-2}$ with an arbitrary choice of initial conditions. Using $d_1 = 0, d_2 = 1$, we get the following 24-element m-sequence over GF (5):

{0 1 1 4 2 4 0 2 2 3 4 3 0 4 4 1 3 1 0 3 3 2 1 2}

The mapping law for the code is based on the {4, 1, 0} ordinary cyclic difference set by placing +1 only in the positions where the original m-sequence has a zero and at those positions that belong to some {4, 1, 0} difference set. A possible {4, 1, 0} CDS is the set {1}; thus, we get the binary sequence given by

{1 1 1 −1 −1 −1 1 −1 −1 −1 −1 −1

1 −1 −1 1 −1 1 1 −1 −1 −1 1 −1}

Taking the inverse DFT of the reciprocal DFT of the binary code yields the following reference code:

$$\{5 \quad 11 \quad 11 \quad -7 \quad -7 \quad -7 \quad 5 \quad -7 \quad -7 \quad -7 \quad -7 \quad -7 \quad 5$$
$$-7 \quad -7 \quad 11 \quad -7 \quad 11 \quad 5 \quad -7 \quad -7 \quad -7 \quad 11 \quad -7\}$$

with a three-element alphabet. The mismatch loss is only 1.061 (0.28 dB).

6.6 OPTIMAL FILTERS FOR SIDELOBE SUPPRESSION

Whereas the Ipatov code shows a way of designing code pairs with perfect periodic cross-correlation and minimal mismatch loss, the method described here shows for any given transmitted code, a way of finding a suboptimal reference code, yielding lower aperiodic correlation sidelobes. Assume that the transmitted signal is defined by an M-element code $\{s_m\}$. We are interested in designing a suboptimal filter defined by a $K \geq M$ element code $\{q_k\}$ such that the cross-ambiguity function between the two signals gives minimal ambiguity function values for a specific selection of points on the delay–Doppler grid (for brevity, we use *ambiguity* instead of *cross-ambiguity*). We start by describing a simplified problem where we wish to control only the cross-correlation function on grid points given by integer multiples of the chip duration (Griep et al., 1995).

Define a filter vector $\mathbf{q}$ (with length K) and a signal vector $\mathbf{s}$ (with length M) such that the elements of $\mathbf{q}$ and $\mathbf{s}$ are given by

$$\mathbf{q} = [q_1 \quad q_2 \quad \cdots \quad q_K], \qquad \mathbf{s} = [s_1 \quad s_2 \cdots \quad s_M] \qquad (6.15)$$

The cross-correlation function for integer delay l between the pulse coded by $\{s_m\}$ and the reference pulse defined by $\{q_m\}$ is given by

$$R_{sq}[l] = \sum_{m=1}^{K} s_m q_{m+l}^* \qquad (6.16)$$

where $-(K-1) \leq l \leq (K-1)$. Define the correlation vector $\mathbf{y}$, where $y_m = R_{sq}[m]$; then (6.16) can be represented in matrix form as $\mathbf{y} = \mathbf{q}^* \mathbf{\Lambda}$, and

$$\mathbf{\Lambda} = \begin{bmatrix} s_K & \cdots & s_2 & s_1 & 0 & \cdots & 0 \\ 0 & s_K & & s_2 & & \ddots & \vdots \\ \vdots & & \ddots & & \vdots & & s_1 & 0 \\ 0 & \cdots & 0 & s_K & \cdots & s_2 & s_1 \end{bmatrix} \qquad (6.17)$$

Define a $P \times P$ weight diagonal matrix $\mathbf{F}$ such that E, the total weighted energy of $\mathbf{y}$, is given by

$$E = \mathbf{y}\mathbf{F}\mathbf{y}^H = (\mathbf{q}^*\mathbf{\Lambda})\mathbf{F}(\mathbf{q}^*\mathbf{\Lambda})^H = \mathbf{q}^*(\mathbf{\Lambda}\mathbf{F}\mathbf{\Lambda}^H)\mathbf{q}^T = \mathbf{q}^*\mathbf{A}\mathbf{q}^T \qquad (6.18)$$

The vector $\mathbf{q}$ that minimizes E under the restriction that $\mathbf{sq}^H = \mathbf{ss}^H$ (same response for zero Doppler and zero delay), and that $\mathbf{A} = \mathbf{\Lambda F \Lambda}^H$ is not singular, is given by $\mathbf{q} = \mathbf{sA}^{-1}(\mathbf{ss}^H)/(\mathbf{sA}^{-1}\mathbf{s}^H)$. The minimal value of E is given by $(\mathbf{ss}^H)^2/(\mathbf{sA}^{-1}\mathbf{s}^H)$.

The example below shows the design of a minimum integrated sidelobe level (ISLL) reference code in the case where the code used for the transmitted pulse is the five-element Barker code. Griep et al. (1995) also give additional methods for designing reference codes for a minimum peak sidelobe level and minimum cross-correlation peak power signals. We start with a Barker binary code of length $M = 5$. Assume that the length K of the reference signal q is also 5. The cross-correlation function between s and q is to be minimized on grid points laying on the zero Doppler axis and given by $\tau = \pm kt_b (1 \leq k \leq 5)$. Define

$$\Lambda = \begin{bmatrix} 1 & -1 & 1 & 1 & 1 & 0 & 0 & 0 & 0 \\ 0 & 1 & -1 & 1 & 1 & 1 & 0 & 0 & 0 \\ 0 & 0 & 1 & -1 & 1 & 1 & 1 & 0 & 0 \\ 0 & 0 & 0 & 1 & -1 & 1 & 1 & 1 & 0 \\ 0 & 0 & 0 & 0 & 1 & -1 & 1 & 1 & 1 \end{bmatrix},$$

$$F = \begin{bmatrix} 1 & 0 & 0 & 0 & 0 & 0 & 0 & 0 & 0 \\ 0 & 1 & 0 & 0 & 0 & 0 & 0 & 0 & 0 \\ 0 & 0 & 1 & 0 & 0 & 0 & 0 & 0 & 0 \\ 0 & 0 & 0 & 1 & 0 & 0 & 0 & 0 & 0 \\ 0 & 0 & 0 & 0 & 0 & 0 & 0 & 0 & 0 \\ 0 & 0 & 0 & 0 & 0 & 1 & 0 & 0 & 0 \\ 0 & 0 & 0 & 0 & 0 & 0 & 1 & 0 & 0 \\ 0 & 0 & 0 & 0 & 0 & 0 & 0 & 1 & 0 \\ 0 & 0 & 0 & 0 & 0 & 0 & 0 & 0 & 1 \end{bmatrix}$$

and get that $\mathbf{q} = [\,0.7692 \quad 1.3462 \quad 0.7692 \quad -1.3462 \quad 0.7692\,]$. The resulting cross-correlation is shown in Fig. 6.25. Note the reduction in the sidelobe level. Using a longer reference signal can yield much lower correlation sidelobes. An example of a longer reference code ($K = 11$) using $\mathbf{q} = [-0.2769 \quad -0.3175 \quad 0.0554 \quad 0.7540 \quad 1.3293 \quad 0.8334 \quad -1.3293 \quad 0.7540 \quad -0.0554 \quad -0.3175 \quad 0.2769]$ is shown in Fig. 6.26. The SNR loss involved with using a suboptimal receiver (nonmatched) is only 0.33 dB with $K = 5$ and only 0.59 dB with $K = 11$.

Note that for a given signal $\{s_m\}$, the ambiguity function and its derivatives of any order for any value of τ and ν are a linear function of $\{q_k\}$. Thus, the procedure described by Griep et al. (1995) can be generalized. We define an output vector $\mathbf{y}$ where each element $y_p (p = 1, 2, \ldots, P)$ represents a point $\langle \tau_p, \nu_p \rangle$ where we wish to impose restrictions on the r_p derivative of the ambiguity function in a direction ε_p relative to the delay axis ($r_p \geq 0$; $r_p = 0$ stands for the value of the ambiguity function itself). Since each element y_p is a linear function of $\{q_k\}$, we can still write $\mathbf{y} = \mathbf{q}^* \Lambda$, where Λ is a matrix with K rows and P columns and the elements of Λ are a function of $\mathbf{A}, \tau_p, \nu_p, r_p$, and ε_p.

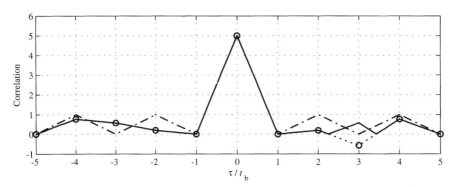

FIGURE 6.25 ACF of the five-element Barker and cross-correlation with the minimum ISLL reference of length $K = 5$.

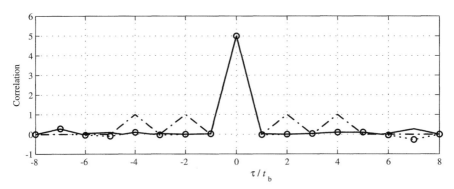

FIGURE 6.26 ACF of the five-element Barker and cross-correlation with the minimum ISLL reference code of length $K = 11$.

6.7 HUFFMAN CODE

Huffman (1962) considered the idealized compressed pulse that contains no sidelobes except the unavoidable sidelobes at the two ends of the compressed pulse. The end sidelobes are unavoidable because at the corresponding delay only one chip remains in the integrated product of the received and reference signals (see Fig. 6.27). The end sidelobe level is a design parameter. It turns out that the resulting pulses vary in amplitude as well as in phase. Because of the amplitude variations the waveforms were not very practical for radar applications. However, with the use of solid-state transmitters, they may become more attractive.

FIGURE 6.27 Reason for the edge sidelobe.

HUFFMAN CODE

The Huffman code is derived simply by describing the time sequence using polynomials. Accordingly, the transmitted sequence polynomial $S(z)$ is given by

$$S(z) = s_1 + s_2 z^1 + s_3 z^2 + \cdots + s_M z^{M-1} \qquad (6.19)$$

where $\{s_m\}$ is the complex-valued code. Huffman showed that to obtain the desired criteria, the roots of the $S(z)$ polynomial should lie in the z-plane at intervals of $2\pi/(M-1)$ on two circles whose radii R and R^{-1} are given by

$$\left[\left|\frac{1}{2a}\right| \pm \left(\frac{1}{4a^2} - 1\right)^{1/2}\right]^{1/(M-1)} \qquad (6.20)$$

The design of the Huffman code consists of specifying the number of code elements M, the edge sidelobe level a, and the particular choice of z-plane zeros which results from choosing one zero from each of the $M-1$ pairs. The design of a Huffman sequence with length $M = 6$ is given below.

We chose to design a Huffman code with peak-to-peak sidelobe level of $a = \frac{1}{6}$. The roots of the $S(z)$ polynomial lie in the z-plane at intervals of $2\pi/5$ on two circles whose radii R and R^{-1} are 1.423 and 0.703. We calculate the $S(z)$ polynomial by selecting one zero from each of the five pairs. For example,

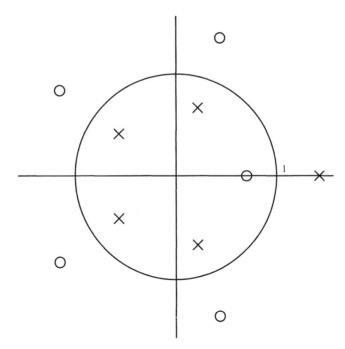

FIGURE 6.28 Zero location of a possible Huffman code polynomial.

selecting all the inner zeros that lie on radii $R = 0.703$ gives the trivial solution of $S(z) = z^5 - 0.1716$, implying the code $s = [1 \quad 0 \quad 0 \quad 0 \quad 0 \quad -0.1716]$.

A different selection of zeros is shown in Fig. 6.28. The resulting polynomial is $S(z) = z^5 - 0.7198z^4 + 1.024z^3 + 1.4569z^2 + 2.0727z - 2.8796$, implying the code $\{1 \quad -0.7198 \quad 1.0240 \quad 1.4569 \quad 2.0727 \quad -2.8796\}$. The aperiodic autocorrelation of the Huffman code of length $M = 6$ is shown in Fig. 6.29. Selecting the zero patterns in a random manner generally results in codes that vary considerably in amplitude from bit to bit. This represents a loss in terms of the power that could be transmitted at the maximum level.

An example of a longer Huffman code ($M = 23$) is given in Figs. 6.30 to 6.32. Figure 6.30 displays the real envelope of the Huffman signal. The peak-to-mean envelope power ratio (PMEPR) in this signal is 3.8. Methods to reduce the PMEPR of Huffman-coded signals were outlined by Ackroyd (1970). Figure 6.31 displays the autocorrelation function (in dB). The peak ACF sidelobe, which is a design parameter, is -63 dB. Figure 6.32 displays the partial ambiguity function. Clearly evident are the sidelobe-free zero-Doppler cut and the fast sidelobe buildup with Doppler.

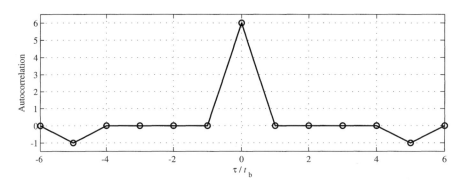

FIGURE 6.29 ACF of the six-element Huffman code.

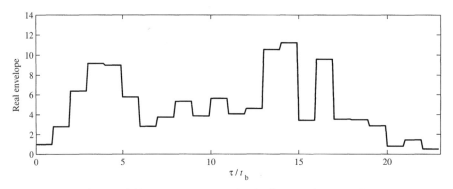

FIGURE 6.30 Real envelope of a Huffman code ($M = 23$).

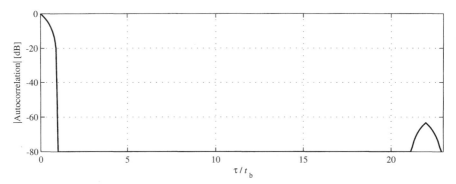

FIGURE 6.31 Autocorrelation function of a Huffman code ($M = 23$).

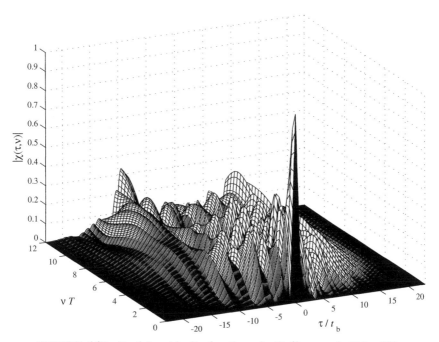

FIGURE 6.32 Partial ambiguity function of a Huffman code ($M = 23$).

6.8 BANDWIDTH CONSIDERATIONS IN PHASE-CODED SIGNALS

The instantaneous phase switching and amplitude rise time in Barker and other phase-coded signals causes extended spectral sidelobes. For example, the autocorrelation function (ACF) and spectrum (both in dB) of an ideal Barker 13 signal appear in Fig. 6.33. Note the expected ACF sidelobe peaks of $20 \log(\frac{1}{13}) = -22.3$ dB. The spectral sidelobes, referred to as spectral *skirt*, decay rather slowly, at a rate of 6 dB/octave. Of special interest is the first null at $f = 13/T = 1/t_b$, namely at the inverse of the subpulse "bit" duration. The spectral skirt poses

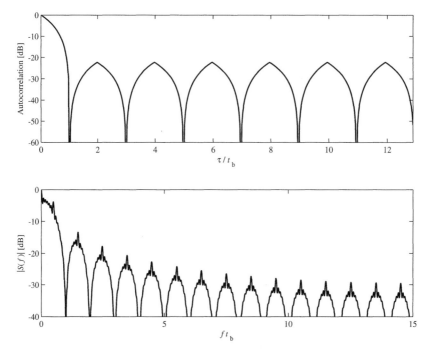

FIGURE 6.33 ACF (top) and spectrum (bottom) of an ideal Barker 13 signal.

problems at both the transmitter and the receiver. The signal transmitted may not meet spectral emission regulations. It may also exceed the bandwidth of the final transmitting stages (e.g., the antenna).

In a practical receiver (see Fig. 6.34), the matched filter will be implemented digitally; in a correlator, that follows an analog-to-digital (A/D) converter. To limit the noise power reaching the A/D, an analog narrowband filter is likely to precede it. That filter will cut off the spectrum tail. In some receivers (Farnett and Stevens, 1990; Taylor, 1990) the bandpass filter matches (or nearly matches) the subpulse, and the sample interval equals the subpulse duration. That approach is not likely to produce an output that is the exact ACF. For example, the highly spaced samples could miss the output's peak, causing a "straddle" or "cusping" loss of 2.3 dB (Klein and Fujita, 1979).

Rather than allow the bandpass filter to influence the delay response, the spectrum tail can be narrowed by modifying the signal in the transmitter (e.g., by slowing the phase switching rate). Figure 6.35 shows the phase evolution of a Barker 13 signal in which the phase transition occupies one-fifth of the bit duration. The resulted ACF and spectrum appear in Fig. 6.36. The ACF peak sidelobe increased only slightly (to -20 dB), while the -30 dB spectral width was nearly halved.

A more drastic modification to biphase coding is the biphase-to-quadriphase (BTQ) transformation (Taylor and Blinchikoff, 1988) that yields the quadriphase

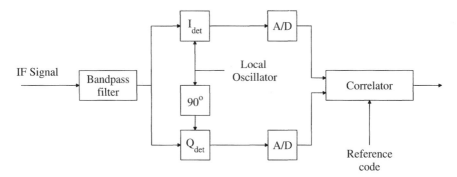

FIGURE 6.34 Matched-filter implementation for phase-coded signal.

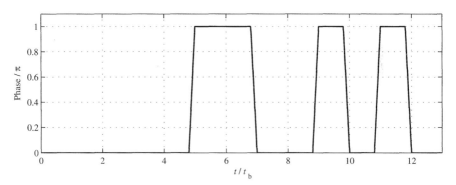

FIGURE 6.35 Phase evolution of a practical Barker 13 signal (rise time is one-fifth of the bit duration).

code. The complex envelope of a quadriphase code, which follows a biphase code with elements W_k, is

$$u(t) = \sum_{k=1}^{M} j^{s(k-1)} W_k p(t - kt_b) \qquad (6.21)$$

where the subpulse $p(t)$ is a half-cycle of a cosine wave of length $2t_b$, namely,

$$p(t) = \cos\frac{\pi t}{2t_b}, \qquad -t_b \le t \le t_b \qquad (6.22)$$

The constant coefficient s can be either 1 or -1. The quadriphase radar signal is similar to the minimum shift keying (MSK) digital modulation technique. The magnitude $a(t)$ and phase $\phi(t)$ of the complex envelope

$$u(t) = a(t)\exp[j\phi(t)]$$

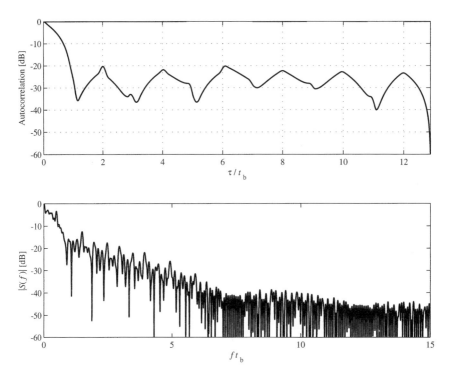

FIGURE 6.36 ACF (top) and spectrum (bottom) of the practical Barker 13 signal.

are continuous functions. If the original sequence of phases is θ_k, $k = 1, \ldots, M$, the transformed phases, at multiples of the subpulse duration t_b, are given by

$$\phi(kt_b) = \begin{cases} 0, & k = 0 \\ s(k-1)\pi/2 + \theta_k, & k = 1, \ldots, M \\ 0, & k = M + 1 \end{cases} \quad (6.23)$$

In between multiples of t_b, the phase is given by straight-line segments, connecting the values at multiples of t_b. For Barker 13 the phases derived from (6.23) appear as shown in Table 6.11.

The magnitude $a(t)$, which is a rectangle in the biphase case, is given in the quadriphasic case as

$$a(t) = \begin{cases} A\sin(2\pi t/4t_b), & 0 \leq t \leq t_b \\ A, & t_b \leq t \leq Mt_b \\ A\cos[2\pi(t - Mt_b)/4t_b], & Mt_b \leq t \leq (M+1)t_b \end{cases} \quad (6.24)$$

Figure 6.37 shows the magnitude, phase, and frequency evolution of the quadriphase signal that corresponds to Barker 13. The frequency subplot indicates a frequency switching between $-1/(4t_b)$ and $+1/(4t_b)$, except at the edge bits, where the frequency of the complex envelope is zero.

TABLE 6.11 Biphase-to-Quadriphase Transformation ($s = +1$) of a Barker 13 Sequence

k	θ_k	ϕ_k mod 2	k	θ_k	ϕ_k mod 2
0	—	0	8	0	$3\pi/2$
1	0	0	9	0	2π
2	0	$\pi/2$	10	π	$3\pi/2$
3	0	π	11	0	π
4	0	$3\pi/2$	12	π	$\pi/2$
5	0	2π	13	0	0
6	π	$3\pi/2$	14	—	0
7	π	2π			

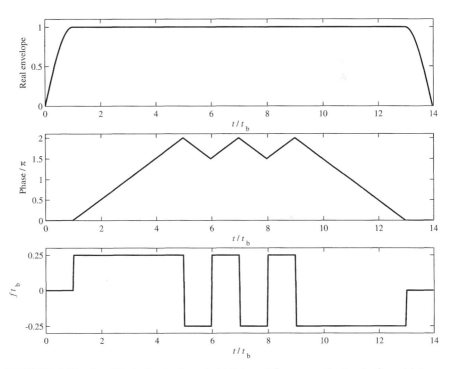

FIGURE 6.37 Amplitude (top), phase (middle), and frequency (bottom) of quadriphase code 13.

The ACF and spectrum of quadriphase code 13 are shown in Fig. 6.38, and should be compared to the ACF and spectrum of Barker 13 (Fig. 6.33). The ACF is very similar to Barker 13. However, the spectral skirt falloff rate has doubled (12 dB/octave). The first null is now at $f \approx 1/t_b\sqrt{2}$ and the -30 dB point is now at $f \approx 1/t_b$. Although the ACF changed little from Barker 13 to quadriphase 13, the AF (Fig. 6.39) off the zero-Doppler axis is quite different (Levanon and

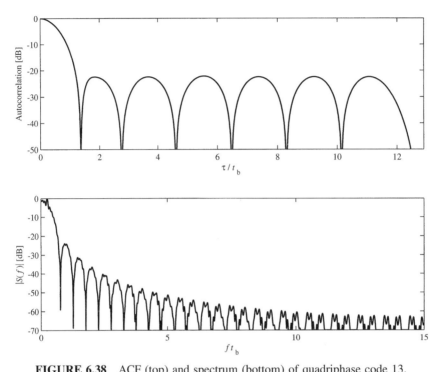

FIGURE 6.38 ACF (top) and spectrum (bottom) of quadriphase code 13.

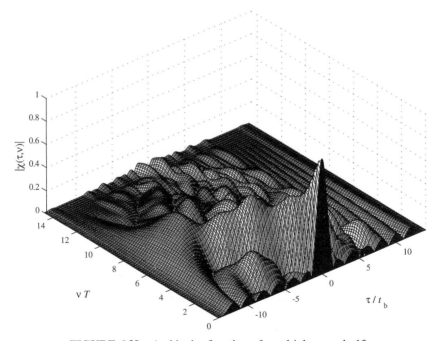

FIGURE 6.39 Ambiguity function of quadriphase code 13.

Freedman, 1989) from that of Barker 13 (Fig. 6.6). Quadriphase code 13 was generated using the MATLAB code in Appendix 6B and the ambiguity plotting codes.

Chen and Cantrell (2002) suggested another bandlimiting technique, in which the rectangular shape of the phase bits is replaced by the shape of a Gaussian-windowed sinc function. A generalization of their signal, which allows polyphase coding (rather than just binary), will define the complex envelope of a single bit as

$$u_m(t) = \exp(j\phi_m)\frac{\sin(\pi t/t_b)}{\pi t/t_b}\exp\left[\frac{-t^2}{2(\sigma t_b)^2}\right],$$

$$-2t_b \leq t \leq 2t_b, \quad \text{zero elsewhere} \tag{6.25}$$

A pulse constructed from M bits is defined as

$$u(t) = \sum_{m=1}^{M} u_m[t - (m-1)t_b] \tag{6.26}$$

It is important to note that the bit duration now extends over $4t_b$, yet consecutive bits are spaced only t_b apart. Hence there is an overlap of $3t_b$. This also implies that the duration of the entire pulse is $T = (M+3)t_b$ and not Mt_b, as in a conventional phase-coded signal. The magnitude of a bit is shown in Fig. 6.40

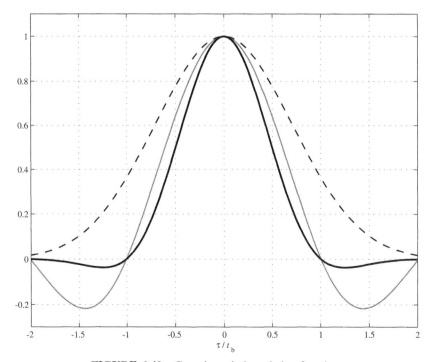

FIGURE 6.40 Gaussian-windowed sinc function.

(solid), which also displays its two components: the sinc (dotted) and the Gaussian (dashed). σ = 0.7 was used in (6.25).

The use of this bit shape will first be demonstrated on a Barker 13 signal. The magnitude and phase of the signal are shown in Fig. 6.41. The variable amplitude points to the main difficulty with this signal—the requirement for linear power amplification. The autocorrelation function and spectrum are shown in Fig. 6.42. They should be compared to Figs. 6.33 and 6.38. The ACF (top subplot of Fig. 6.42) peak sidelobe level is not much different than in the conventional Barker or the quadriphase Barker. However, the spectrum (lower subplot of Fig. 6.42) is dramatically narrower, with a negligible spectral skirt. The ambiguity function (AF) is plotted in Fig. 6.43 and is very similar to the AF of a conventional Barker 13 signal.

The second example will be a 25-element P4 signal utilizing a Gaussian-windowed sinc bit shape. In the P4 case the overlap of 3 bits will cause a variable phase within a bit. The amplitude and phase of the signal are given in Fig. 6.44. σ = 0.7 is still used in (6.25). The need for a linear power amplifier is obvious from the top subplot. The phase, which is a smoothed P4 phase, begins to resemble the phase behavior of a linear-FM signal. The ACF and spectrum of the modified 25-element P4 signal are plotted in Fig. 6.45 and should be compared to those of a conventional 25-element P4 (Fig. 6.46). The modified P4

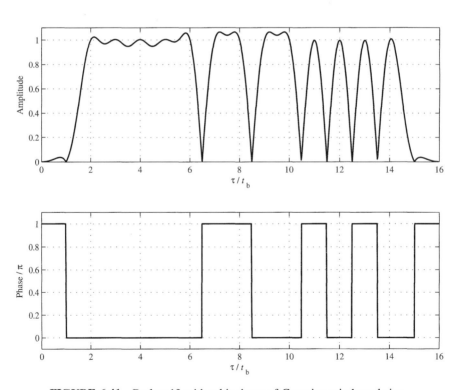

FIGURE 6.41 Barker 13 with a bit shape of Gaussian-windowed sinc.

BANDWIDTH CONSIDERATIONS IN PHASE-CODED SIGNALS 153

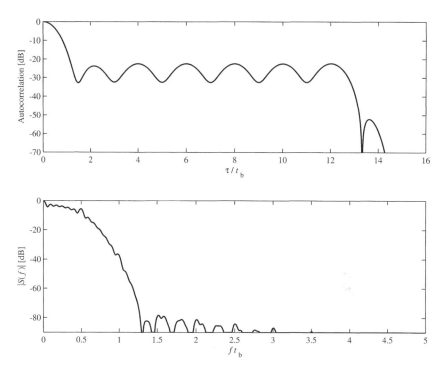

FIGURE 6.42 ACF and spectrum of Barker 13 with a bit shape of Gaussian-windowed sinc.

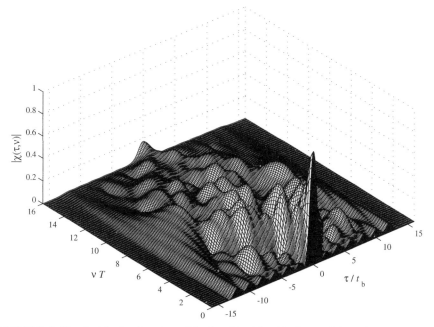

FIGURE 6.43 Ambiguity function of Barker 13 with a bit shape of Gaussian-windowed sinc.

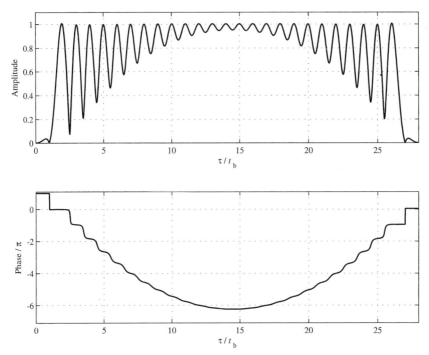

FIGURE 6.44 P4 (25-bit) with a bit shape of Gaussian-windowed sinc.

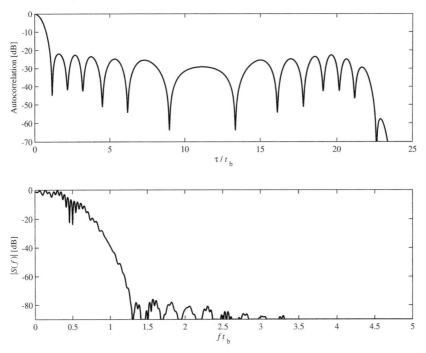

FIGURE 6.45 ACF and spectrum of 25-element P4 with a bit shape of Gaussian-windowed sinc.

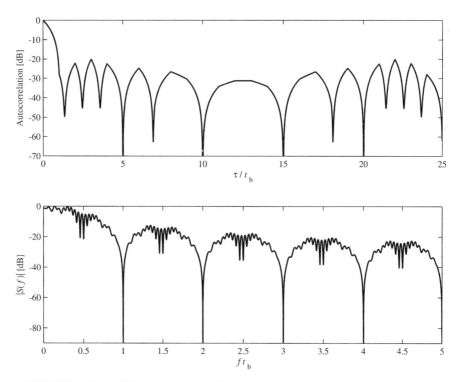

FIGURE 6.46 ACF and spectrum of 25-element P4 with a rectangular bit shape.

exhibits a small reduction of ACF sidelobes and a dramatic (almost complete) reduction of the spectral sidelobes. The examples of phase-coded signal with a Gaussian-windowed sinc bit waveform in this chapter were generated using the MATLAB program in Appendix 6C.

Although not shown, it should be pointed out that the ideal periodic autocorrelation of a conventional P4 signal is lost in the modified P4 signal. In contrast, while Barker 13 PACF has rather high sidelobes, its quadriphase modification turns out to be very nearly ideal, with a PACF sidelobe level near -60 dB. This property is not repeated in the quadriphase signal based on Barker 7.

6.9 CONCLUDING COMMENTS

We have discussed some sets of codes that can be operated either periodically or aperiodically under various conditions and optimization rules. Most of the codes described here are used in operational radar systems. Many other codes were not presented in detail. Among these codes we choose to mention the primitive root and quadratic residue codes (Schroeder, 1986), generalized Frank codes (Suehiro and Hatori, 1988a; Kretschmer and Gerlach 1991), N-shift cross-orthogonal sequences (Suehiro and Hatori, 1988b), reciprocal perfect phase codes (Kretschmer and Gerlach, 1991), generalized P4/Zadoff–Chu codes (Kretschmer and Gerlach, 1991; Popovic, 1992), index codes (Kretschmer and Gerlach, 1991), and binary alexis sequences (Luke, 2001). Many other codes have been published. Finally, we note the theoretical work initiated by Welch (1974) and Sarwate (1979) putting bounds on the cross-correlation and autocorrelation of sets of periodic and aperiodic sequences.

APPENDIX 6A: GALOIS FIELDS

A Galois field is an algebraic identity having a finite number of members and obeying a set of roles. The simplest and smallest example of such a field is the set containing the elements 1 and 0. We denote this field as GF(2). Galois fields play a fundamental role in the theory and application of error-control coding, multiuser communication, cryptography, and digital signal processing. This appendix is not a replacement for a good textbook on the subject or a dedicated course. Instead, it can serve as a quick tutorial for those who are not familiar with the subject at all, or as a quick reference for those who know the basics of the subject but do not practice it. We start by defining some basic algebraic identities.

Definition: A *field* is a set $\mathcal{F}$ of elements in which it is possible to add, subtract, multiply, and divide (except that division by 0 is not defined). Addition and multiplication must satisfy the commutative, associative, and distributive laws. That is, for all elements x, y, and z of $\mathcal{F}$:

1. $x + y = y + x$ (addition is commutative).
2. $(x + y) + z = x + (y + z)$ (addition is associative).
3. $x \cdot y = y \cdot x$ (multiplication is commutative).
4. $x \cdot (y \cdot z) = (x \cdot y) \cdot z$ (multiplication is associative).
5. $x \cdot (y + z) = x \cdot y + x \cdot z$ and $(x + y) \cdot z = x \cdot z + y \cdot z$ (the distributive law).

The field must also contain elements 0, 1, $-x$, and x^{-1} with the properties that $x + 0 = x$, $1 \cdot x = x$, $x + (-x) = 0$, and $x \cdot x^{-1} = 1$ for all elements x of $\mathcal{F}$.

A *finite field* (or Galois field) contains a finite number of elements, this number being called the *order* of the field. For example, let p be a prime number. Then the integers modulo p form a field of order p, denoted by GF(p). The elements of GF(p) are $\{0, 1, 2, \ldots, p-1\}$ and $+, -, \cdot,$ and $\div$ are carried out mod p. GF(3) is the ternary field $\{0, 1, 2\}$ with $1 + 2 = 3 = 0$ (mod 3), $2 \cdot 2 = 4 = 1$ (mod 3), $1 - 2 = -1 = 2$ (mod 3), $1 \div 2 = 4 \div 2 = 2$ (mod 3), and so on.

The multiplication and addition tables for the GF(3) are given by

+	0	1	2
0	0	1	2
1	1	2	0
2	2	0	1

·	0	1	2
0	0	0	0
1	0	1	2
2	0	2	1

Finite fields of order p^m exist for all p and m where p is a prime and m is some integer. Furthermore, all fields of the same order are equivalent. An example of a field with $2^2 = 4$ elements is now given. Each element of GF(4) can be represented by two elements of GF(2). This gives the four elements given by 00, 01, 10, and 11.

Any two elements of GF(4) can clearly be added by adding the elements in the corresponding location using the addition rule in GF(2) (i.e., modulo 2). To define multiplication we associate with each element a polynomial in α:

Vector representation	Polynomial representation
00	0
10	1
01	α
11	$1 + \alpha$

The multiplication of two elements in the field is defined as the multiplication of the two polynomials. Since simple multiplication of any two polynomials in the field does not always give a polynomial with degree ≤ 1, we must also agree that α will satisfy a certain fixed equation of degree 2. A suitable equation is $\pi(\alpha) = \alpha^2 + \alpha + 1 = 0$. For the multiplication to have an inverse, which it must if our system is to be a field, $\pi(x)$ must be irreducible over GF(2).

Definition: A *primitive element* of GF(p^m) is a cyclic generator of the group of nonzero elements of GF(p^m). This means that every nonzero element of the field can be expressed as the primitive element raised to some integer power. Usually, the primitive element is denoted by α. In the example given before for the construction of GF(4), note that the "10" element is $1 = \alpha^0$, "01" is $\alpha = \alpha^1$, and "11" is $1 + \alpha = \alpha^2$, since $\pi(\alpha) = \alpha^2 + \alpha + 1 = 0$.

Definition: A *primitive polynomial* for GF(p^m) is the minimal polynomial of some primitive element of GF(p^m). That is, it is the p-ary coefficient polynomial of smallest nonzero degree having a certain primitive element as a root in GF(p^m). Consequently, a primitive polynomial has degree m and is irreducible (i.e., it is not the product of two polynomials of lower degree in the field). For example, the polynomial $\pi(x) = x^2 + x + 1$ has degree 2 and is irreducible over GF(2) since $\pi(1) = 1^2 + 1 + 1 = 1$ and $\pi(0) = 0^2 + 0 + 1 = 1$.

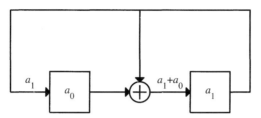

FIGURE 6A.1 Shift register implementation for multiplication by α in GF(4), where $\pi(\alpha) = \alpha^2 + \alpha + 1 = 0$.

Using the definition of the primitive polynomial and primitive element, we can now generalize the example given before for the construction of Galois fields of order p^m.

Definition: Suppose that $\pi(x)$ is irreducible over GF(p) and has degree m. Then the set of all polynomials in x of degree $\leq m - 1$ and with coefficients from GF(p), with calculations performed modulo $\pi(x)$, form a field of order p^m. The prime number p is called the *characteristics* of the field.

Computation in a finite field with characteristics 2 is easily manipulated using digital circuits or in a binary computer. An element of GF(2^m) is stored in a row of m binary storage elements (a register). To multiply by α where $\pi(\alpha) = 0$, we use a linear feedback shift register with the taps connected according to the coefficients of the primitive polynomial $\pi(x)$. For example, if the shift register in Fig. 6A.1 initially contains $a_0, a_1 \leftrightarrow a_0 + a_1\alpha$, one time instant later it contains $a_0\alpha + a_1\alpha^2 = a_0\alpha + a_1(\alpha + 1) = a_1 + (a_0 + a_1)\alpha$.

If the initial condition is not 00 and α is primitive, the output of the circuit of Fig. 6A.1 is periodic with period $2^2 - 1 = 3$. This is the maximum possible period with two storage elements. The resulting sequence of 0's and 1's is called a *maximal-length shift register code*, or *m-sequence*. In a similar way, circuits with ternary or p-ary elements can be implemented, forming *m*-sequences over a p-ary alphabet.

APPENDIX 6B: QUADRIPHASE BARKER 13

The following MATLAB program generates a numerical quadriphase Barker 13 signal, which is then used with the ambiguity function plot programs in Appendix 3A to study its performances. After running the code given here, the workspace includes the variables u_amp1 and f_basic1, which define the quadriphase Barker 13 signal complex envelope. Use u_amp=u_amp1 and f_basic =f_basic1 and run the ambiguity function code of Appendix 3A to plot the ambiguity function of the code.

```
% quadriphase_barker13.m - Barker 13 using Taylor's quadriphase modification
% written by Nadav Levanon

nn=14;

mm=100;
mmm=1:mm;
amp_rise=sin(2*pi/4/mm*mmm);
amp_fall=fliplr(amp_rise);
a1 =ones(1,mm);
b1 =-a1;
c1=zeros(1,mm);

% barker13   0 0 0 0 0 1 1 0 0 1 0 1 0

f_basic1=0.25/mm*[c1 a1 a1 a1 a1 b1 a1 b1 a1 b1 b1 b1 b1 c1];
u_amp1=[amp_rise a1 a1 a1 a1 a1 a1 a1 a1 a1 a1 a1 a1 amp_fall];

figure(10)
plot( u_amp1,'+')
grid
axis([0 mm*nn -.1 1.1])
```

APPENDIX 6C: GAUSSIAN-WINDOWED SINC

The following MATLAB program generates several phase-coded pulses in which the bit is a Gaussian-windowed sinc extended over four code bits. After running the code given here, the workspace includes the variables u_amp1 and f_basic1, which define the Gaussian-windowed sinc signal complex envelope. Use u_amp =u_amp1 and f_basic =f_basic1 and run the ambiguity function code of Appendix 3A to plot the ambiguity function of the code.

```
% gaussian_sinc.m - Gaussian windowed sinc waveform
% written by Nadav Levanon

% single bit shape
nn=201;   nn2=(nn-1)/2; nn22=nn2/2;
small=.00000001;
nnn=-nn2:nn2;
arg_bit=small+4*pi/nn*nnn;
amp_bit=sin(arg_bit)./arg_bit;

s_gauss=.7;
gauss_weight=exp(-0.5*(nnn./(nn/4*s_gauss)).^2);
ab_a=amp_bit.*gauss_weight;

sig_type=input(' Barker13 =1, P4 =2, single bit =3,  = ? ');

if sig_type==1
    phase_vec=pi*[0 0 0 0 0 1 1 0 0 1 0 1 0];   % Barker 13
elseif sig_type==2
    mm=input(' No. of elements of P4 signal = ? ');
    m=1:mm;
    phase_vec=pi*(1/mm*(m-1).^2-(m-1));  % P4
```

```
elseif sig_type==3
    phase_vec=[0 0 0 0 0 0 0 0 0];
end

lb=length(phase_vec);
vec_length=(lb+3)*nn22+1;
ab=zeros(lb,vec_length);
bpv=ones(1,nn); % bit phase vector

for k=1:lb
  ab(k,:)=[zeros(1,(k-1)*nn22), ab_a.*exp(j*phase_vec(k)*bpv),
    zeros(1,(lb-k)*nn22)];
end

u_amp_complex=sum(ab);

u_amp1=abs(u_amp_complex);
u_phase1=1/pi*angle(u_amp_complex);
t_axis=nnn/nn22;

figure(10)
plot(u_amp1,'k','linewidth',1.5)
grid

figure(11), clf, hold off
plot( t_axis,ab_a,'k','linewidth',2.5)
hold on
plot(t_axis,gauss_weight,'k--','linewidth',1.5)
plot(t_axis,amp_bit,'k:','linewidth',1.5)
grid
axis([ -inf inf -.3 1.1])
xlabel('τ/t_b')
```

PROBLEMS

6.1 Equivalent codes

Show that every polyphase code is equivalent to one that begins $\{\phi_1 = 0, \phi_2 = 0, \phi_3, \ldots\}$, where $0 \leq \phi_3 \leq \pi$. What does the equivalence imply regarding:

(a) The autocorrelation function of the code?

(b) The cross-correlation function of the code with a different code?

(c) The frequency spectrum of the code?

6.2 Cyclic shift operation

Show that cyclically shifting a code preserves the periodic autocorrelation function. What happens to the periodic cross-correlation when cyclically shifting one of the codes?

6.3 Decimation

Show that decimating a sequence by d, which is relatively prime to the code length M, results with decimating the periodic autocorrelation function.

6.4 Symmetric aperiodic ACF sidelobe property of a perfect code
Show that for a perfect code, the aperiodic correlation function has sidelobes symmetric around $\tau = Mt_b/2$.

6.5 New perfect codes
Consider a new code obtained by multiplying the elements of a Frank code and a Chu code of the same odd length.
(a) Calculate the code elements for a length $M = 25$.
(b) Show that the new code is also a perfect code.
(c) Prove that all codes formed in the same way are perfect.

6.6 Code comparison I
For the Barker, polyphase Barker, Chu, P3, and P4 codes with $M = 13$ elements, calculate and plot:
(a) The phase history.
(b) The aperiodic autocorrelation function.
(c) The periodic correlation function.
(d) The power spectrums.

What can be said regarding the periodic and aperiodic correlation functions of codes from observing the power spectrums?

6.7 Code comparison II
Calculate the aperiodic autocorrelation function of a minimum peak sidelobe code of length $M = 6$. Give a four-phase code of length 6 with a peak sidelobe of 2, using progressive multiplication.
(a) Plot the autocorrelation function of the new codes and compare to the MPS code obtained using Table 6.3.
(b) Compare the two codes' structure, integrated sidelobe levels, frequency spectrums, and ambiguity functions.

6.8 Code comparison III
For a P4, palindromic P4, and P(4,k_{opt}) codes of length $M = 64$, compare:
(a) The aperiodic and periodic mainlobe width.
(b) Autocorrelation peak response and autocorrelation peak sidelobe.

Use the results to define scenarios or applications, for which one of the three types of signals is more suitable than the others.

6.9 Code comparison IV
(a) Calculate the phase elements of an m-sequence of length $M = 15$ (use the shift register initial condition where all states are "1").
(b) Plot the aperiodic and periodic autocorrelation functions of the resulting code.
(c) Calculate Golomb's biphase code based on the m-sequence just calculated and plot the aperiodic and periodic autocorrelation function of the Golomb code.

(d) Compare the resulting periodic autocorrelation and aperiodic autocorrelation with a globally optimal Ipatov binary code of length $M = 15$.

6.10 Code comparison V
(a) Calculate an Ipatov binary code of length $M = 63$ (use Table 6.10).
(b) Compare a code with an m-sequence and a Golomb biphase code of length $M = 63$.
(c) Compare the codes to Frank and Px codes of length $M = 64$.

6.11 Suboptimal reference code
(a) For a Barker code of length $M = 13$, calculate an optimal filter with length 13 and 27 such that the integrated cross-correlation sidelobe level is minimized.
(b) What is the SNR loss when using the longer reference code compared to a shorter code and compared to the matched-filter case?

6.12 Primitive root code
The M code words of the primitive root code are defined as $u_m = \exp[-j2\pi/(M+1)\alpha^{m-1}]$, where $M-1$ must be a prime number and α is a primitive root modulo $M+1$.
(a) Calculate and plot the phase and autocorrelation functions of the primitive root codes of length $M = 100$ (101 is prime) with $\alpha = 2$ and 3.
(b) Plot the aperiodic cross-correlation between the two codes.
The results are typical for the cross-correlations between two primitive root codes of the same length or number of elements; the sidelobes are down from the peak by approximately the pulse compression ratio.

6.13 Quaternary residue code
We introduce the Legendre symbol $\{q/p\}$. This symbol is defined for all q that are not divisible by p; it is equal to 1 if q is a quadratic residue of p and is equal to -1 otherwise. Note that q is a quadratic residue of p if the congruence $z^2 = q \bmod p$ has a solution. The code is defined as $u_m = \{(m-1)/M\}$, where M is a prime number of the form $4N-1$. Note that we define $\{0/M\} = 1$.
(a) Calculate the quaternary residue codes of length $M = 11$ and $M = 19$.
(b) Plot the aperiodic autocorrelation functions of the two codes.

6.14 Suboptimal filtering of nested codes
A nested Barker code of length 65, shown below, was created from an outer Barker code of length 5 in which each element is a Barker code of length 13.

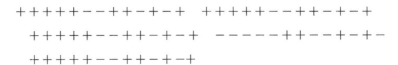

(a) Design a filter of length 31 with impulse response H31 that will yield a minimum integrated sidelobe for Barker code of length 13.

(b) Perform convolution between H31 and the entire code of length 65. The product will be labeled OUT1. Plot OUT1 using a linear scale.

(c) Design a filter of length 17 with impulse response H17 that will yield a minimum integrated sidelobe for a Barker code of length 5.

(d) Extend H17 by inserting 12 zero elements after each element of H17. The new impulse response, of length $13 \times 17 = 221$, will be labeled H17W.

(e) Perform convolution between OUT1 and H17W. The overall convolution product will be labeled OUT. Plot OUT in decibels.

6.15 Quadriphase code I
Derive a quadriphase signal based on Barker 11. Calculate and plot:
(a) Its amplitude, phase, and frequency evolution.
(b) Its autocorrelation function (in dB).
(c) Its spectrum (in dB).
(d) Its ambiguity function.

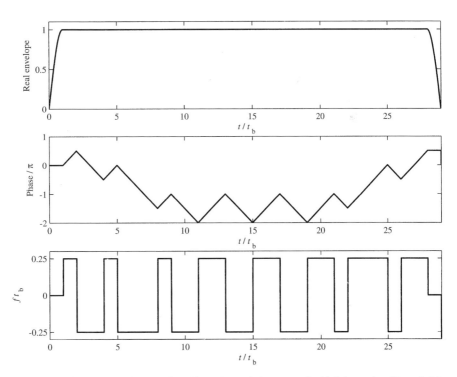

FIGURE P6.16 Quadriphase signal based on minimum peak sidelobe code of length 28.

6.16 Quadriphase code II
Show that the minimum peak sidelobe code of length 28 (see Table 6.3) can be modified to the quadriphase signal shown in Fig. P6.16. Calculate and plot:

(a) Its autocorrelation function (in dB).

(b) Its spectrum (in dB).

(c) Its ambiguity function.

REFERENCES

Ackroyd, M. H., The design of Huffman sequences, *IEEE Transactions on Aerospace and Electronic Systems*, vol. AES-6, November 1970, pp. 790–796.

Antweiler, M., and L. Bomer, Merit factor of Chu and Frank sequences, *Electronics Letters*, vol. 26, no. 25, 6 December 1990, pp. 2068–2070.

Barker, R. H., Group synchronization of binary digital systems, in Jackson, W. (ed.), *Communication Theory*, Academic Press, London, 1953, pp. 273–287.

Boehmer, A. M., Binary pulse compression codes, *IEEE Transactions on Information Theory*, vol. IT-13, no. 2, April 1967, pp. 156–167.

Bomer, L., and M. Antweiler, Polyphase Barker sequences, *Electronics Letters*, vol. 25, no. 23, November 1989, pp. 1577–1579.

Brenner, A. R., Polyphase Barker sequences up to length 45 with small alphabets, *Electronics Letters*, vol. 34, no. 16, August 1998, pp. 1576–1577.

Chen, R., and B. Cantrell, Highly bandlimited radar signals, *Proceedings of the 2002 IEEE Radar Conference*, Long Beach, CA, April 22–25, 2002, pp. 220–226.

Chu, D. C., Polyphase codes with good periodic correlation properties, *IEEE Transactions on Information Theory*, vol. IT-18, no. 4, July 1972, pp. 531–532.

Cohen, M. N., J. M. Baden, and P. E. Cohen, Biphase codes with minimum peak sidelobes, *Proceedings of the IEEE National Radar Conference*, 1989, pp. 62–66.

Cohen, M. N., M. R. Fox, and J. M. Baden, Minimum peak sidelobes pulse compression codes, *Proceedings of the IEEE International Radar Conference*, Arlington, VA, May 1990, pp. 633–638.

Coxson, G. E., A. Hirschel, and M. N. Cohen, New results on minimum-PSL binary codes, *Proceedings of the 2001 IEEE Radar Conference*, Atlanta, GA, May 2001, pp. 153–156.

Ein-Dor, L., I. Kanter, and W. Kinzel, Low autocorrelated multiphase sequences, *Physical Review E (Statistical, Nonlinear, and Soft Matter Physics)*, vol. 65, no. 2, February 2002, pp. 020102/1–020102/4

Eliahou, S., and M. Kervaire, Barker sequences and difference sets, *L'Enseignement Mathematique*, vol. 38, no. 3–4, 1992, pp. 345–382.

Farnett, E. C., and G. H. Stevens, Pulse compression radar, in Skolnik, M. I. (ed.), *Radar Handbook*, 2nd ed., McGraw-Hill, New York, 1990, Chap. 10.

Felhauer, T., Design and analysis of new P(n, k) polyphase pulse compression codes, *IEEE Transactions on Aerospace and Electronic Systems*, vol. 30, no. 3, July 1994, pp. 865–874.

REFERENCES

Frank, R. L., Phase coded communication system, U.S. patent 3,099,795, July 30, 1963a.

Frank, R. L., Polyphase codes with good non-periodic correlation properties, *IEEE Transactions on Information Theory*, vol. IT-9, no. 1, January 1963b. pp. 43–45.

Frank, R. L., Comments on polyphase codes with good correlation properties, *IEEE Transactions on Information Theory*, vol. IT-19, no. 2, March 1973, p. 244.

Frank, R. L., and S. A. Zadoff, Phase shift pulse codes with good periodic correlation properties, *IRE Transactions on Information Theory*, vol. IT-8, no. 6, October 1962, pp. 381–382.

Frank, R. L., and S. A. Zadoff, Phase coded signal receiver, U.S. patent 3,096,482, July 2, 1963.

Friese, M., Polyphase Barker sequences up to length 36, *IEEE Transactions on Information Theory*, vol. 42, no. 4, July 1996, pp. 1248–1250.

Friese, M., and H. Zottmann, Polyphase Barker sequences up to length 31, *Electronics Letters*, vol. 30, no. 23, November 1994, pp. 1930–1931.

Gerlach, K., and F. F. Kretschmer, Reciprocal perfect and asymptotically perfect periodic radar waveforms and their aperiodic properties, NRL Report 9059, Naval Research Laboratory, Washington, DC, May 1989.

Golomb, S. W., Two-valued sequences with perfect periodic autocorrelation, *IEEE Transactions on Aerospace and Electronic Systems*, vol. 28, no. 2, March 1992, pp. 383–386.

Golomb, S. W., and R. A. Scholtz, Generalized Barker sequences, *IEEE Transactions on Information Theory*, vol. IT-11, no. 4, October 1965, pp. 533–537.

Griep, K. R., J. A. Ritcey, and J. J. Burlingame, Poly-phase codes and optimal filters for multiple user ranging, *IEEE Transactions on Aerospace and Electronic Systems*, vol. 31, no. 2, April 1995, pp. 752–767.

Heimiller, R. C., Phase shift pulse codes with good periodic correlation properties, *IRE Transactions on Information Theory*, vol. IT-7, no. 6, October 1961, pp. 254–257.

Heimiller, R. C., Author's comment, *IRE Transactions on Information Theory*, vol. IT-8, 1962, p. 382.

Huffman, D. A., The generation of impulse-equivalent pulse trains, *IRE Transactions on Information Theory*, vol. IT-8, September 1962, pp. S10–S16.

Ipatov, V. P., *Periodic Discrete Signals with Optimal Correlation Properties*, Radio I Svyaz, Moscow, 1992. English translation, ISBN 5-256-00986-9, available from the University of Adelaide, Australia.

Kerdock, A. M., R. Mayer, and D. Bass, Longest binary pulse compression codes with given peak sidelobe levels, *Proceedings of the IEEE*, vol. 74, no. 2, February 1986, p. 366.

Klein, A. M., and M. T. Fujita, Detection performance of hard-limited phase-coded signals, *IEEE Transactions on Aerospace and Electronic Systems*, vol. AES-15, no. 6, November 1979, pp. 795–802.

Kretschmer, F. F., and K. Gerlach, Low sidelobe radar waveforms derived from orthogonal matrices, *IEEE Transactions on Aerospace and Electronic Systems*, vol. 27, no. 1, January 1991, pp. 92–102.

Kretschmer, F. F., and B. L. Lewis, Doppler properties of poly-phase coded pulse compression waveforms, *IEEE Transactions on Aerospace and Electronic Systems*, vol. AES-19, no. 4, July 1983, pp. 521–531.

Levanon, N., and A. Freedman, Ambiguity function of quadriphase coded radar pulse, *IEEE Transactions on Aerospace and Electronic Systems*, vol. 25, no. 6, November 1989, pp. 848–853.

Lewis, B. L., and F. F. Kretschmer, A new class of polyphase pulse compression codes and techniques, *IEEE Transactions on Aerospace and Electronic Systems*, vol. AES-17, no. 3, May 1981a, pp. 364–372.

Lewis, B. L., and F. F. Kretschmer, Corrections to: A new class of polyphase pulse compression codes and techniques, *IEEE Transactions on Aerospace and Electronic Systems*, vol. AES-17, no. 5, September 1981b, p. 726.

Lewis, B. L., and F. F. Kretschmer, Linear frequency modulation derived polyphase pulse compression codes, *IEEE Transactions on Aerospace and Electronic Systems*, vol. AES-18, no. 5, September 1982, pp. 637–641.

Lindner J., Binary sequences up to length 40 with best possible autocorrelation function, *Electronics Letters*, vol. 11, no. 21, 16 October 1975, p. 507.

Luke, H. D., Binary Alexis sequences with perfect correlation, *IEEE Transactions on Communications*, vol. 49, no. 6, June 2001, pp. 966–968.

Maschietti, A., Difference sets and hyperovals, *Designs, Codes and Cryptography*, vol. 14, no. 1, April 1998, pp. 89–98.

Popovic, B. M., Generalized chirp-like polyphase sequences with optimum correlation properties, *IEEE Transactions on Information Theory*, vol. 38, no. 4, July 1992, pp. 1406–1409.

Popovic, B. M., M. Antweiler, and L. Bomer, Comment on: Merit factor of Chu and Frank sequences (and reply), *Electronics Letters*, vol. 27, no. 9, April 25, 1991, pp. 776–778.

Rapajic, P. B., and R. A. Kennedy, Merit factor based comparison of new polyphase sequences, *IEEE Communication Letters*, vol. 2, no. 10, October 1998, pp. 269–270.

Sarwate, D. V., Bounds on the crosscorrelation and autocorrelation of sequences, *IEEE Transactions on Information Theory*, vol. IT-25, no. 6, November 1979, pp. 720–724.

Schroeder, M. R., Synthesis of low-peak factor signals and binary sequences with low autocorrelation, *IEEE Transactions on Information Theory*, vol. IT-16, no. 1, January 1970, pp. 85–89.

Schroeder, M. R., *Number Theory in Science and Communications*, 2nd ed., Springer-Verlag, New York, 1986.

Suehiro, N., and M. Hatori, Modulatable orthogonal sequences and their application to SSMA systems, *IEEE Transactions on Information Theory*, vol. 34, no. 1, January 1988a, pp. 93–100.

Suehiro, N., and M. Hatori, N-shift cross-orthogonal sequences, *IEEE Transactions on Information Theory*, vol. 34, no. 1, January 1988b, pp. 143–146.

Taylor, J. W., Jr., Receivers, in Skolnik, M. I. (ed.), *Radar Handbook*, 2nd ed., McGraw-Hill, New York, 1990, Chap. 3.

Taylor, J. W., Jr., and H. J. Blinchikoff, Quadriphase code: a radar pulse compression signal with unique characteristics, *IEEE Transactions on Aerospace and Electronic Systems*, vol. 24, no. 2, March 1988, pp. 156–170.

Turyn, R., On Barker codes of even length, *Proceedings of the IEEE*, vol. 51, no. 9, September 1963, p. 1256.

Wai Ho Mow and S. Y. R. Li, A-periodic autocorrelation and crosscorrelation of polyphase sequences, *IEEE Transactions on Information Theory*, vol. 43, no. 3, May 1997, pp. 990–1007.

Welch, L. R., Lower bounds on the maximum crosscorrelation of signals, *IEEE Transactions on Information Theory*, vol. IT-20, no. 3, May 1974, pp. 397–399.

Welti, G. R., Quaternary codes for pulsed radar, *IRE Transactions on Information Theory*, vol. IT-6, June 1960, pp. 400–408.

Zadoff, S. A., Phase coded communication system, U.S. patent 3,099,796, July 30, 1963.

Zhang, N, and S. W. Golomb, Sixty-phase generalized Barker sequences, *IEEE Transactions on Information Theory*, vol. 35, no. 4, July 1989, pp. 911–912.

Zhang, N., and S. W. Golomb, Polyphase sequences with low autocorrelation, *IEEE Transactions on Information Theory*, vol. 39, no. 3, May 1993, pp. 1085–1089.

7

COHERENT TRAIN OF LFM PULSES

Good Doppler resolution requires a long coherent signal. To minimize eclipsing requires a short transmission time (unless the radar is especially designed to transmit and receive simultaneously, as in CW radar). A solution that meets both requirements is a coherent train of pulses. The basic type of such a signal was discussed in Sections 3.6 and 4.3, where the train was constructed from identical constant-frequency pulses. In many practical cases the pulses are modulated and are not identical. Modulation produces wider bandwidth, hence pulse compression. The identicalness is violated by even the simple introduction of interpulse weighting, used to lower Doppler sidelobes. In some signals, significant diversity is introduced between the pulses in order to obtain additional advantages, such as lower delay sidelobes or lower recurrent lobes.

In this and the following chapters we extend the discussion in two directions: adding modulation and adding diversity. Adding modulation but keeping the pulses identical still allows us to use the theoretical results regarding the periodic ambiguity function (Section 3.6) and to obtain analytic expression for the ambiguity function (within the duration of one pulse, i.e., $|\tau| \leq T$). Adding diversity in amplitude (i.e., weighting) or by different modulation in different pulses within the coherent train will usually require numerical analysis, except for a few simple cases in which theoretical analysis is available. Such is the case for the subject of this chapter—a coherent train of LFM pulses—probably the most popular radar signal in airborne radar (Rihaczek, 1969; Stimson, 1983; Nathanson et al., 1991).

Radar Signals, By Nadav Levanon and Eli Mozeson
ISBN 0-471-47378-2 Copyright © 2004 John Wiley & Sons, Inc.

7.1 COHERENT TRAIN OF IDENTICAL LFM PULSES

A train of identical linear-FM pulses provides both range resolution and Doppler resolution—hence its importance and popularity in radar systems. Its ambiguity function still suffers from significant sidelobes, both in delay (range) and in Doppler. Thus, modifications are usually applied to reduce these sidelobes. In this section we consider the basic signal without modifications. The coherency is reflected in the expression of the real signal, given by

$$s(t) = \text{Re}[u(t)\exp(j2\pi f_c t)] \tag{7.1}$$

where the complex envelope is a train of N pulses with pulse repetition period T_r,

$$u(t) = \frac{1}{\sqrt{N}} \sum_{n=1}^{N} u_n[t - (n-1)T_r] \tag{7.2}$$

The uniformity of the pulses is expressed by assuming that $u_n(t) = u_1(t)$. The LFM nature of a pulse of duration T is expressed in its complex envelope,

$$u_1(t) = \frac{1}{\sqrt{T}} \text{rect}\left(\frac{t}{T}\right) \exp(j\pi k t^2), \qquad k = \pm \frac{B}{T} \tag{7.3}$$

An example of a real signal is shown in Fig. 7.1, where all the pulses begin with the same initial phase. This is not a mandatory requirement for coherence. Coherency can be maintained as long as the initial phase of each pulse transmitted is known to the receiver.

Changes in phase from pulse to pulse will be expressed in the complex envelope of the nth pulse as

$$u_n(t) = \frac{1}{\sqrt{T}} \text{rect}\left(\frac{t}{T}\right) \exp[j(\phi_n + \pi k t^2)], \qquad k = \pm \frac{B}{T} \tag{7.4}$$

As long as $T < T_r/2$ (which will be assumed henceforth), the additional phase element has no effect on the ambiguity function for $|\tau| \leq T$. It will only affect the recurrent lobes of the AF: namely, over $|\tau \pm nT_r| \leq T$ ($n = 1, 2, \ldots$). The

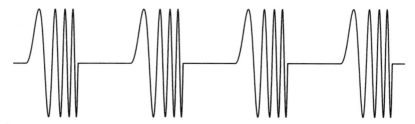

FIGURE 7.1 Coherent train of identical LFM pulses.

additional phase term can already be considered as some sort of diversity, but one that affects only recurrent lobes.

To get an analytic expression for the partial ambiguity function (AF) of our signal, we start with the AF of a constant-frequency pulse, apply AF property 4 to it, and obtain the AF of a single LFM pulse (as done in Section 4.2):

$$|\chi(\tau, \nu)| = |\chi_T(\tau, \nu)| = \left| \left(1 - \frac{|\tau|}{T}\right) \frac{\sin \alpha}{\alpha} \right|, \qquad |\tau| \leq T \quad \text{(single pulse)} \quad (7.5)$$

where

$$\alpha = \pi T \left(\nu \mp B\frac{\tau}{T}\right) \left(1 - \frac{|\tau|}{T}\right) \qquad (7.6)$$

The first equality $|\chi(\tau, \nu)| = |\chi_T(\tau, \nu)|$ stems from the fact that $T < T_r/2$. To describe the AF of the train, for the limited delay $|\tau| \leq T$, we now apply the relationship of the periodic AF:

$$|\chi(\tau, \nu)| = |\chi_T(\tau, \nu)| \left| \frac{\sin N\pi\nu T_r}{N \sin \pi\nu T_r} \right|, \qquad |\tau| \leq T \quad \text{(train of pulses)} \quad (7.7)$$

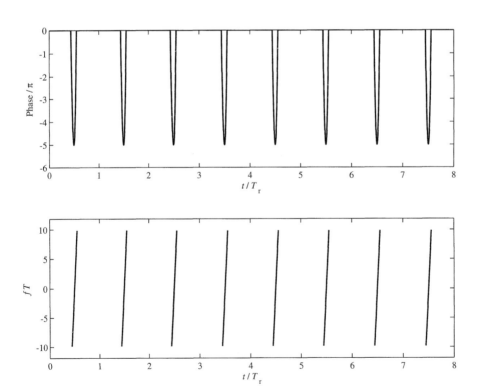

FIGURE 7.2 Phase (top) and frequency (bottom) of a coherent train of eight LFM pulses ($BT = 20$).

Using (7.5) in (7.7) yields the ambiguity function of a train of N identical LFM pulses:

$$|\chi(\tau, \nu)| = \left|\left(1 - \frac{|\tau|}{T}\right)\frac{\sin \alpha}{\alpha}\right|\left|\frac{\sin N\pi\nu T_r}{N \sin \pi\nu T_r}\right|, \qquad |\tau| \leq T \qquad (7.8)$$

We will demonstrate the signal and its AF using a train of eight LFM pulses. The time–bandwidth product of each pulse is 20 and the duty cycle is $T/T_r = \frac{1}{9}$. The phase and frequency history are given in Fig. 7.2. The partial ambiguity function, plotted in Fig. 7.3, is restricted in delay to ± 1 pulse width, and in Doppler to $10/8 = 1.25$ the pulse repetition frequency (PRF, $1/T_r$). Note the first null in Doppler that occurs at $\nu = 1/(NT_r) = 1/(8T_r) = 1/T_c$, where $T_c = 8T_r$ is the coherent processing time. This improved Doppler resolution is the main contribution of the pulse train. Note also the first recurrent Doppler peak at $\nu = 1/T_r$. It is difficult to note from the plot, but the recurrent Doppler lobe is slightly lower and slightly delayed compared to the main lobe. The zero-Doppler cut of Fig. 7.3 (i.e., the magnitude of the normalized autocorrelation function) is identical to the cut that would have been obtained with a single LFM pulse. This is a property of all trains of *identical* pulses. An AF plot extending beyond the first delay recurrent lobe is plotted in Fig. 7.4, and an AF plot extending over the

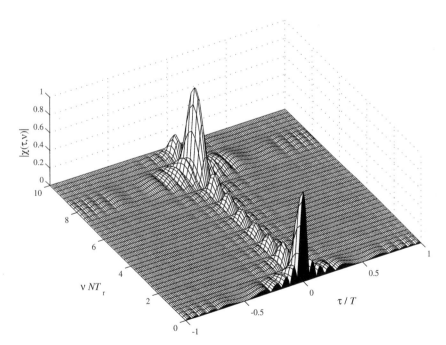

FIGURE 7.3 Partial ambiguity function ($|\tau| \leq T$) of a coherent train of eight LFM pulses ($BT = 20$).

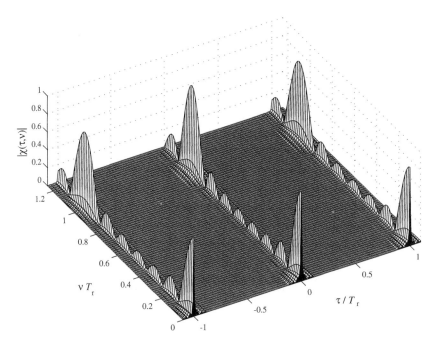

FIGURE 7.4 Partial ambiguity function ($|\tau| \leq T + T_r$) of a coherent train of eight LFM pulses ($BT = 20$).

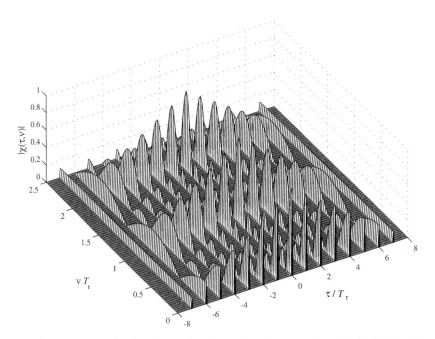

FIGURE 7.5 Ambiguity function ($|\tau| \leq 8T_r$) of a coherent train of eight LFM pulses ($BT = 20$).

entire delay span appears in Fig. 7.5. The Doppler span displayed was doubled in Fig. 7.5, showing two Doppler recurrent lobes.

7.2 FILTERS MATCHED TO HIGHER DOPPLER SHIFTS

The ambiguity function displays the response of a filter matched to the original signal, without Doppler shift. As shown in Fig. 7.3, a coherent pulse train achieves good Doppler resolution, and a matched filter will produce an output only when the Doppler shift of the received signal is within the Doppler resolution. A typical radar processor is likely to contain several filters, matched to several different Doppler shifts. In a coherent pulse train, implementing such a processor is relatively simple, especially if the intrapulse modulation is Doppler tolerant, as LFM is.

The principle of Doppler filter implementation is summarized in Figs. 7.6 and 7.7. Figure 7.6 shows that each pulse is processed by a zero-Doppler matched filter. The N outputs from the N pulses are then fed to an FFT. The first output of the FFT is equivalent to a zero-Doppler filter. For that first output the FFT coherently sums the N inputs. For the second FFT output, the nth pulse $n = 0, 1, \ldots, N - 1$ is first multiplied by the complex coefficient $\exp(j2\pi n/N)$, before being added to the outputs of the other N processed pulses. This complex coefficient is equivalent to a phase shift,

$$\phi_n = 2\pi \frac{n}{N} = 2\pi \nu n T_r \Rightarrow \nu = \frac{1}{NT_r} \qquad (7.9)$$

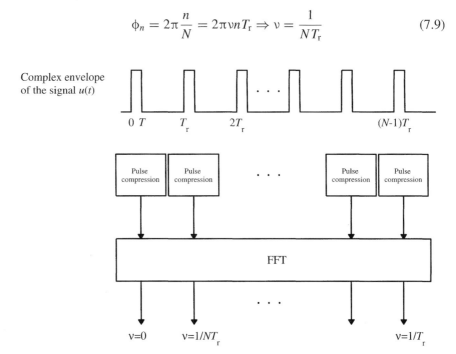

FIGURE 7.6 Implementing several Doppler filters using FFT.

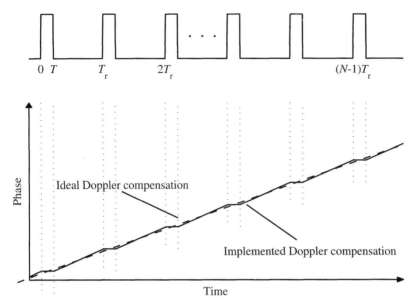

FIGURE 7.7 Interpulse phase compensation for a higher Doppler filter.

The phase shift ϕ_n, added to the nth pulse which is delayed by nT_r, is equal to the phase shift that would have been accumulated after a delay of nT_r by a Doppler shift of $\nu = 1/NT_r$. We thus see that the second output of the FFT is effectively a Doppler filter matched to the Doppler frequency at which the first Doppler filter produces a null response. (The first Doppler filter is centered at zero Doppler.) The $(k+1)$ FFT filter sums the pulses after multiplying the nth pulse by the complex coefficient $\exp(j2\pi kn/N)$, and is therefore matched to a Doppler frequency $\nu = k/NT_r$ or because of Doppler ambiguity (mod $1/T_r$), to $\nu = -(N-k)/NT_r$.

The Doppler compensation achieved by this method is represented by the solid line in Fig. 7.7. The ideal Doppler compensation is represented by the dashed line. The difference is that the FFT implementation lacks Doppler compensation within each pulse. If the output of a compressed pulse is relatively insensitive to Doppler shift (like LFM), the performance of the FFT approach is nearly as good as the performance of an ideal Doppler filter.

The expected response of the second Doppler filter (matched to $\nu = 1/NT_r$) of the train of eight LFM pulses studied in Section 7.1 is shown in Fig. 7.8. The response was obtained by calculating a cross-ambiguity function between two signals, one having the normal phase history of a train of identical LFM pulses (Fig. 7.9, top), and the other in which phase steps of $2\pi/N = \pi/4$ were added between pulses (Fig. 7.9, bottom). Note in Fig. 7.8 that the response was zero at zero Doppler and peaked at $\nu = 1/NT_r$, while in Fig. 7.3 the response peaks at zero Doppler and reaches a null at $\nu = 1/NT_r = 1/(8T_r) = 1/T_c$. Note that we denote the cross-ambiguity or delay–Doppler response of a mismatched filter as

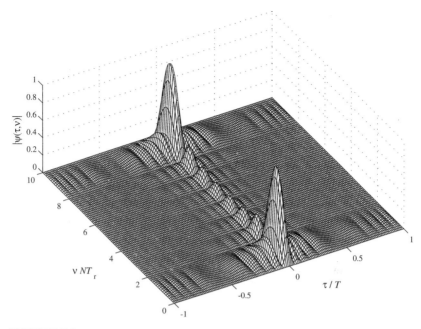

FIGURE 7.8 Delay–Doppler response of the second Doppler filter, $\nu = 1/NT_r$.

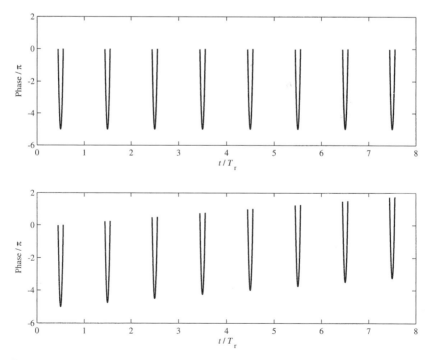

FIGURE 7.9 Phase of the reference signal for the $\nu = 0$ filter (top) and the $\nu = 1/NT_r$ filter (bottom).

$\psi(\tau, \nu)$. As for the notations used for the (auto) ambiguity function in this book, a positive value of ν corresponds to a target with higher closing velocity, and positive delay (τ) implies a target farther away from the radar.

Some targets may have a Doppler value that falls between successive FFT filters. The return from those targets suffers a power loss according to the FFT filter response at the target Doppler. This power loss is usually referred to as *straddle loss*. To minimize straddle loss it is possible to either reduce the frequency separation between consecutive Doppler filters or to widen the filter response such that the overlap between successive filters is higher. To reduce the frequency separation between consecutive Doppler filters and thus achieve overlap between their responses, it is necessary to reduce the phase step between pulses of the reference signal. This could be implemented by adding zero-padded inputs to the FFT. The number of FFT outputs will grow accordingly, adding more Doppler filters. Widening the filter response is achieved by introducing interpulse amplitude modulation, which is discussed in the next section. Finally, note that the imperfection of this kind of FFT Doppler filtering grows as the center frequency of the Doppler filter grows (the LFM Doppler tolerance has a stronger effect on the response).

7.3 INTERPULSE WEIGHTING

Figure 7.3 reveals high delay and Doppler sidelobes. As we learned with respect to the single LFM pulse, the delay sidelobes of the autocorrelation function of a single LFM pulse can be reduced by shaping the spectrum according to a weight window. Because of the linear relationship between frequency and time in an LFM pulse, spectral shaping can be implemented by applying intrapulse amplitude weighting. In a dual way, the Doppler sidelobes, shown in Fig. 7.3, can be reduced by implementing interpulse amplitude weighting.

Figure 7.10 presents two versions of the amplitude of eight transmitted LFM pulses following a square root of a Hamming window. The reference signal in the receiver will also have such a square-root weight window. Their product will yield a true Hamming window. In the top subplot the weight window has eight steps, yielding flat-top pulses. In the bottom subplot the window is smooth over the entire train duration (NT_r), yielding variable-amplitude pulses. As long as the duty cycle is small, there is little difference in performances. Hence, the more practical stepwise window is preferred. For a large duty cycle, and especially in periodic CW signals (duty cycle = 1), the smooth window should be used. The phase and frequency are identical in both signals and remain as in Fig. 7.2. The resulting ambiguity function is shown in Fig. 7.11 and should be compared with Fig. 7.3. The comparison clearly demonstrates the reduction in Doppler sidelobes and a widening of the Doppler mainlobe lobe (at $\nu = 0$) and recurrent lobe (at $\nu = 1/T_r$ or $\nu N T_r = 8$).

The variable amplitude of the transmitted pulses could be a problem to a high-power RF amplifier, hence the readiness to implement the window only on receive, despite the mismatch loss that this creates. Next we will demonstrate that

INTERPULSE WEIGHTING

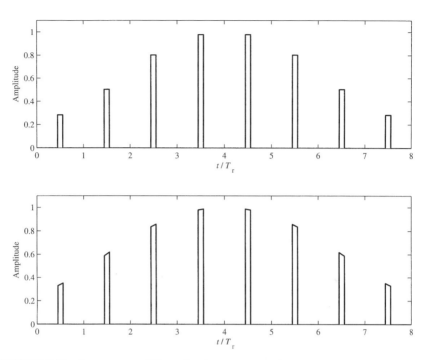

FIGURE 7.10 Square root of Hamming interpulse amplitude weighting: stepwise (top) and smooth (bottom).

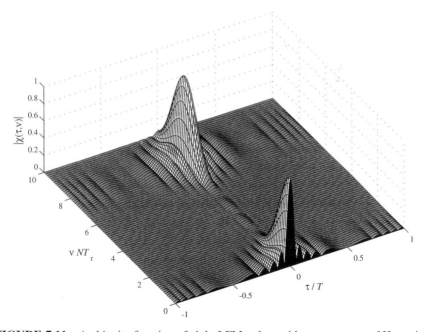

FIGURE 7.11 Ambiguity function of eight LFM pulses with a square root of Hamming interpulse amplitude weighting.

178 COHERENT TRAIN OF LFM PULSES

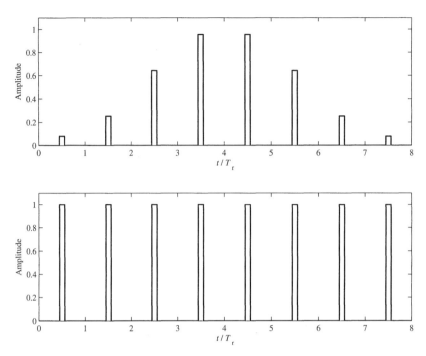

FIGURE 7.12 Pulse amplitude of Hamming-weighted reference signal (top) and transmitted signal (bottom).

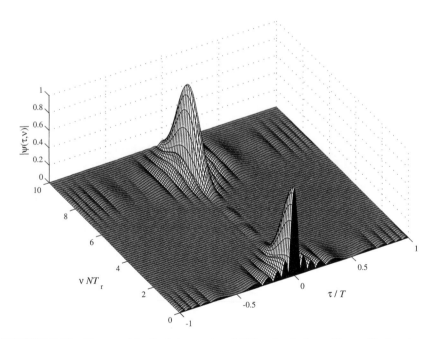

FIGURE 7.13 Cross-ambiguity between two LFM pulse trains with amplitudes shown in Fig. 7.12.

other than SNR loss, implementing the interpulse amplitude weighting only on receive, does not result in a different delay–Doppler response. Note that since the reference signal is not matched to the signal transmitted, we cannot use the phrase *ambiguity function*. Instead, we use the phrases *delay–Doppler response* or *cross-ambiguity* to describe the output of a mismatched receiver. Figure 7.12 displays the amplitude of the eight pulses used in the train transmitted (bottom) and in the reference train (top). The phase and frequency are identical in both signals and were shown in Fig. 7.2. Comparing the top of Fig. 7.12 with Fig. 7.10 shows the difference between a Hamming window and a square root of a Hamming window. The resulting cross-ambiguity is shown in Fig. 7.13. It is practically indistinguishable from Fig. 7.11. We can conclude that with regard to interpulse weighting, there is hardly any difference in delay–Doppler response between matched weighting and mismatched (on receive) weighting. A difference will be seen when intrapulse weight is introduced.

7.4 INTRA- AND INTERPULSE WEIGHTING

Interpulse weighting mitigates Doppler sidelobes, as demonstrated in Fig. 7.11 or 7.13. Intrapulse weighting in LFM mitigates range sidelobes, as was shown in Section 4.2.1. The weightings can be combined to reduce both range and Doppler sidelobes. Again, square-root weight windows can be applied

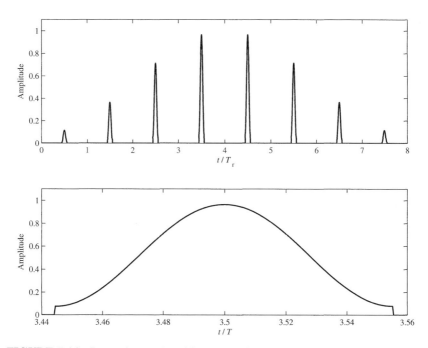

FIGURE 7.14 Interpulse (top) and intrapulse (bottom) Hamming weight windows.

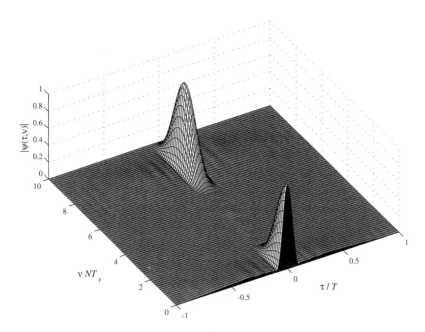

FIGURE 7.15 Delay–Doppler response of eight LFM pulses with inter- and intrapulse Hamming weight windows.

in both transmitter and receiver, maintaining match filtering. However, to allow fixed-amplitude transmission, the weighting can be concentrated in the receiver, at a cost of SNR loss and some difference in delay–Doppler response. Figure 4.15 demonstrated the difference in the zero-Doppler response of a single LFM pulse between matched and mismatched Hamming intrapulse weight windows. That difference remains in the response of a train of LFM pulses with intrapulse weighting.

Figure 7.14 shows the amplitude of a reference signal in a receiver for a train of eight LFM pulses. Hamming weight windows were implemented both between the pulses (see the top) and within each pulse (see the zoom on a single pulse at the bottom). The resulting delay–Doppler response ($BT = 20$) is shown in Fig. 7.15. The SNR loss due to mismatch is rather high in this case. It can be calculated using Section 4.2.2. For simple weight windows such as Hann or Hamming, the delay–Doppler response (plotted in Fig. 7.15) can be calculated analytically. An outline of the calculation is given in the next section.

7.5 ANALYTIC EXPRESSIONS OF THE DELAY–DOPPLER RESPONSE OF AN LFM PULSE TRAIN WITH INTRA- AND INTERPULSE WEIGHTING

A simple analytic expression is presented here for the delay–Doppler response of a coherent pulse-train radar waveform consisting of LFM pulses. The receiver differs from a matched filter by intra- and interpulse weighting. The response differs from the ambiguity function of an unweighted pulse train by widened lobes and reduced sidelobes, due to weightings, and by SNR loss due to the mismatch. The analysis follows Getz and Levanon (1995). The analysis will be limited in delay to the pulse width T: namely, $|\tau| \leq T$. This will allow us to use the periodic ambiguity function, which is simpler to develop and which is equal to the ambiguity function over that delay range. Recall from Section 3.6 that the transmitted pulse train is not bounded in time. Only the reference signal in the receiver is limited to N pulses.

7.5.1 Ambiguity Function of N LFM Pulses

The complex envelope of a single LFM pulse that starts at $t = 0$ can be described as

$$u_T(t) = \frac{1}{\sqrt{T}} \exp\left[j\pi k \left(t - \frac{1}{2}T\right)^2\right], \qquad 0 \leq t \leq T \qquad (7.10)$$

where T is the pulse duration and k is the frequency sweep rate. The product $|kT^2| = TB$ is known as the *time–bandwidth product* or *compression ratio*. The envelope of an infinitely long coherent pulse train is defined as

$$u(t) = \sum_{i=-\infty}^{\infty} u_T(t - iT_r) \qquad (7.11)$$

where T_r is the pulse repetition interval. At the receiver the number of processed pulses will be limited to N. For delay within the pulse duration $|\tau| \leq T$, when the receiver is matched to N pulses, the delay–Doppler response is equal to the periodic ambiguity function of N pulses, which will be written slightly differently from its form in (3.37),

$$|\chi(\tau, \nu)| = \frac{1}{NT_r} \left| \int_0^{NT_r} u(t - \tau) u^*(t) \exp(j2\pi\nu t)\, dt \right|, \qquad |\tau| \leq T \qquad (7.12)$$

As given by (7.6) and (7.8),

$$|\chi(\tau, \nu)| = \left| \left(1 - \frac{|\tau|}{T}\right) \frac{\sin \alpha}{\alpha} \right| \left| \frac{\sin N\pi\nu T_r}{N \sin \pi\nu T_r} \right|, \qquad |\tau| \leq T \qquad (7.13)$$

where

$$\alpha = \pi T \left(\nu \mp B\frac{\tau}{T}\right)\left(1 - \frac{|\tau|}{T}\right) \qquad (7.14)$$

This simple and well-known result will serve as a reference to which we will compare the response of a more general processor that includes weighting.

7.5.2 Delay–Doppler Response of a Mismatched Receiver

Figure 7.16 summarizes the magnitude of the signals involved in a receiver mismatched by introducing weight windows. It contains the periodic (and infinite) signal received, $u(t - \tau)$, the periodic reference signal $r(t)$, which can be more complicated than just the complex conjugate of $u(t)$, and an interpulse weight window, limited to the duration of N pulses:

$$w_p(t) = \begin{cases} w(t), & 0 \leq t \leq NT_r \\ 0, & \text{elsewhere} \end{cases} \quad (7.15)$$

This interpulse weight window will reduce sidelobes along the Doppler axis. The receiver delay–Doppler response is the magnitude of the cross-correlation between the weighted reference and the received signal shifted in delay τ and Doppler v:

$$|\psi(\tau, v)| = \left| \int_{-\infty}^{\infty} u(t - \tau) r(t) w_p(t) \exp(j2\pi vt) \, dt \right| \quad (7.16)$$

Using the well-known property that the Fourier transform of a product is the convolution between the individual Fourier transforms, (7.16) can be written as

$$|\psi(\tau, v)| = \left| \int_{-\infty}^{\infty} u(t - \tau) r(t) \exp(j2\pi vt) \, dt \otimes \int_{0}^{NT} w(t) \exp(j2\pi vt) \, dt \right| \quad (7.17)$$

Since $u(t)$ and $r(t)$ are infinite and periodic with period T, the first transform on the right-hand side of (7.17) can be written as an infinite sum:

$$\int_{-\infty}^{\infty} u(t - \tau) r(t) \exp(j2\pi vt) \, dt = \sum_{n=-\infty}^{\infty} \delta\left(v - \frac{n}{T}\right) g_n(\tau) \quad (7.18)$$

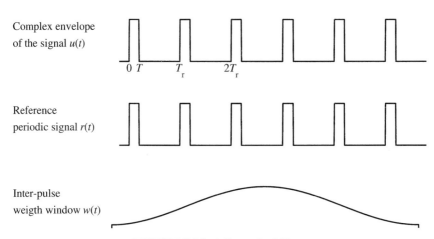

FIGURE 7.16 Mismatched filter.

where

$$g_n(\tau) \triangleq \frac{1}{T} \int_0^T u(t-\tau) r(t) \exp\left(\frac{j2\pi nt}{T}\right) dt \quad (7.19)$$

When $T_r > 2T$, u and r in (7.19) can be replaced by their one-period versions, u_T and r_T, respectively.

The second transform is a Fourier transform of the external weight window

$$W(\nu) = \int_0^{NT} w(t) \exp(j2\pi \nu t) \, dt \quad (7.20)$$

The convolution between (7.18) and (7.20) yields

$$|\psi(\tau, \nu)| = \left| \sum_{n=-N_\infty}^{N_\infty} g_n(\tau) W\left(\nu - \frac{n}{T}\right) \right|, \quad N_\infty \to \infty \quad (7.21)$$

Rigorously, $N_\infty \to \infty$, but for most practical numerical calculations it is sufficient to choose $N_\infty \approx 5$. Before we can obtain from (7.21) the specific delay–Doppler response of the LFM pulses, we need to define the reference periodic signal $r(t)$, which may include an intrapulse weight, and also choose a specific interpulse weight function $w(t)$.

We limit the weight windows to the family containing the uniform, Hann, and Hamming windows. All three can be defined by

$$w_p(t) = \begin{cases} \dfrac{1}{NT_r}\left[1 - \dfrac{1-c}{c} \cos \dfrac{2\pi t}{NT_r}\right], & 0 \leq t \leq NT_r \\ 0, & \text{elsewhere} \end{cases} \quad (7.22)$$

with $c = 1$ for a uniform window, $c = 0.5$ for a Hann window, and $c = 0.53836$ for a Hamming window. Their Fourier transform is given by

$$W(\nu) = \frac{\sin \pi \nu NT_r}{\pi \nu NT_r} \left\{ 1 + \frac{(1-c)(\nu NT_r)^2}{c[1-(\nu NT_r)^2]} \right\} \exp(j2\pi \nu NT_r) \quad (7.23)$$

Note that these are smooth windows that will create variable-top pulses (as in the bottom subplot of Fig. 7.10). For a small duty cycle, the delay–Doppler response of variable-top pulses will be very similar to that of flat-top pulses. Flat-top pulses are easier to implement and process but are more difficult to analyze analytically.

7.5.3 Adding Intrapulse Weighting

When the periodic reference in the receiver is simply the complex conjugate of the complex envelope of the transmitted signal: namely,

$$r(t) = u^*(t) \quad (7.24)$$

then the $g_n(\tau)$ function is found from (7.10) and (7.19) to be (up to a constant)

$$g_n(\tau) = \left(1 - \frac{\tau}{T}\right) \frac{\sin[\pi n (T/T_r)(1 - \tau/T)]}{\pi n (T/T_r)(1 - \tau/T)} \exp\left[j\pi n \frac{T}{T_r}\left(1 + \frac{\tau}{T}\right)\right], \quad 0 \leq \tau \leq T \quad (7.25)$$

During an LFM pulse the frequency is changing linearly with time. Adding an intrapulse weight window with a duration T is equivalent to implementing a weight window in the frequency domain. Such a window reduces sidelobes along the delay axis (Farnett and Stevens, 1990). We will therefore apply an intrapulse weight window:

$$w_{\text{int}}(t) = \begin{cases} \frac{1}{T}\left[1 - \frac{1-c}{c}\cos\frac{2\pi t}{T}\right] = \frac{1}{T}\left\{1 - \frac{1-c}{2c}\left[\exp\left(j\frac{2\pi t}{T}\right) + \exp\left(-j\frac{2\pi t}{T}\right)\right]\right\}, & 0 \leq t \leq T \\ 0, & \text{elsewhere} \end{cases} \quad (7.26)$$

For simplicity the intrapulse weight (7.26) was chosen to be similar to the interpulse weight (7.22). This is not a requirement, however. For example, c in (7.26) could be different from c in (7.22).

With intrapulse weighting, one period of the reference signal becomes the product

$$r_T(t) = u_T^*(t) w_{\text{int}}(t) \quad (7.27)$$

which is clearly zero for $t < 0$ and $t > T$. With $r_T(t)$ defined through (7.10), (7.26), and (7.27), the final expression for the $g_n(\tau)$ function of an LFM pulse, including intrapulse weighting, is

$$g_n(\tau) = g_n^{(0)}(\tau) - \frac{1-c}{2c}[g_n^{(1)}(\tau) + g_n^{(-1)}(\tau)] \quad (7.28)$$

where

$$g_n^{(m)}(\tau) = \left(1 - \frac{\tau}{T}\right) \frac{\sin \beta_n^{(m)}}{\beta_n^{(m)}} \exp\left[j\pi T\left(1 + \frac{\tau}{T}\right)\left(\frac{n}{T_r} + \frac{m}{T}\right)\right], \quad m = -1, 0, 1 \quad (7.29)$$

and

$$\beta_n^{(m)} = \pi T\left(1 - \frac{\tau}{T}\right)\left(\frac{n}{T_r} + \frac{m}{T} - k\tau\right), \quad m = -1, 0, 1 \quad (7.30)$$

The five equations (7.21), (7.23), and (7.28)–(7.30) fully define the delay–Doppler response for $0 \leq \tau \leq T$. For negative delays we can sometimes assume symmetry with respect to the origin, borrowed from the ambiguity function, namely

$$|\psi(\tau, \nu)| = |\psi(-\tau, -\nu)| \quad (7.31)$$

Note that (7.31) was not proved as (3.22) was. Figure 7.8 is an example where there is no symmetry. The number of terms in (7.28) can be increased, and their respective coefficients adjusted, to provide for intrapulse Taylor weighting (Farnett and Stevens, 1990).

7.5.4 Examples

Figure 7.17 presents the first quadrant of the cross-ambiguity function of unweighted processing (top) and Hamming weighting (bottom). Identical types of windows were used for both inter- and intrapulse weighting. The signal parameters were $T_r/T = 5$, $TB = 20$, and $N = 16$ pulses. Note from Fig. 7.17 that at a Doppler of $v/T_r = 1$, the recurrent peak is shifted in delay relative to the peaks at zero Doppler. Note also that the direction of the shift agrees with our notations of the ambiguity function, where positive delay corresponds to a target farther from the radar than the reference ($\tau = 0$). Indeed, using an up-chirp LFM (as used for Fig. 7.17) implies that a target with negative Doppler $1/T_r$ would appear as if the target location is farther from the radar than the true target location. The shift (typical of LFM) is

$$\frac{\tau}{T} = \frac{1}{TB(T_r/T)} \tag{7.32}$$

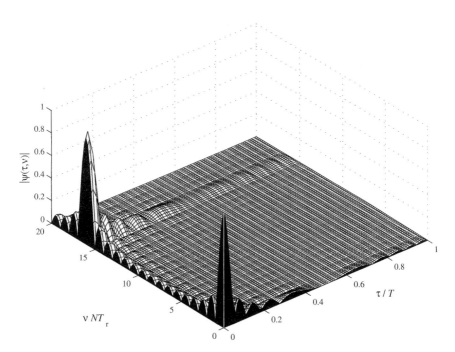

FIGURE 7.17 Cross-ambiguity function of 16 LFM pulses, with no weighting (above) and Hamming weighting (next page).

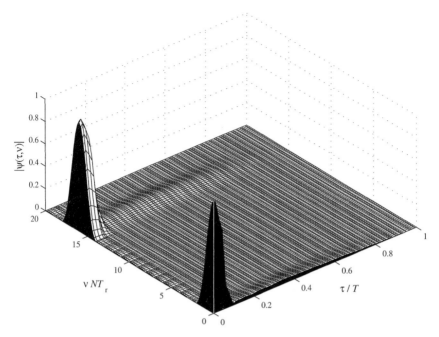

FIGURE 7.17 (*continued*)

The corresponding cuts along the Doppler axis (in dB) are presented in Fig. 7.18. The top plot (uniform window) behaves according to $|(\sin N\pi\nu T)/(N\sin \pi\nu T)|$. The bottom plot demonstrates the flat sidelobes level typical of a Hamming window.

The corresponding zero-Doppler cuts (in dB) are presented in Fig. 7.19. The top part (no weighting) is the well-known zero-Doppler AF cut of an unweighted single LFM pulse [see equation (4.10)]. The lower plot (Hamming weighting) was calculated numerically using equations (7.28)–(7.30) for $n = 0$. The relatively high delay sidelobes (≈ -30 dB), despite the Hamming weighting, will drop with higher time–bandwidth product. An example for $TB = 100$ is given in Fig. 7.20, where the sidelobe level dropped to -42 dB.

Note the identity of Fig. 7.19 (bottom) and Fig. 4.15 (bottom). They both describe the zero-Doppler cut of the delay–Doppler response of mismatched (only on receive) Hamming-weighted LFM pulse. However, Fig. 7.19 (bottom) was based on a numerical calculation of equations (7.28)–(7.30), whereas Fig. 4.15 (bottom) was based on numerical cross-correlation between the (unweighted) transmitted signal and the (weighted) reference signal. This could serve as some sort of confirmation of the analysis in Section 7.5.

Finally, we should point out that the analysis of Section 7.5 assumed that the intrapulse sampling rate is higher than twice the pulse bandwidth, meeting the Nyquist criteria. In some very high bandwidth LFM pulses, hardware limitations may not allow high-enough sampling rate. Using a lower sampling rate will

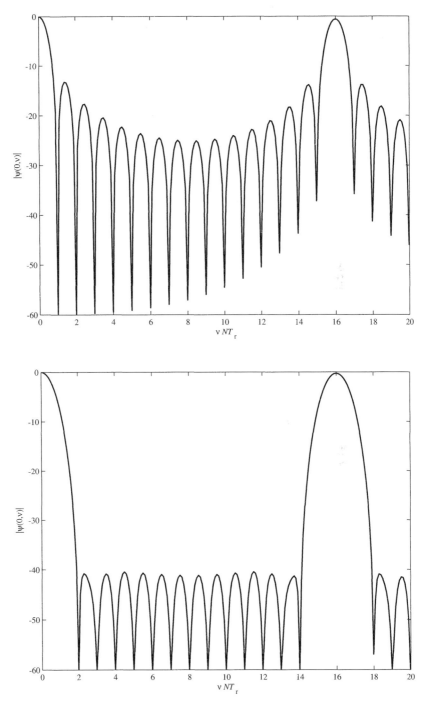

FIGURE 7.18 Zero-delay cut of the cross-ambiguity function of 16 LFM pulses, with no weighting (top) and Hamming weighting (bottom).

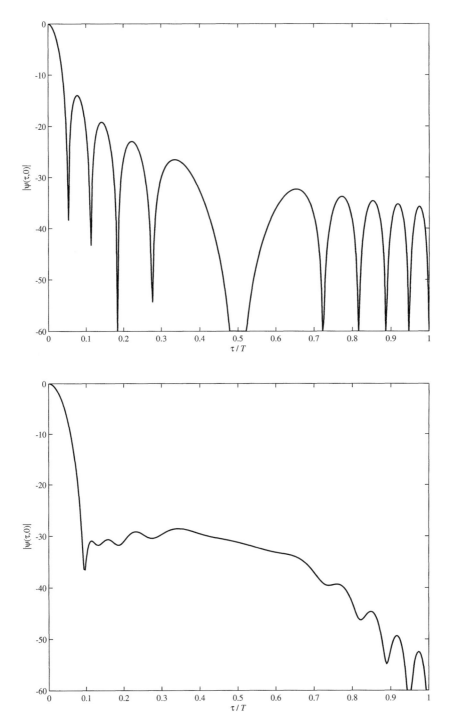

FIGURE 7.19 Zero-Doppler cut of the cross-ambiguity function of 16 LFM pulses ($TB = 20$), with no weighting (top) and Hamming weighting (bottom).

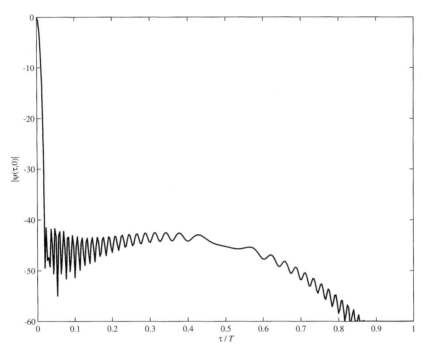

FIGURE 7.20 Zero-Doppler cut of the cross-ambiguity function of 16 LFM pulses with Hamming weighting ($TB = 100$).

create spurious peaks in delay. This issue was discussed by Abatzoglou and Gheen (1998).

PROBLEMS

7.1 Doppler tolerance

Consider a pulse train of $N = 8$ pulses, pulse duration $T = 100\,\mu\text{s}$, and repetition period $T_r = 10T$.

(a) What is the Doppler resolution (first null)?

(b) For constant-frequency pulses, if the Doppler filter is matched to a higher Doppler shift, at what (matched to) Doppler shift will the peak response drop to $\frac{1}{2}$ (relative to the peak response of a zero-Doppler filter)?

(c) Repeat part (b) for the case of LFM pulses with $TB = 20$.

7.2 SNR loss due to weighing

Calculate the SNR loss for a train of eight LFM pulses when Hamming intrapulse and stepwise interpulse weightings are implemented only on receive.

7.3 Opposite LFM slope doublet

(a) In a train of two identical LFM pulses, what is the relative peak height (in dB relative to the peak at the origin) of the zero-Doppler response at the recurrent lobe?

(b) Repeat part (a) for the case when the two pulses are of opposite LFM slopes and $TB = 20$.

7.4 Frequency spectrum of a train of identical pulses

Consider a coherent train of eight identical pulses with $T_r = 4T$.

(a) For a constant-frequency pulse, compare, discuss, and explain the similarity between the ambiguity function's zero-delay cut (in dB) with the energy spectral density of the signal (in dB) over a Doppler or frequency span of $0 \leq f \leq 6/T_r = 1.5/T$.

(b) Qualitatively compare the energy spectral densities of the signal in part (a) with a similar train in which the pulses are LFM with $TB = 10$. Consider both the fine details of the spectrum plot in dB (over the frequency span $0 \leq f \leq 6/T_r = 1.5/T$) and the overall spectral shape (over the frequency span $0 \leq f \leq 60/T_r = 15/T$).

7.5 AF of a train of LFM pulses

Consider a coherent train of eight identical (unweighted) LFM pulses with $T_r = 4T$ and $TB = 50$. Plot the ambiguity function with the following zooms:

(a) $|\tau| \leq T, 0 \leq \nu \leq 9T/8$.

(b) $|\tau| \leq T/40, 0 \leq \nu \leq 9T/8$.

Using the latter zoom, find how much has the first recurrent Doppler peak (at $\nu = 1/T_r$) shifted in delay from $\tau = 0$? Confirm your result with equation (7.32).

REFERENCES

Abatzoglou, T. J., and G. O. Gheen, Range, radial velocity, and acceleration MLE using radar LFM pulse train, *IEEE Transactions on Aerospace and Electronic Systems*, vol. 34, no. 4, October 1998, pp. 1070–1084.

Farnett, E. C., and G. H. Stevens, Pulse compression radar, in Skolnik, M. I. (ed.), *Radar Handbook*, 2nd ed., McGraw-Hill, New York, 1990, Chap. 10.

Getz, B., and N. Levanon, Weight effects on the periodic ambiguity function, *IEEE Transactions on Aerospace and Electronic Systems*, vol. 31, no. 1, January 1995, pp. 182–193.

Nathanson, F. E., J. P. Reilly, and M. N. Cohen, *Radar Design Principles: Signal Processing and the Environment*, 2nd ed., McGraw-Hill, New York, 1991.

Rihaczek A. W., *Principles of High Resolution Radar*, McGraw-Hill, New York, 1969.

Stimson, G. W., *Introduction to Airborne Radar*, Hughes Aircraft Co., El Segundo, CA, 1983.

8

DIVERSE PRI PULSE TRAINS

In Chapter 7, a coherent train of identical pulses was used to improve Doppler resolution. Using a train of identical pulses allowed for the effective use of FFT in the receiving end. However, the main drawback of the identical pulse train is the ambiguity in range and Doppler, which is a by-product of using both identical pulses and equal spacing between pulses. In this chapter we introduce a method for diversifying the identical pulse train by using different pulse separation periods. Diversifying the pulse train by using different pulses is addressed in Chapter 9.

Examples of *pulse repetition interval* (PRI) diversity will be given in the form of *pulse-to-pulse* PRI staggering, usually found in *moving-target indicator* (MTI) radars, and *batch-to-batch* PRI diversity in the form of a *medium pulse repetition frequency* (MPRF) radar, maximizing the radar's unambiguous range and Doppler while minimizing blind range and speed (maximizing target visibility). Methods are discussed for selecting the medium PRF set, including minor–major PRF selection. Using binary integration (M-out-of-N) for target detection and for solving ambiguities will also be addressed and demonstrated.

8.1 INTRODUCTION TO MTI RADAR

There are many radar applications in which the clutter is stationary relative to the radar but the targets are moving. An example is air traffic control (ATC) radar. In such radars, most of the clutter power can be removed if the near-zero

Radar Signals, By Nadav Levanon and Eli Mozeson
ISBN 0-471-47378-2 Copyright © 2004 John Wiley & Sons, Inc.

192 DIVERSE PRI PULSE TRAINS

Doppler spectrum is filtered out. Techniques for implementing clutter filtering are the basis of MTI radars. The DFT processor discussed in Chapter 7 could serve as an MTI filter by separating the output of the first (zeroth) filter from all the others. However, the DFT processor utilizes $N \gg 1$ pulses, whereas MTI could be implemented using as few as two pulses (Schleher, 1991).

8.1.1 Single Canceler

The simplest form of an MTI filter is the single canceler shown in Fig. 8.1. *Single* stands for a single delay line. The delay T must be equal to the pulse interval T_r. The filter was drawn purposely following a phase detector to emphasize the advantage of coherent detection prior to filtering. Intuitively, it is clear that the filter output at a given instant t is constructed from the input signal at t, minus the signal at $t - T$. In other words, the present pulse return from a given range is compared with the previous pulse return from the same range. Only the difference between the two appears at the output. A coherent detector can yield differences as big as twice the signal amplitude, due to Doppler-induced phase differences from pulse to pulse.

The MTI filter can be described in many forms (e.g., as a first-order difference equation, using Z transforms, etc.). Maybe the simplest way of describing it is as a FIR (finite impulse response) filter. In the case of the single canceler of Fig. 8.1, the FIR filter is of length 2 and the impulse response (FIR coefficients) are simply $+1, -1$. The description method selected does not necessarily imply that the filter is implemented in digital form. Indeed, as will also be discussed later, traditional MTI filters are implemented using analog circuits, which can sometimes be followed by a sampling circuit for the purpose of increasing the receiver dynamic range.

In Chapter 7 it was shown that the periodic ambiguity function (PAF) can be used to describe the matched filter response. In the case shown in Fig. 8.1, the output of the phase detector is the response of a filter matched to zero Doppler and is described by the periodic ambiguity function of a train of identical pulses $\chi(\tau, \nu)$.

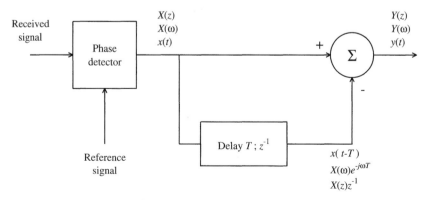

FIGURE 8.1 Single canceler.

Single-canceler output as a function of the returned signal delay and Doppler can be written as

$$\chi_{SC}(\tau, \nu) = \chi_{NT}(\tau, \nu) - \chi_{NT}(\tau - T, \nu) \tag{8.1}$$

where the periodic ambiguity function of a train of N identical pulses $\chi_{NT}(\tau, \nu)$ has the shape of a bed of nails. Each "nail" is shaped as the single pulse ambiguity function multiplied by a relative phase term which is a function of target Doppler.

Denote the single-pulse ambiguity function by $\chi_1(\tau, \nu)$; then the periodic ambiguity function of a train of N identical pulses is given by

$$\chi_{NT}(\tau, \nu) = \sum_{p=-\infty}^{\infty} \exp(j2\pi\nu p T_r)\chi_1(\tau - pT_r, \nu) \tag{8.2}$$

Using (8.2) in (8.1), changing variables, and using the fact that $T = T_r$ gives the single-canceler ambiguity function as

$$\chi_{SC}(\tau, \nu) = \sum_{p=-\infty}^{\infty} \exp(j2\pi\nu p T_r)\chi_1(\tau - pT_r, \nu)$$

$$- \sum_{p=-\infty}^{\infty} \exp(j2\pi\nu p T_r)\chi_1(\tau - T - pT_r, \nu)$$

$$\underset{T=T_r}{=} \sum_{p=-\infty}^{\infty} \exp(j2\pi\nu p T_r)\chi_1(\tau - pT_r, \nu)$$

$$- \exp(-j2\pi\nu T_r) \sum_{p'=-\infty}^{\infty} \exp(j2\pi\nu p' T_r)\chi_1(\tau - p'T_r, \nu)$$

$$= [1 - \exp(-j2\pi\nu T_r)]\chi_{NT}(\tau, \nu) \tag{8.3}$$

where the first term is the frequency response of the FIR filter and the second term is the periodic ambiguity function of the identical-pulse train. The magnitude of the frequency response of the single canceler is given by

$$|H(\nu)| = |1 - \exp(-j2\pi\nu T_r)| = 2|\sin(\pi\nu T_r)| \tag{8.4}$$

A plot of the frequency response of a single canceler appears in Fig. 8.2 (solid line). The response has a periodicity of $1/T_r$. The rectified sine shape has a finite slope at $f = 0$, yielding too fast a rise of the response. For some applications, the resulting attenuation of the low frequencies may not be enough. Cascading two single cancelers will improve the attenuation at low frequencies.

8.1.2 Double Canceler

A double canceler is effectively two cascaded single cancelers. Two implementations of a double canceler are shown in Fig. 8.3. Mathematically, the two filters are identical. In practice, however, the first implementation is less sensitive to

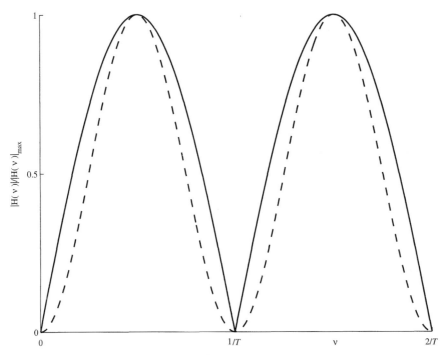

FIGURE 8.2 Frequency response of a single canceler (solid line) and a double canceler (dashed line).

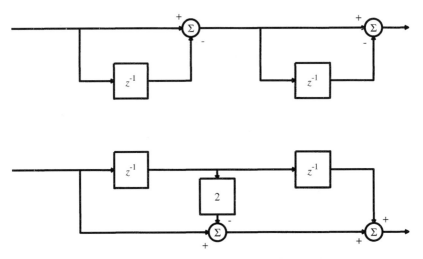

FIGURE 8.3 Two versions of a double canceler.

poor matching of delay and attenuation in the two delay lines. The frequency response of a double canceler is simply the square power of the response of a single canceler. The response of a double canceler also appears in Fig. 8.2 (dashed line).

Single- and double-canceler filters belong to a family of filters called *feedforward filters, nonrecursive filters, finite-memory filters, taped delay-line filters, moving-average* (MA) *filters, transversal filters,* or *finite-impulse-response* (FIR) *filters*. Additional degrees of freedom, and therefore improved response, could be achieved by introducing feedback. Introducing feedback converts the filter to a recursive type. Recursive filters, of which there are many configurations (Linden and Steinberg, 1957; Urkowitz, 1957), allow for more control of the shape of the nulls at $\pm n/T_r$.

8.2 BLIND SPEED AND STAGGERED PRF FOR AN MTI RADAR

The various cancelers discussed in Section 8.1 differ in the shape of the response between multiples of $1/T_r$. However, all the cancelers exhibited nulls at multiples of $1/T_r$. These nulls create total attenuation of Doppler frequencies equal to multiples of the pulse repetition frequency (PRF). Returns from targets with corresponding velocities will be strongly attenuated. In other words, the MTI filter will render the radar blind to such velocities—hence the name *blind speed*.

Note that signals' blind speeds can be visualized intuitively by use of the periodic ambiguity function. We have already seen that the periodic ambiguity function of a train of identical pulses is in the form of a bed of nails, and the nails are spaced one PRI (T_r) in delay and a single PRF ($1/T_r$) in Doppler. Thus, once a matched filter (*phase detector*) was used prior to an MTI filter, all signal returns for targets having the same range (modulo the PRI) and Doppler (modulo the PRF) are ambiguous. Specifically, if a MTI filter is implemented to filter out the clutter return (zero Doppler) for all ranges, all targets having zero Doppler (modulo PRF) are also filtered out. The blind-speed problem can be alleviated by modulating the interpulse period and by using corresponding delays in the canceler filter. This technique is called *staggered PRF*.

8.2.1 Staggered-PRF Concept

The staggered-PRF concept will be explained through the simple double canceler whose frequency response is given by

$$|H(\nu)| = |[1 - \exp(-j2\pi\nu T_r)][1 - \exp(-j2\pi\nu T_r)]|$$
$$= |1 - 2\exp(-j2\pi\nu T_r) + \exp(-j2\pi\nu 2T_r)| \qquad (8.5)$$

We should note here that the filter and the response to be developed in this section do not apply directly to the staggered-MTI radar. However, they allow us to understand the concepts involved. The actual response is discussed in Section 8.2.2.

We note that the concept of the double canceler is a special case of a more general nonrecursive filter; that is,

$$|H(\nu)| = \left| \sum_{n=0}^{N-1} C_n \exp(-j2\pi\nu\tau_n) \right| \qquad (8.6)$$

The double canceler can be obtained from (8.6) by choosing $N = 3$, $C_0 = C_2 = 1$, $C_1 = -2$, $\tau_0 = 0$, $\tau_1 = T_r$, and $\tau_2 = 2T_r$. Constant-PRF MTI filters are characterized by the quality $\tau_n - \tau_{n-1} = T_r$. However, delays (τ_n) that are not multiples of T_r could be utilized in (8.6) as well as weights (C_n) that are not integers.

Before discussing any general rules for choosing the delays and the weights, let us look at a very elementary modification of the double canceler, involving only one change, $\tau_1 = aT_r$. The normalized frequency response of the new filter reduces to

$$|H(\nu)| = \tfrac{1}{4}[(1 - 2\cos 2\pi\nu aT_r + \cos 2\pi\nu 2T_r)^2 \\ + (-2\sin 2\pi\nu aT_r + \sin 2\pi\nu 2T_r)^2]^{1/2} \qquad (8.7)$$

The normalization in (8.7) is with respect to the maximum gain of the filter (the maximum gain is equal to the sum of the absolute values of all the weights). Normalization with respect to the maximum gain aligns the peak of the response with the 0-dB line. A plot of (8.7) in decibels for $a = \tfrac{3}{4}$ is given in Fig. 8.4. We note that the first complete null appears at $\nu = 4/T_r$. From (8.7) we can deduce that the first null will appear at the lowest frequency that is a solution to the two equations

$$a\nu T_r = n_1, \qquad 2\nu T_r = n_2 \qquad (8.8)$$

where n_1 and n_2 are any integers.

From (8.8) we can construct a table of the frequency (normalized with respect to T_r) of the first complete null as a function of a (see Table 8.1). In addition to a, the table also presents the ratio between the two delay lines, which is called the *stagger ratio*. In other words, if

$$a = \frac{m-1}{m} \qquad (8.9)$$

then

$$\text{stagger ratio} = \frac{a}{2-a} = \frac{m-1}{m+1} \qquad (8.10)$$

For odd m the blind speed will effectively be located at one-half the blind speed for even m:

$$f_{\text{blind}} = \begin{cases} \dfrac{m}{T_r}, & m \text{ even} \\ \dfrac{m}{2T_r}, & m \text{ odd} \end{cases} \qquad (8.11)$$

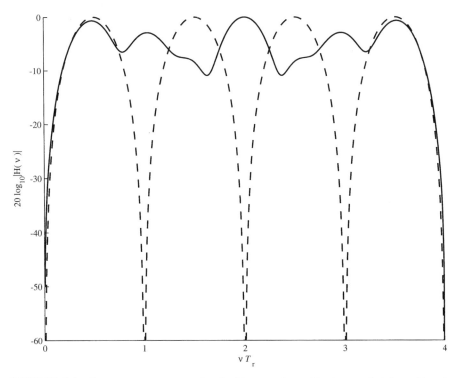

FIGURE 8.4 Frequency response of a double canceler with staggered delays, T and $3T/4$.

TABLE 8.1 First Complete Null of a Staggered Double Canceler

a		1	1/2	2/3	3/4	4/5	5/6	⋯	63/64	⋯
Stagger Ratio $= a/(2-a)$		1	1/3	1/2	3/5	2/3	5/7	⋯	63/65	⋯
$f_{\text{complete null}} T_r$		1	2	1.5	4	2.5	6	⋯	64	⋯

A plot of the response of a staggered double canceler with a stagger ratio 63/65 is given in Fig. 8.5. Note that the frequency axis extends as far as $\nu = 32/T_r$ (half the blind speed for this case). Figure 8.5 emphasizes the passband of the staggered filter, where the staggering has improved the response considerably. However, comparing the responses at the stopband (near zero) reveals a deterioration in performance. Such a comparison appears in Fig. 8.6, where the solid curve represents the response of the staggered (63/65 ratio) double canceler, and the dashed curve represents the nonstaggered double canceler.

The response at the stopband can be corrected by changing the weights from their nominal $\{1, -2, 1\}$ sequence. It is possible to show that if we use the

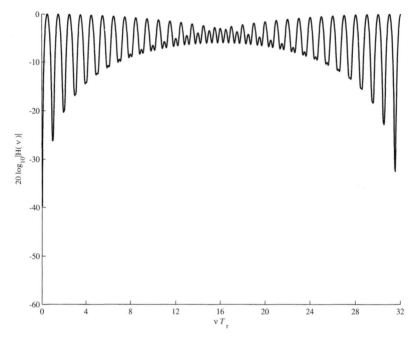

FIGURE 8.5 Frequency response of a double canceler with a 63/65 stagger ratio.

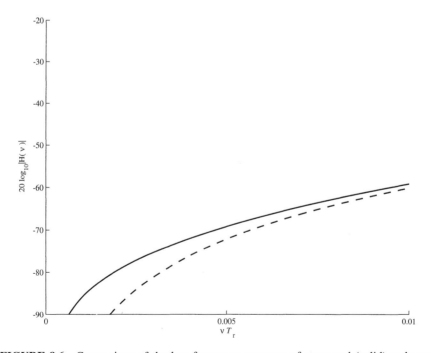

FIGURE 8.6 Comparison of the low-frequency response of staggered (solid) and non-staggered (dash) double cancelers.

coefficients given by

$$C_0 = 1 \tag{8.12a}$$

$$C_1 = \frac{2}{a-2} = -\frac{2m}{m+1} \tag{8.12b}$$

$$C_2 = \frac{a}{2-a} = -\frac{m-1}{m+1} = \text{stagger ratio} \tag{8.12c}$$

the stopband response of the staggered filter with the modified weights becomes identical to the nonstaggered filter with the "binomial" weights. For the stagger ratio 63/65, the response is identical to the dashed curve of Fig. 8.6.

8.2.2 Actual Frequency Response of Staggered-PRF MTI Radar

For a staggered-PRF waveform the transmitted pulses have staggered repetition periods. The signal received from a given range window becomes a set of signal samples at nonuniform intervals. The reference signal used in the matched filter (phase detector) is usually a single pulse. A single canceler will be implemented by subtracting the preceding sample from the present one (both from the same range). Similarly, a double canceler will subtract twice the preceding sample from the present sample and add the sample before the preceding one. It is important to understand that the number of different interpulse periods is not limited by the order of the canceler.

Figure 8.7 presents an example of a digital MTI with a single canceler and a staggered PRF with three different interpulse periods. The z^{-1} block does not represent a given delay but a separation of one sample. The transfer function of a first-order $(1 - z^{-1})$ digital filter operating on a sequence of samples

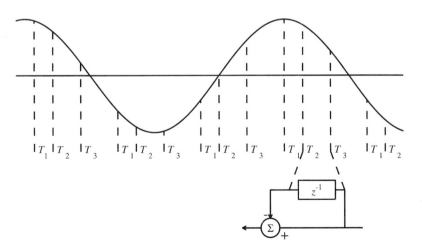

FIGURE 8.7 Single canceler with staggered PRF.

(pulses) with N interpulse periods is discussed in (Roy and Lowenschuss, 1970) and (Thomas and Abram, 1976). Since the set of N periods can begin at any phase $\varphi > 0$ of a received sine-wave signal, the transfer function was defined (Roy and Lowenschuss, 1970) as the ratio of the mean-squared output (MSO) to the mean-squared input (MSI). The averaging is done on all possible values of $\varphi > 0$ that is distributed uniformly between 0 and 2π. The transfer function of a single canceler with N different interpulse periods T_n was derived explicitly by Roy and Lowenschuss (1970) and is given by

$$|H_N(\nu)|^2 = 1 - \frac{1}{N}\sum_{k=1}^{N} \cos 2\pi\nu T_k \tag{8.13}$$

where the response (squared) is normalized with respect to the mean-squared gain of the filter, which is equal to the sum of the squares of all the weights (for a single canceler that sum is equal to $1^2 + 1^2 = 2$).

For $N = 1$, equation (8.13) reduces to (8.4), which is the response of the non-staggered single canceler. The "normalized with respect to the mean-squared gain" response of a single canceler having two interpulse periods with a stagger ratio of $T_2/T_1 = 63/65$ is shown in Fig. 8.8. Note that the actual response

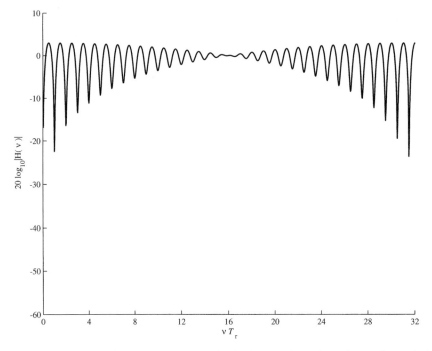

FIGURE 8.8 Actual frequency response of a single canceler and two interpulse periods with $T_1/T_2 = 63/65$.

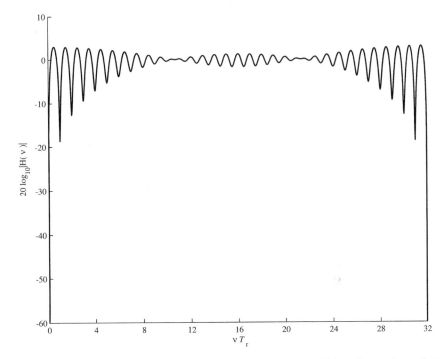

FIGURE 8.9 Actual frequency response of a single canceler and three interpulse periods with a stagger ratio of 31/32/33.

as defined in (8.13) and plotted in Fig. 8.8 is different from the response of an analog filter with two (staggered) delay lines, which was plotted in Fig. 8.5. The difference is not only in the form of a bias due to the different normalization definitions, but also in essence. The normalized response of a single canceler with three interpulse periods and a stagger ratio of 31/32/33 is shown in Fig. 8.9.

The filter included in Fig. 8.7 was a single-canceler filter, but higher-order filters can be used. The normalized response of a second-order filter (double canceler) with N interpulse periods was given in (Roy and Lowenschuss, 1970) as

$$|H_N(\nu)|^2 = 1 - \frac{2}{3(N-1)}(\cos 2\pi\nu T_1 + \cos 2\pi\nu T_N) - \frac{4}{3(N-1)}\sum_{k=2}^{N-1}\cos 2\pi\nu T_k$$

$$+ \frac{1}{3(N-1)}\sum_{k=1}^{N-1}\cos[2\pi\nu(T_k + T_{k+1})] \qquad (8.14)$$

A general expression for the nonnormalized response of higher-order filters can be found in (Thomas and Abram, 1976). Higher-order cancelers are useful for their effect in the stopband. However, the actual response of double (or higher-order) cancelers with PRF staggering suffers from the same deterioration of the stopband response, as discussed in Section 8.2.1. Here, too, the nonstaggered stopband response can be restored by modifying the weights. However, optimal weights modification must be a function of the last two (or more) interpulse periods. Hence, a system of time-varying weights is required (Shrader, 1970).

From a discussion of the various filters, it seems that there is a trade-off between the order of the filter, number of interpulse periods, introduction of feedback, time-varying weights, and between these and the ever-present consideration of the instrument complexity. To deal with this trade-off, we need a quantitative measure of the filter performance. Such a figure of merit is described in the next section.

8.2.3 MTI Radar Performance Analysis

There are several figures of merit that allow us to make a comparative performance analysis among the various MTI filters. The first basic criterion determines the clutter attenuation by the filter; that is,

$$\text{CA} = \text{clutter attenuation} = \frac{\text{clutter power at the filter input}}{\text{clutter power at the filter output}} \quad (8.15)$$

If $S_c(v)$ is the clutter power spectral density and $|H(v)|$ is the magnitude of the filter response, the clutter attenuation will be given by

$$\text{CA} = \frac{\int_{-\infty}^{\infty} S_c(v) \, dv}{\int_{-\infty}^{\infty} S_c(v) |H(v)|^2 \, dv} \quad (8.16)$$

When the clutter spectrum is symmetrical (which is not always the case), both integrations in (8.16) could begin from 0 rather than $-\infty$. With regard to the upper boundary of the integral, it should be recalled that in a fixed-PRF radar, with a repetition period T_r, the clutter spectrum and the filter response fold over at $v = 1/(2T_r)$. Thus, the upper integration limit should be $1/(2T_r)$ rather than ∞. In staggered PRF, T_r should be replaced with the extended repetition period.

At the same time that the filter attenuates the clutter, it may attenuate or amplify the signal. Because various signals may have different spectrums, it is customary to assume that all Doppler frequencies are equally likely and to average the filter gain up to the highest Doppler frequency. The improvement factor, I, is defined as the output target-to-clutter ratio divided by the input target-to-clutter ratio. Hence

$$\text{I} = \text{improvement factor} = \text{CA} \cdot G \quad (8.17)$$

where G is the *average power gain* over all possible target Doppler values.

Another measure of filter performance is *subclutter visibility* (SCV), which is the ratio between the input clutter power and the input target power that yields equal powers at the output of the canceler filter. It can be shown that when the target spectrum can also be approximated by a normal distribution and the filter response at the target Doppler frequencies has a unit magnitude, the CA and SCV criteria are equivalent.

For a Gaussian clutter power spectral density (see Box 8A), the improvement factors of single and double cancelers are given approximately by

$$I \cong \frac{2}{(T\sigma_\omega)^2} \quad \text{(single canceler)} \tag{8.18a}$$

$$I \cong \frac{2}{(T\sigma_\omega)^4} \quad \text{(double canceler)} \tag{8.18b}$$

where σ_ω is the clutter power spectral density standard deviation (in rads) and where we also assumed $\sigma_\omega T \ll 0.5$ rad. For larger values of $\sigma_\omega T$ the exact expressions of I should be used; that is,

$$I = \frac{1}{1 - \exp[-(T\sigma_\omega)^2/2]} \quad \text{(single canceler)} \tag{8.19a}$$

$$I = \frac{1}{1 - \frac{4}{3}\exp[-(T\sigma_\omega)^2/2] + \frac{1}{3}\exp[-(2T\sigma_\omega)^2/2]} \quad \text{(double canceler)} \tag{8.19b}$$

The exact expressions of the improvement factor are developed in Box 8A using a simple procedure that utilizes the clutter autocorrelation function. Plots of (8.19a) and (8.19b) are given in Fig. 8.10.

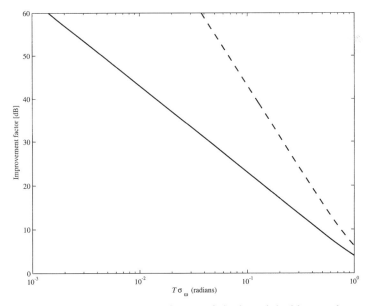

FIGURE 8.10 Improvement factors of single and double cancelers.

BOX 8A: Improvement Factor Introduced through the Autocorrelation Function

We assume a zero-centered clutter spectral density given by

$$S_C(\omega) = \frac{P_c}{\sqrt{2\pi\sigma_\omega^2}} \exp\left(-\frac{\omega^2}{2\sigma_\omega^2}\right) \tag{8A.1}$$

where P_c is the clutter power and σ_ω is the standard deviation of the spectrum. The clutter autocorrelation function is obtained by the inverse Fourier transform of the spectral density, yielding a normalized correlation function

$$\rho(\tau) = \frac{R_c(\tau)}{R_c(0)} = \exp\left(-\frac{\tau^2}{2\sigma_\tau^2}\right) = \exp\left(-\frac{\tau^2 \sigma_\omega^2}{2}\right) \tag{8A.2}$$

where $\sigma_\tau = 1/\sigma_\omega$. If the clutter input signal is $x(t)$, then

$$R_c(\tau) = \overline{x(t)x(t-\tau)} \tag{8A.3a}$$

and

$$R_c(0) = \overline{x^2(t)} \tag{8A.3b}$$

We will now use the correlation function to obtain the single- and double-canceler improvement factors. The single-canceler output signal $y(t)$ is obtained from the input signal $x(t)$ as follows:

$$y(t) = x(t) - x(t-T) \tag{8A.4}$$

The mean-squared output is given by

$$\overline{y^2(t)} = \overline{[x(t)-x(t-T)]^2} = \overline{x^2(t)} - \overline{2x(t)x(t-T)} + \overline{x^2(t-T)}$$
$$= 2R_c(0) - 2R_c(T) \tag{8A.5}$$

The clutter attenuation (CA) was defined as the clutter input power divided by the clutter output power; hence,

$$CA = \frac{\overline{x^2(t)}}{\overline{y^2(t)}} = \frac{R_c(0)}{2[R_c(0) - R_c(T)]} = \frac{1}{2[1 - \rho(T)]} \tag{8A.6}$$

The improvement factor is obtained by multiplying the CA by the average power gain G. The single canceler has $G = 2$ (sum of all the coefficients squared). Thus,

$$I = G \cdot CA = \frac{1}{1 - \rho(T)} = \frac{1}{1 - \exp(-T^2/2\sigma_\tau^2)}$$

$$= \frac{1}{1 - \exp(-T^2\sigma_\omega^2/2)} \tag{8A.7}$$

The output signal of a double canceler is given by

$$y(t) = x(t) - 2x(t - T) + x(t - 2T) \tag{8A.8}$$

Hence, the mean-squared output will be

$$\overline{y^2(t)} = \overline{x^2(t)} + 4\overline{x^2(t - T)} + 4\overline{x^2(t - 2T)} - 4\overline{x(t)x(t - T)}$$
$$+ 2\overline{x(t)x(t - 2T)} - 4\overline{x(t - T)x(t - 2T)} \tag{8A.9}$$

or

$$\overline{y^2(t)} = 6R_c(0) - 8R_c(T) + 2R_c(2T) \tag{8A.10}$$

The clutter attenuation will therefore be

$$CA = \frac{\overline{x^2(t)}}{\overline{y^2(t)}} = \frac{R_c(0)}{6\left[R_c(0) - \frac{4}{3}R_c(T) + \frac{1}{3}R_c(2T)\right]} \tag{8A.11}$$

The double-canceler average power gain is 6 ($= 1^2 + 2^2 + 1^2$), hence

$$I = \frac{1}{1 - \frac{4}{3}\rho(T) + \frac{1}{3}\rho(2T)} = \frac{1}{1 - \frac{4}{3}\exp(-T^2\sigma_\omega^2/2) + \frac{1}{3}\exp(-4T^2\sigma_\omega^2/2)} \tag{8A.12}$$

The improvement factor depends mostly on the stopband response of the MTI filter. In staggered PRF we showed a considerable deterioration in the stopband response unless the binomial weights are modified. This deterioration is reflected in the improvement factor. In (8.12b) and (8.12c) we suggested weights that will restore the nonstaggered performance at the stopband. Having the improvement factor criteria as performance measures, it is possible to seek weights that will not only restore but maximize I (Capon, 1964). An exhaustive search reported by Hsiao (1974) indicated that the highest improvement factor of a staggered PRF lies between two bounds. The lower bound is the improvement factor of an MTI with a fixed PRF equal to the lowest PRF of the staggered MTI. The upper bound is the improvement factor of a fixed PRF equal to the highest PRF of the staggered system. An example of optimizing the MTI weights is given in Box 8B.

BOX 8B: Optimal MTI Weights

Nonstaggered MTI FIR filters were developed by cascading single cancelers. This method led to binomial weights: {1 −2 1}, {1 −3 3 −1}, and so on. However, now having quality criteria such as the improvement factor, we can check if these weights are indeed optimal. Let us rewrite the output of the double canceler with the second weight as a parameter:

$$y(t) = x(t) + C_2 x(t - T) + x(t - 2T) \tag{8B.1}$$

The corresponding improvement factor becomes

$$I = \left\{ 1 + \frac{4C_2}{2 + C_2^2} \exp\left[-\frac{(T\sigma_\omega)^2}{2} \right] + \frac{2}{2 + C_2^2} \exp\left[-\frac{(2T\sigma_\omega)^2}{2} \right] \right\}^{-1} \tag{8B.2}$$

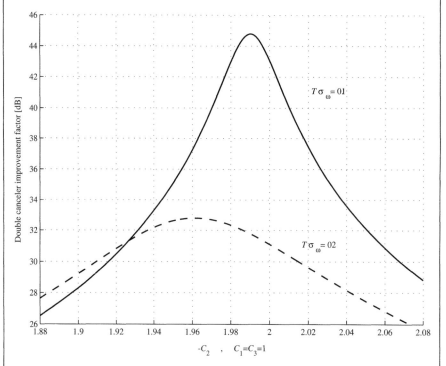

FIGURE 8.11 Improvement factor of a double canceler as a function of the central weight.

A plot of I as function of C_2 for two values of $T\sigma_\omega$ is given in Fig. 8.11. Figure 8.11 shows that the optimal weight of the middle tap (C_2) is lower than 2 and that this optimal value is a function of the clutter spectral width σ_ω. In a double canceler, only one weight (C_2) needs to be optimized and the optimal value can be obtained using several simple methods.

What about longer MTI filters? It turns out that the optimal filter coefficients can be obtained easily for any filter length, using eigenvalues and eigenvectors (Capon, 1964; Hsiao, 1974). As pointed out in Box 8A, the normalized correlation function of clutter with zero-centered Gaussian distribution is given by

$$\rho(\tau) = \exp\left[-\frac{(\tau\sigma_\omega)^2}{2}\right], \qquad \rho(0) = 1 = \text{average input power} \quad (8\text{B}.3)$$

For equally spaced pulses, T seconds apart, the clutter samples covariance matrix is

$$\mathbf{A} = \begin{bmatrix} 1 & \rho & \rho^4 & \rho^9 & \cdots & \rho^{(n-1)^2} \\ \rho & 1 & \rho & \rho^4 & \cdots & \rho^{(n-2)^2} \\ \rho^4 & \rho & 1 & \rho & \ddots & \vdots \\ \rho^9 & \rho^4 & \rho & 1 & \ddots & \rho^4 \\ \vdots & \vdots & \ddots & \ddots & \ddots & \rho \\ \rho^{(n-1)^2} & \rho^{(n-2)^2} & \cdots & \rho^4 & \rho & 1 \end{bmatrix} \quad (8\text{B}.4)$$

where $\rho = \rho(T)$. The elements a_{ij} of $\mathbf{A}$ are real and positive (A is a Toeplitz matrix). Let the filter coefficients be described by the vector

$$\mathbf{c} = [\,c_1 \quad c_2 \quad \cdots \quad c_N\,]^\mathrm{T} \quad (8\text{B}.5)$$

The clutter average power at the output of the filter is proportional to

$$\frac{P_{c_\text{out}}}{P_{c_\text{in}}} = \sum_i \sum_j c_i c_j^* a_{ij} \quad (8\text{B}.6)$$

Since a_{ij} are real and, as will be shown, the filter coefficients are also real, the average clutter output power can also be represented by the inner product

$$\frac{P_{c_\text{out}}}{P_{c_\text{in}}} = \langle \mathbf{c}, \mathbf{A}\mathbf{c} \rangle \quad (8\text{B}.7)$$

If the vector of filter coefficients $\mathbf{c}$ is an eigenvector of $\mathbf{A}$, corresponding to an eigenvalue λ (a scalar, not to be confused with wavelength), then, by definition,

$$\mathbf{Ac} = \lambda \mathbf{c} \tag{8B.8}$$

Using (8B.8) in (8B.7) yields

$$\frac{P_{c_{out}}}{P_{c_{in}}} = \langle \mathbf{c}, \lambda \mathbf{c} \rangle = \lambda \langle \mathbf{c}, \mathbf{c} \rangle = \lambda \sum_{n=1}^{N} c_n^2 = \lambda G \tag{8B.9}$$

Assuming that the Doppler signal is distributed uniformly at all frequencies, the signal power gain is G; hence for the given coefficient vector, the improvement factor is given by

$$I = \frac{G}{P_{c_{out}}/P_{c_{in}}} = \frac{1}{\lambda} \tag{8B.10}$$

The improvement factor will be maximum when the eigenvalue λ is the smallest of the eigenvalues of $\mathbf{A}$. Hence the optimal filter coefficients are given by the eigenvector of $\mathbf{A}$ that corresponds to the minimum eigenvalue of $\mathbf{A}$.

For example, we calculate the optimal coefficients of a canceler of length 5 when $T\sigma_\omega = 0.1$. First, the value of $\rho = \rho(T) = 0.995$ is obtained using (8B.3). Next, the 5×5 matrix $\mathbf{A}$ is obtained [see equation (8B.4)] and the eigenvalues of $\mathbf{A}$ are calculated. The eigenvector corresponding to the minimal eigenvalue (3.4×10^{-9}) is normalized such that the leading element is 1, yielding the normalized vector of coefficients given by $\mathbf{c}/c_1 = [\,1.0000 \quad -3.9409 \quad 5.8821 \quad -3.9409 \quad 1.0000\,]^T$. The corresponding improvement factor is obtained by $10\log_{10}(1/\lambda_{min}) = 84.7\,\text{dB}$.

For comparison, we calculate the improvement factor of the filter with binomial coefficients $\mathbf{c} = [\,1 \quad -4 \quad 6 \quad -4 \quad 1\,]$. First we find that for this specific selection of filter coefficients, the inner product in equation (8B.7) is $\mathbf{c}^T\mathbf{Ac} = 1.02 \times 10^{-6}$. The average gain is calculated from the sum of the filter coefficients squared as $\mathbf{c}^T\mathbf{c} = 70$. Finally, the improvement factor is obtained from $10\log_{10}(70/1.02 \times 10^{-6}) = 78.4\,\text{dB}$, which is about 6.3 dB less than the improvement factor of the optimal five-element FIR filter.

Figure 8.12 displays the magnitude of the frequency response of a five-element FIR MTI filter with the optimal coefficients (solid line) compared to a five-element filter with binomial coefficients (dashed line). Both filters were normalized to give the same average gain. The dotted line presents a clutter spectrum with a frequency spread ($T\sigma_\omega = 0.1$) to which the filter coefficients were optimized. The higher improvement factor is probably due to the extended bandstop section in the frequency response of the optimized filter.

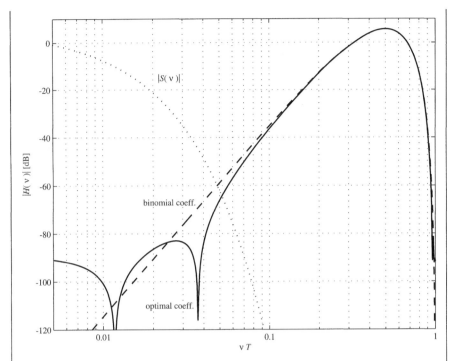

FIGURE 8.12 Frequency responses of five-element MTI filters with optimal and binomial coefficients.

In the introduction to this chapter we pointed out that the DFT processor, discussed in Chapter 7, can also serve as an MTI filter, by avoiding the output of the zeroth filter. The MTI improvement factor of DFT processors is discussed in (Ziemer and Ziegler, 1980). The various weight functions (Hamming, Hann, etc.) strongly affect the results. The performance of a DFT processor depends on the analog-to-digital (A/D) conversion preceding it. If the initial signal-to-clutter ratio is very small, the dynamic range of the A/D may not be able to process the signal properly. In that case it is possible to insert a canceler filter in front of the DFT. The canceler will attenuate the clutter and thus reduce the dynamic range of the input to the DFT.

The MTI techniques discussed so far utilized the pulse-to-pulse changes of a coherently detected return from a moving target compared to the relatively fixed return of the clutter. The clutter return is fairly constant not only from pulse to pulse but also from batch to batch (or "burst" to "burst"). This fact suggests a postdetection reduction of clutter returns by binary integration of the returns from different bursts (performed at different PRIs), which is the topic of the next section.

8.3 DIVERSIFYING THE PRI ON A DWELL-TO-DWELL BASIS

The pulse-to-pulse PRI diversity discussed in Section 8.2 has the clear advantage of receiver simplicity (MTI processing) while also providing the required filtering at zero Doppler. In some applications the loss involved in using MTI processing or the requirement to have better Doppler resolution between targets implies using the constant-PRI pulse train described in Chapter 7. Each pulse train having a constant PRF is usually referred to as a *burst* or a *dwell*; threshold crossings based on the processing of a single dwell are named *hits*, while target *detection* is made on a multiple-dwells basis (and in some cases even on a scan-to-scan basis).

8.3.1 Single-PRF Pulse Train Blind Zones and Ambiguities

The unambiguous range (in km) that can be observed using a specific PRF (in kHz) is given by 150/PRF. The unambiguous velocity (in m/s) is related to the PRF (in kHz) as 150 PRF/f, where f is the carrier frequency (in GHz). In general, we would like to be able to choose the PRF to provide the desired unambiguous range and Doppler bandwidth simultaneously. However, unambiguous range increases with decreasing PRF, while unambiguous Doppler increases with increasing PRF (see Fig. 8.13).

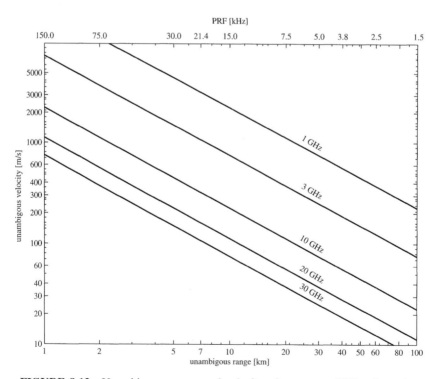

FIGURE 8.13 Unambiguous range and velocity of a constant-PRF pulse train.

In dwell-to-dwell PRF stagger, a dwell (coherent processing interval) of N pulses is transmitted at a fixed PRF. A second dwell is then transmitted at a different fixed PRF. Because the blind areas are different for each PRF used, a target that falls in a blind area of one PRF will be visible in the others. Radar systems utilizing bursts of constant PRF for separating radar echoes using their Doppler properties are often referred to as *pulse Doppler systems*. The PRF values that give both unambiguous range and Doppler are referred to as *medium PRF*, as opposed to *high PRF values* that give highly ambiguous range but unambiguous Doppler, or *low PRF values* producing unambiguous range but highly ambiguous Doppler. Note that one PRF could be considered medium for one application but low or high for others (depending on the expected target parameters).

Figure 8.14 shows an example of the blind zone areas of a fixed PRF pulse train. The vertical lines at integer multiples of the PRI represent the timing where the receiver is blocked due to the transmitting pulses (*eclipsing*). The horizontal lines at integer multiples of the PRF represent Doppler cells that are filtered out due to clutter or interference at a specific frequency. The figure also shows the masking caused by a point target at a velocity equivalent to 2.5 the PRF and in a range corresponding to 1.25 the PRI (marked with a circle). The single point target masks other targets separated by integer multiples of the PRI and integer multiples of the PRF. The blind zone footprint is a function of the interfering target signal power, the signal to interference power required for detection, range,

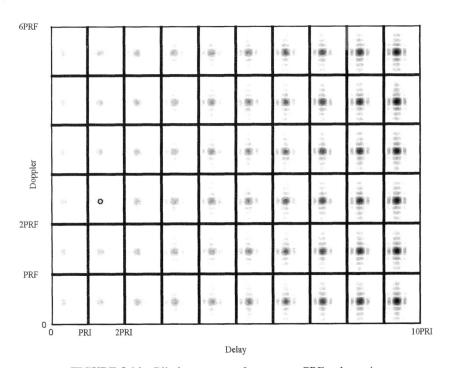

FIGURE 8.14   Blind zone map of a constant-PRF pulse train.

and Doppler. In the example shown in Fig. 8.14, the interference was assumed to be a point target with nonzero range and Doppler sidelobes (e.g., due to pulse compression). The targets of interest were assumed to have power range dependence on $1/R^4$. The gray level indicated the level of interference. In the area where the gray level is identical to the one observed at the interfering target position, the signal-to-interference level would be 0 dB for targets with RCS identical to the RCS of the interfering target. Where the gray level is higher (e.g., higher range), the RCS required for detection will increase. Under other assumptions on the interference (e.g., mainlobe clutter, altitude line, sidelobe clutter, jamming) and signal (e.g., the single pulse ambiguity function shape, constant target power at the receiver, etc.), different blind zone footprints will have to be considered.

8.3.2 Solving Range–Doppler Ambiguities

The process of solving the range and Doppler ambiguities involves sequential measurement of the ambiguous range and Doppler in each of the PRF. The process involves comparison of the measurements to eliminate ambiguities. In a multiple-target environment, the process of solving range and Doppler ambiguities can yield false solutions (usually referred to as *ghosts*). An example of a two-PRF waveform having both range and Doppler ambiguities is shown in Fig. 8.15.

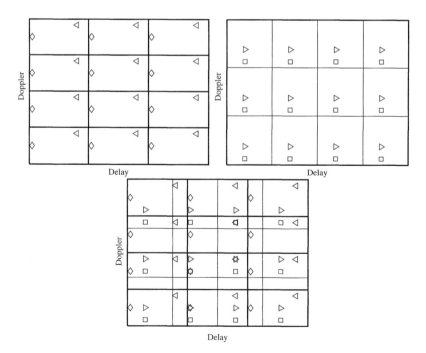

FIGURE 8.15 Solving range and Doppler ambiguities using a two-PRF waveform.

The two PRFs used are $3f$ and $4f$, where f is the LCM of the two PRFs, implying a maximal unambiguous "range" of $1/f = R_u$ and unambiguous Doppler of $f_u = 12f$. The top left subplot shows the range–Doppler ambiguous hits of first PRF dwell (unambiguous range of $R_u/3$ and unambiguous Doppler of $3f$), while the top right subplot shows the range Doppler ambiguous hits of the second PRF dwell (unambiguous range of $R_u/4$ and unambiguous Doppler of $4f$). The two subplots show hits caused by two targets. One target is marked using triangles (facing right on one dwell and facing left on the other dwell), and the second target is marked using the square (first dwell) and diamond (second dwell) shapes.

The lower subplot shows the two subplots overlaid. The place where the two triangles intersect is the true location of the first target, while the position where the square and diamond intersect is the true location of the second target. The cross intersections (first target from first dwell with second target from second dwell, and the other way around) cause false detections (ghosts). Note that the ghosts are a result of the specific selection of PRF and target position; as the targets change their positions, the ghosts may disappear (or reappear elsewhere) and real targets can be identified. Note also that the ghosts problem is greatly minimized when the target is detected using three PRFs instead of two.

The procedure described using Fig. 8.15 is usually referred to as the *coincidence algorithm*. The coincidence algorithm is computationally intensive even for a small number of targets but involves few constraints on the PRF selection. As demonstrated, the algorithm operates by taking the target returns in a single PRI and repeating them until the maximum range may be present. The process is repeated for all the visible PRIs and the results overlaid. If a true target is present, it will appear in the same position in all visible PRIs. Similarly, the true Doppler may be resolved in the frequency domain. When accounting for range and Doppler, the process can be performed with a two-dimensional map in range–Doppler space as demonstrated in Fig. 8.15.

An older technique to resolve the ambiguities is based on the Chinese remainder theorem. The Chinese remainder theorem employs a PRI of an integer number of range cells and subsequent modulo mathematics that is sufficiently simple to enable a hardware solution. However, integer mathematics imposes limitations on the number of suitable PRFs and does not address the minimization of blind zones. The procedure of solving a target range using the Chinese remainder theorem is demonstrated using a two-PRF example. Assume that the target range R corresponds to a true target delay T_t. Assume that the desired system unambiguous range yields an unambiguous delay T_{un}. We select $PRI_1 = 2T_{un}/N$ and $PRI_2 = 2T_{un}/(N+1)$, where N is an integer. Thus, T_t can be expressed as a function of PRI_1, PRI_2, and the ambiguous range measurements T_1 and T_2 as

$$T_t = n\text{PRI}_1 + T_1 \qquad (8.20a)$$

$$T_t = m\text{PRI}_2 + T_2 \qquad (8.20b)$$

where $n, m = 0, 1, 2, \ldots$. Ideally, the two equations could be solved simultaneously to yield an expression for T_t in terms of the measurements T_1, T_2 and the known quantities N and T_{un}. Substituting the values of PRI_1 and PRI_2 in (8.20) gives

$$NT_t = 2nT_{un} + NT_1 \tag{8.21a}$$

$$(N+1)T_t = 2mT_{un} + (N+1)T_2 \tag{8.21b}$$

Subtracting the two equations gives

$$T_t = 2(m-n)T_{un} + N(T_2 - T_1) + T_2 = kT_{un} + N(T_2 - T_1) + T_2 \tag{8.22}$$

where $k = m - n = 0, \pm 2, \pm 4, \ldots$. The equation still holds after taking the modulo of both sides; thus,

$$T_t = kT_{un} + N(T_2 - T_1) + T_2 \pmod{T_{un}} \tag{8.23a}$$

or

$$T_t = N(T_2 - T_1) + T_2 \pmod{T_{un}} \tag{8.23b}$$

Thus, the unambiguous range (modulo T_{un}) is the remainder of $N(T_2 - T_1) + T_2$ after division with T_{un}. In the practical case, the delay measurements are corrupted by interference and noise. A dominant factor in the performance of the range resolver is the transmitted pulse width causing eclipsing and the possible target straddle between range gates. For this matter it is possible to select PRIs that are spaced not by one pulse width count but by two counts. The measurement errors may result in different values of T_t for integer values of n and m when used in (8.20). A value of T_t that minimizes the difference is then chosen. A simple algorithm could be to compute trial values of T_t using (8.20) for various values of m and n, and to select the true value of T_t as the one that satisfies, with minimum error, both equations of the set.

8.3.3 Selection of Medium-PRF Sets

Selecting the set of medium-PRF values used in a medium-PRF burst waveform is strongly connected to the application. Maybe the oldest method is the one utilizing major and minor PRFs. *Major PRFs* are used to extend the unambiguous Doppler. Each major PRF may be accompanied by one or two additional *minor PRFs* to resolve range ambiguities. A different method for finding the PRF values is the *M*-out-of-*N* selection method. The objective is to select a set of *N* PRFs such that the target will be in the clear Doppler region for at least M, often $M + 1$, PRFs. Detection on at least M PRFs is required for declaring a target. Any M of the PRFs may be used for range resolving; that is, none is designed as major or minor. A third option of designing the PRF set is to use an exhaustive search over all possible PRF sets maximizing the clear area in the range–Doppler map.

The exhaustive search could in many cases be too heavy to implement. Davis and Hughes (2002) have demonstrated evolutionary algorithms that give a better solution than the major–minor or M-out-of-N methods in a reasonable time.

For all methods the PRF design is subject to constraints based on the minimal average power sufficient to achieve the required detection range, the maximal allowable duty cycle, the maximal allowable pulse width required for minimal detection range, and so on. We first describe the ground rules for a typical scenario. Later, the selection of various PRF sets using the various methods will be described and compared. We use the scenario used by Davis and Hughes (2002) involving an X-band radar (3-cm wavelength) using a 1-μs pulse width (150-m range bin) and allowable PRF values between 10 and 20 kHz. The scenario also assumes blinding 11 range cells at each repetition of the PRI (starting at a zero range) due to eclipsing (one cell) and sidelobe clutter (10 additional cells). The mainlobe clutter notch is assumed to blind $\Delta v = 3.4$ kHz (± 25.5 m/s) at every repetition of the PRF and corresponds to ± 17 cells at multiples of each PRF. The region of interest is limited in range to 1000 range cells (150 km) and 200 velocity cells of 100 Hz each (20 kHz, 300 m/s).

We also assume that all PRIs are integer multiple of the pulse width (one range cell). The upper and lower bounds on the available PRI set are in this case 96 and 50 pulse widths, respectively. Thus, each PRF is one of 47 discrete values. Selecting a set of eight PRFs out of the available 47 will result in an exhaustive search over a space of more than 3×10^8 possible sets. Furthermore, to allow range and Doppler ambiguities to be resolved properly, the PRF set chosen must be fully decodable using the methods described in Section 8.3.2.

8.3.3.1 Major–Minor Selection of Medium-PRF Sets In the major–minor PRF set selection method, the major PRFs are selected first, ensuring that a high percentage of the spectral region between PRF lines is visible (e.g., free of mainlobe clutter). The minor PRF values that accompany the major PRFs are selected to ensure good range resolve performance. For the X-band radar scenario described above, we select the first major PRF as the lowest available PRF ($N_1 = 96$, $PRI_1 = 96$ μs, $PRF_1 = 10.4$ kHz). Selecting a minor PRF associated with the first major PRF is based on range-resolving considerations. The PRI as measured in number of range gates should have factors that are relatively prime to ensure that the system's unambiguous range after resolving is satisfactory. The range gate counts N_2 and N_3 are selected as $N_2 = N_1 - 2 = 94$ and $N_3 = N_2 - 2 = 92$. A difference of 2 was selected to avoid problems due to range straddle while maintaining the unambiguous range after resolving of at least 150 km. The lowest common multiple of N_1, N_2, and N_3 is 103,776, which translates to T_{un} of $103{,}776 \cdot 1$ μs ($= 15{,}566$ km). However, the large blind zone in range (11 cells) could yield detection in only two of the three PRFs. Thus, the maximal detectable range for all ranges is based on the minimal lowest common multiple of any of the pairs $N_1; N_2$, $N_1; N_3$, and $N_2; N_3$, which is only 2208 (331 km).

The PRF values are $PRF_2 = 10.6\,kHz$ and $PRF_3 = 10.8\,kHz$, which are both very close to PRF_1, thus cover almost the same area in Doppler. Note that selecting a higher count difference between the PRFs (e.g., $N_2 = N_1 - 3 = 93$, $N_3 = N_2 - 3 = 90$) would still give the desired unambiguous range (216 km) but will widen the Doppler notch and could require using additional PRFs for detection. The second major PRF (PRF_4) is selected such there is no overlap between the blind Doppler zones in both major PRFs. Thus, PRF_4 is selected as the lowest one that exceeds $PRF_1 + \Delta v = 13.8\,kHz$. Using the constant range gate interval of 1 μs, the integer N_4 should be less than 72.4. We select $N_4 = 72$ ($PRF_4 = 13.9\,kHz$). Note that selecting lower PRF could yield a blind Doppler zone in the overlapping of the Doppler blind zones of the first and second major PRFs. The minor PRFs that accompany the second major PRF are selected such that $N_5 = N_4 - 2 = 70$ ($PRF_5 = 14.3\,kHz$) and $N_6 = N_5 - 2 = 68$ ($PRF_6 = 14.7\,kHz$). The minimal lowest common multiple of the pairs $N_4;N_5$, $N_4;N_6$, and $N_5;N_6$ is 1224, which translates to T_{un} of $1224 \cdot 1\,\mu s$ (=183.6 km).

The first major PRF covers unambiguously 10.4 kHz (104 Doppler cells), while the second major PRF covers 139 Doppler cells. To allow minimal unambiguous Doppler cover of 200 cells for all Doppler values, a third major PRF is required (a target in the blind zone of one of the PRFs cannot be resolved in Doppler using only one PRF). The third major PRF is selected as the lowest one that exceeds $PRF_4 + \Delta v = 17.3\,kHz$. Using the constant range gate interval of 1 μs, the integer N_7 should be less than 57.8. We select $N_4 = 57$ ($PRF_7 = 17.5\,kHz$). The minor PRFs that accompany the third major PRF are selected such that $N_8 = N_7 - 2 = 55$ ($PRF_8 = 18.2\,kHz$) and $N_9 = N_8 - 2 = 53$ ($PRF_9 = 18.9\,kHz$). The minimal lowest common multiple of the pairs $N_7;N_8$, $N_7;N_9$, and $N_8;N_9$ is 2915, which gives an unambiguous range of 437 km.

The resulting set of 9 PRFs is {10.4 10.6 10.8 13.9 14.3 14.7 17.5 18.2 18.9} (in kHz) can be also written in terms of cell counts as {96 94 92 72 70 68 57 55 53}. Figure 8.16 shows the blind zone map of the nine major–minor PRFs. The dark areas mark regions that cannot be detected using three different PRF values. The requirement for detection in three different PRFs is to allow for proper range and Doppler resolving in the presence of noise and to lower the number of ghosts significantly. The total number of blind range Doppler cells in Fig. 8.16 is 41,583. Within the designed area (150 km, 300 m/s) only 19,387 cells are blind. Excluding the zero Doppler and minimum range areas (19,013 cells), only 374 blind cells remain (0.1%). The total antenna dwell time must be divided among the ensemble of PRFs. If the number of PRFs can be reduced by an improved PRF selection strategy without increasing the blind zone, more time can be allotted to each PRF. The resulting blind zone map when using only the highest eight PRFs of the nine major–minor PRFs is shown in Fig. 8.17. Note the blind area increase to 1706 range Doppler cells.

8.3.3.2 M-out-of-N Selection of Medium-PRF Sets In the major–minor PRF set the PRFs were divided in groups where each group allows for a different

DIVERSIFYING THE PRI ON A DWELL-TO-DWELL BASIS

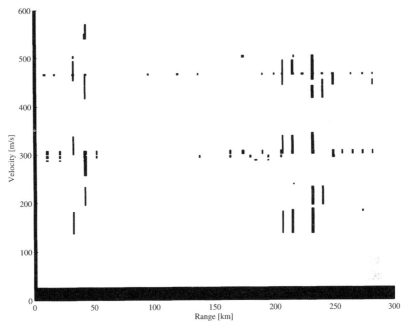

FIGURE 8.16 Minimum three PRF detect and resolve blind zone map for the major–minor nine-PRF set.

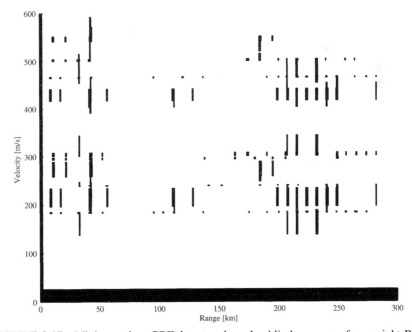

FIGURE 8.17 Minimum three PRF detect and resolve blind zone map for an eight-PRF set based on the major–minor nine-PRF set (lowest PRF excluded).

Doppler cover and the members of the group were selected to allow range resolve. In the M-out-of-N strategy, the different PRFs are distributed evenly such that their Doppler blind area partly overlaps but each area can be detected by at least M PRFs.

Consider the scenario described before but instead of using nine PRFs as in the major–minor strategy, we use only eight PRFs. We still require detection in three of the eight PRFs. The required change between PRFs is $\Delta v/4 = 0.85\,\text{kHz}$. While this ensures that all Doppler frequencies are covered by at least four PRFs, the coverage of these PRFs in range is found only by trial and error. We start with the highest available PRF (20 kHz) working down and write the initial set of PRFs (in kHz) as { 20 19.15 18.3 17.45 16.6 15.75 14.9 14.05 }. The next step is to adjust the PRFs downward to provide an integer number of range bins. The PRFs become { 20 19.2 18.5 17.5 16.7 15.9 14.9 14.1 } in kHz and the counts are { 50 52 54 57 60 63 67 71 }. The blind zone map for the first guess of three-out-of-eight PRF is shown in Fig. 8.18. Note the relatively high blind zone area of 7092 cells (3.5%) within the required 150-km range and 300-m/s velocity (minimum range and zero Doppler excluded).

A modified approach would be to divide the entire PRF zone uniformly to eight PRFs, resulting in the PRF counts { 52 55 59 64 70 76 84 94 }.

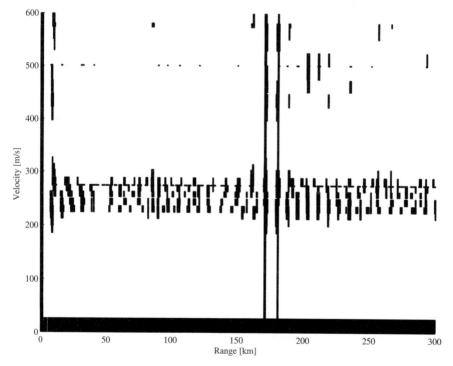

FIGURE 8.18 Minimum three-PRF detect and resolve blind zone map for the maximal average PRF three-out-of-eight set { 50 52 54 57 60 63 67 71 }.

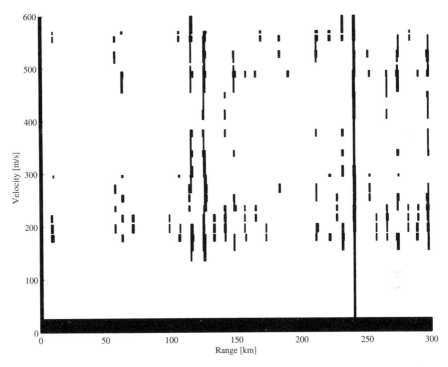

FIGURE 8.19 Minimum three-PRF detect and resolve blind zone map for the uniformly spaced three-out-of-eight PRF set { 52 55 59 64 70 76 84 94 }.

Figure 8.19 shows the blind zone map of the uniform PRF set selection. The blind zone area within the required boundaries (150 km, 300 m/s) was reduced to 2765 (1.4%). Note that while this selection gives better coverage in terms of the blind zone, it can result in lower detection ranges due to a lower average duty cycle. Proper selection of M and N such that the overall P_d is maximized and the overall P_{fa} is minimized is based on the expected P_d and P_{fa} for a single burst. The trade-offs and mathematics involved in selecting the proper value of M for a given set N are described in Box 8C.

8.3.3.3 Medium-PRF Set Selected Using an Evolutionary Algorithm Davis and Hughes (2002) reported the results of finding an optimal PRF set to the X-band radar problem described in previous sections. The resulting PRF set using the evolutionary algorithm reported is based on the counts { 51 53 60 63 67 84 89 93 }. The resulting blind zone map is shown in Fig. 8.20. Note the blind area reduced to only 519 range–Doppler cells, which is almost the same as with the nine major–minor PRFs (achieved here with only eight PRFs).

BOX 8C: Binary Integration

For the multiple-dwell pulse–Doppler waveforms, we repeat the entire detection process N times for any given range–Doppler cell and get N binary decisions. To improve the reliability of our detection decision, we require that a target be detected on some number M out of the N decisions before we finally accept it as valid target detection. This process is called *binary integration, M-out-of-N detection,* or *coincidence detection.*

To analyze binary integration, we begin by assuming a nonfluctuating target so that the probability of detection, P_D, is the same for each of N threshold tests. Then the probability of *not* detecting an actual target (i.e., the probability of a miss) on one trial is $1 - P_D$. If we have N independent trials, the probability of missing the target on all N trials is $(1 - P_D)^N$. Thus, the probability of detecting the target on at least one of N trials, which we denote as the *cumulative probability* P_{CD}, is $1 - (1 - P_D)^N$.

It can easily be verified that the required single-trial probability of detection achieving high P_{CD} in a 1-out-of-N decision rule is relatively low. The trouble with the 1-out-of-N rule is that it "works" for the probability of false alarms also. The cumulative probability of false alarms is the probability of at least one false alarm in N trials. Assuming that $P_{FA} \ll 1$, it is possible to show that the 1-out-of-N rule increases P_{FA} by a factor of approximately N. A binary integration rule that reduces the required single-trial P_D but increases P_{FA} does not accomplish the goal of decreasing the required single-sample SNR to achieve a specified P_D and P_{FA}. What is needed is a binary integration rule that increases P_{CD} compared to P_D while leaving P_{CFA} equal to, or less than, P_{FA}.

An M-out-of-N strategy provides better results. If we repeat the probability calculations for 2-out-of-4 detection rule, it is easy to see that it results in a cumulative false alarm probability that is less than the single-trial P_{FA}, as desired. The 2-of-4 rule not only reduces the probability of false alarms, it also increases the probability of detection as long as the single-trial P_D is reasonably high. This example illustrates the characteristics required of an M-of-N rule. Comparing all possible M-out-of-4 detection rules, the 2-out-of-4 rule appears to be the best choice.

A previously suboptimal set reported by Simpson (1988) which gives a blind area of 2001 range–Doppler cells is described by the counts {51 57 63 66 69 78 90 96}. The blind area map for Simpson's PRF set is shown in Fig. 8.21.

Finally, we note that each time the PRF is changed, a *dead time* must be allowed, equal to the round-trip propagation time to the maximum range of significant clutter. During this dead time, some fill pulses are transmitted such

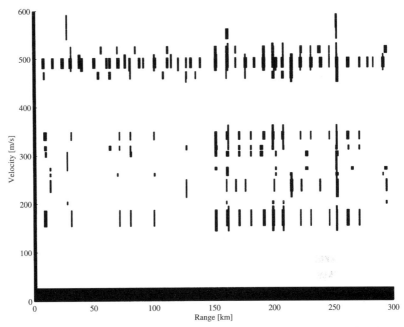

FIGURE 8.20 Minimum three-PRF detect and resolve blind zone map for the evolutionary algorithm three-out-of-eight PRF set {51 53 60 63 67 84 89 93}.

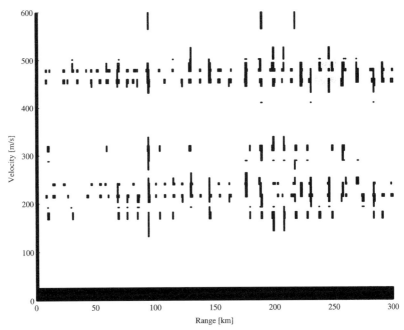

FIGURE 8.21 Minimum three-PRF detect and resolve blind zone map for Simpson's suboptimal three-out-of-eight PRF set {51 57 63 66 69 78 90 96}.

that the receiver "sees" an infinite signal return from the area of interest, and the PAF and PACF can be used to describe the receiver time–Doppler response.

PROBLEMS

8.1 Optimal double canceler

A coastal radar operates near the sea, where the wind behavior is different, such that the expected clutter spreading in the sea area is $T\sigma_w = 0.15$ but is only $T\sigma_w = 0.05$ for the land area.

(a) What should be the optimal double-canceler coefficients for the different conditions, and what are the improvement factors?

(b) Assume that MTI coefficients optimal to $T\sigma_w = 0.15$ are implemented. What is the improvement factor over land?

(c) Assume that MTI coefficients optimal to $T\sigma_w = 0.05$ are implemented. What is the improvement factor over sea?

(d) Two strategies of implementation are tested. The first is to use MTI coefficients optimal to the average clutter conditions ($T\sigma_w = 0.1$). The second is to use average MTI coefficients. Compare the improvement factors obtained for the two strategies.

(e) Assume that the clutter level at the land area is 10 dB higher than the clutter level of the sea area. Give an alternative method of selecting the MTI coefficients to minimize the clutter return for all directions. How will the radar's azimuth beamwidth and sidelobe level affect the calculations?

8.2 Single-canceler staggered-PRF first incomplete null

A staggered-PRF radar system employs a single canceler and two interpulse periods, T_1 and T_2.

(a) Show that the first minimum (incomplete null) of the actual response occurs at

$$f_1 = \frac{2}{T_1 + T_2}$$

(b) Show that the value of the response is given by

$$|H(f_1)|^2 = \frac{1}{2}\left[\frac{2\pi(T_2 - T_1)}{T_2 + T_1}\right]^2$$

and calculate the response in decibels at that frequency for the stagger ratio $T_1/T_2 = 63/65$.

(c) Repeat parts (a) and (b) for a single canceler and three interpulse periods. Calculate the first minimum for the stagger ratio 31/32/33. Compare the results to the value obtained in part (b).

PROBLEMS

8.3 Single canceler and double canceler

For staggered PRF with two interpulse periods and stagger ratio 63/65, plot and compare the actual response of a double canceler and a single canceler.

8.4 Improvement factor of nth-order canceler

Show that the improvement factor of an nth-order canceler (assuming a Gaussian clutter spectrum) is approximately

$$I \cong \frac{2n}{(T\sigma_\omega)^{2n} n!}$$

8.5 Improvement factor of third-order canceler

(a) Develop the exact expression for the improvement factor of a third-order canceler in the presence of clutter with Gaussian power spectral density.

(b) Compare the result to the approximation obtained in Problem 8.4 for $n = 3$ as a function of the clutter spectrum. Explain.

8.6 Improvement factor in matrix form

(a) Show that the results obtained for the single canceler and the two-pulse canceler in Box 8A can be written in matrix form as

$$CA = \frac{2\sigma_w^2}{\mathbf{w}^T \mathbf{M}_c \mathbf{w}^*} \qquad I = 2\sigma_w^2 \frac{\mathbf{w}^T \mathbf{w}^*}{\mathbf{w}^T \mathbf{M}_c \mathbf{w}^*}$$

where $\mathbf{c}^T$ is the vector of successive samples separated at a distance T, $\mathbf{M}_c$ the clutter covariance matrix, and $\mathbf{w}^T$ a column vector of weights applied to the successive samples.

(b) Use the expressions in part (a) to calculate the limit of the clutter attenuation when successive clutter samples become uncorrelated. Explain.

(c) How are the results in part (a) modified for non-Gaussian clutter?

8.7 Factors limiting MTI performance

Consider a radar system with unstable pulse amplitude having a mean value A and variance σ^2.

(a) Calculate the single-canceler output power for a stationary target.

(b) Calculate the canceler input power and the maximum achievable cancellation ratio.

(c) Repeat parts (a) and (b) where the pulse amplitude is fixed but the phase of the transmitted signal is not stable.

8.8 Implementation of a two-pulse canceler

Consider the two implementations of a two-pulse canceler shown in Fig. 8.3.

(a) Compare the frequency response of the two implementations where the delay lines have a delay error of ΔT_1 and ΔT_2.

(b) Calculate the frequency response of the two implementations where the delay lines have a gain of $1 + k_1$ and $1 + k_2$ ($|k_1|, |k_2| << 1$).

8.9 Two PRF ranging

Assume a staggered-PRF waveform having a constant pulse width of $T = 5\,\mu s$ and a waveform clock of $1\,\mu s$ (i.e., all PRIs are integer multiples of the clock duration).

(a) Assume that the radar employs a set of $n = 2$ PRIs given by PRI1 $= 30\,\mu s$ and PRI2 $= 70\,\mu s$. Find the radar first blind range.

(b) Design a set of $n = 2$ PRIs of total duration $T_r = $ PRI1 + PRI2 $= 100\,\mu s$ such that the first blind range is maximized. What is the value of this range?

(c) Repeat part (a) for $n = 3$. What is the value of the unambiguous range now?

(d) Define the visibility, k, as the minimal number of PRIs in which a target can be detected at a given range. Design a set of n PRIs of total burst duration $T_r = 100\,\mu s$ such that there is maximal visibility up to the range found in part (b). What is the value of n, and what is the maximal visibility k? Plot the visibility up to the maximal range with the maximal visibility.

(e) Assume that an m-out-of-n decision logic is used with the n PRIs found in part (d). Find the m that will give the highest minimal P_d over all ranges.

8.10 Two-PRF ranging using the Chinese remainder theorem

The Chinese remainder theorem affords a convenient method for decoding the true range from several ambiguous measurements. We demonstrate the principle of the method for a two-PRF ranging system at X-band.

(a) Assume a maximal target expected range of 100 km. Chose two PRF values having a common submultiple frequency f such that the interpulse period is equal to the maximal expected target range.

(b) Verify that the minimum value of the PRF was chosen to allow a clear Doppler region free of clutter in which to make target detection (assume a clutter spread of 5 m/s) and that the average PRF value is the lowest possible (maximizing duty cycle).

(c) Adjust the PRF values such that the ratio of the PRF to the ranging frequency is set equal to an integer.

(d) Assume a target at 80 km. Determine the ambiguous range to the target first in PRF_1 and then for PRF_2. By taking a time comparison of the two PRFs, obtain a coincidence pulse that is a measure of the true range.

REFERENCES

Capon, J., Optimum weighting functions for the detection of sampled signals in noise, *IEEE Transactions on Information Theory*, vol. IT-10, no. 2, April 1964, pp. 152–159.

Davis, P. G., and E. J. Hughes, Medium PRF set selection using evolutionary algorithm, *IEEE Transactions on Aerospace and Electronic Systems*, vol. 38, no. 3, July 2002, pp. 933–939.

Hsiao, J. K., On the optimization of MTI clutter rejection, *IEEE Transactions on Aerospace and Electronic Systems*, vol. AES-10, no. 5, September 1974, pp. 622–629.

Linden, D. A., and B. D. Steinberg, Synthesis of delay line networks, *IRE Transactions on Aeronautical and Navigational Electronics*, vol. 4, no. 1, January 1957, pp. 34–39.

Morris, G. V., *Airborne Pulsed Doppler Radar*, Artech House, Norwood, MA, 1988.

Roy, R., and O. Lowenschuss, Design of MTI detection filters with nonuniform interpulse periods, *IEEE Transactions on Circuit Theory*, vol. CT-17, no. 4, November 1970, pp. 604–612.

Schleher, D. C., MTI and Pulsed Doppler Radar, Artech House, Norwood, MA, 1991.

Shrader, W. W., MTI radar, in Skolnik, M. I. (ed.), *Radar Handbook*, 2nd ed., McGraw-Hill, New York, 1970, Chap. 17.

Simpson, J., PRF set selection for pulse Doppler radars, Proceedings of the IEEE Region 5 Conference, 1988, pp. 38–44.

Thomas, H. W., and T. M. Abram, Stagger period selection for moving target radars, *Proceedings of the IEE*, vol. 123, no. 3, March 1976, pp. 195–199.

Urkowitz, H., Analysis and synthesis of delay line periodic filters, *IRE Transactions on Circuit Theory*, vol. 4, 1957, pp. 41–53.

Ziemer, R. E., and J. A. Ziegler, MTI improvement factors for weighted DFTs, *IEEE Transactions on Aerospace and Electronic Systems*, vol. AES-16, no. 3, May 1980, pp. 393–397.

9

COHERENT TRAIN OF DIVERSE PULSES

The diversity introduced in Chapters 7 and 8 was caused primarily by interpulse amplitude weighting, designed to reduce Doppler sidelobes, or from staggering the PRF to alleviate the problem of blind speed. More meaningful diversities are discussed in this chapter. Pulse-to-pulse diversity can achieve several different properties, some of which are considered here. One important goal could be a reduction in the height of the recurrent (range) lobes of the autocorrelation function (ACF): namely, around $\tau = nT_r$, $n = \pm 1, \pm 2, \ldots$. Another goal could be a reduction of the near-range sidelobes: namely, around $|\tau| \leq T$. A third goal would be an increase in the overall bandwidth of the signal while maintaining relatively narrow instantaneous bandwidth. An important signal that achieves all three goals is the stepped-frequency signal. It is discussed in detail in this chapter, as is a simple processor that is often used with it—the stretch processor.

9.1 DIVERSITY FOR RECURRENT LOBES REDUCTION

In a coherent pulse train with a fixed pulse repetition interval (PRI), if the pulses are identical, the ACF will exhibit strong peaks around multiples of the PRI (i.e., around $\tau = nT_r$, $n = \pm 1, \pm 2, \ldots$). Even simple diversity can drastically reduce some or all of these recurrent lobes. A simple example is a coherent train of seven fixed-frequency pulses with simple polarity reversals (see Fig. 9.1). This particular polarity reversal (+ + + − − + −), which follows a Barker sequence

---

Radar Signals, By Nadav Levanon and Eli Mozeson
ISBN 0-471-47378-2 Copyright © 2004 John Wiley & Sons, Inc.

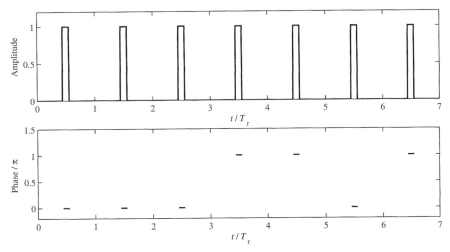

FIGURE 9.1 Pulse train with polarity reversals.

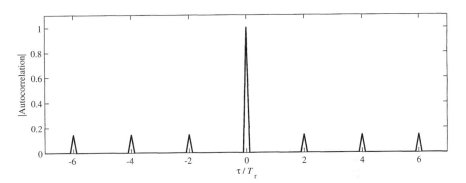

FIGURE 9.2 Aperiodic ACF of the pulse train with polarity reversals.

of length $N = 7$, will lower the ACF recurrent lobes to $\frac{1}{7}$ or 0, as in Barker 7 (see Fig. 9.2).

However, the ambiguity function (Fig. 9.3) shows that the lowered or missing recurrent ACF lobes moved from the zero-Doppler axis to the higher Doppler values. Note that both Figs. 9.2 and 9.3 show the aperiodic ACF and aperiodic AF. In most cases it will be more relevant to plot the periodic ACF and periodic AF since the pulse train is transmitted periodically. The mainlobe area of the AF (and ACF) and the periodic AF (and periodic ACF) are identical. The recurrent lobes are, however, different. Although the Barker code seems a better choice in the aperiodic case, the perfect codes (heaving an ideal PACF) usually is a much better choice in the periodic case.

The same polarity reversal approach will also work with modulated pulses (e.g., LFM pulses). However, modulated pulses provide more possibilities for pulse-to-pulse diversity. Vannicola et al. (2000) described LFM pulse trains in which

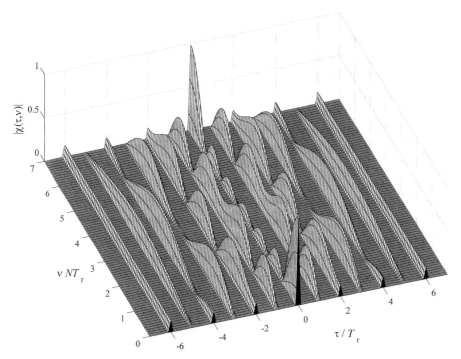

FIGURE 9.3 Aperiodic AF of the pulse train with polarity reversals ($N = 7$).

diversity was obtained by different frequency slopes in the different pulses of the train.

9.2 DIVERSITY FOR BANDWIDTH INCREASE: STEPPED FREQUENCY

An effective way to increase the bandwidth of a coherent pulse train is to add a frequency step Δf between consecutive pulses. A large Δf implies a large total bandwidth, hence improved range resolution. A stepped-frequency pulse train is an efficient method to achieve large overall bandwidth while maintaining narrow instantaneous bandwidth. One advantage of such an approach is that it allows using the interval between pulses to adjust the center frequency of other narrowband components of the radar system (e.g., a phased array).

Frequency steps can be added to a train of unmodulated pulses as well as to a train of modulated (e.g., LFM) pulses. An analytic expression of the ambiguity function is developed in Section 9.2.1. It turns out to be fairly simple to calculate the AF for the more general case, which is a stepped-frequency train of LFM pulses (see Fig. 9.4). The special case of a stepped-frequency train of unmodulated pulses will be derived from the general expression by setting the LFM slope to zero.

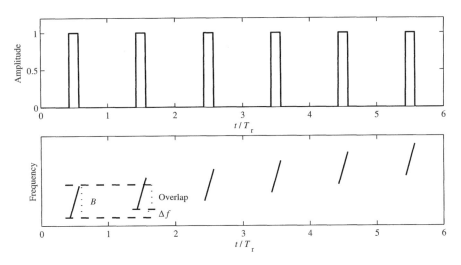

FIGURE 9.4 Stepped-frequency train of LFM pulses.

9.2.1 Ambiguity Function of a Stepped-Frequency Train of LFM Pulses

We begin with the ambiguity function of a single LFM pulse of duration T whose complex envelope $u_1(t)$ has unit energy. The complex envelope of such a pulse can be written as

$$u_1(t) = \frac{1}{\sqrt{T}} \text{rect}\left(\frac{t}{T}\right) \exp(j\pi k t^2) \tag{9.1}$$

where k is the frequency slope, related to the bandwidth B_1 (> 0) of the single pulse according to

$$k = \pm \frac{B_1}{T} \tag{9.2}$$

where a plus sign stands for a positive frequency slope and a minus sign stands for a negative frequency slope. Henceforth we will assume a positive frequency slope ($k > 0$), but the results apply to a negative slope as well. Note that k and B_1 are not the ultimate values of the single-pulse slope and bandwidth. For the sake of brevity we will also use the notation sinc(x), which stands for $(\sin \pi x)/\pi x$.

The ambiguity function of such a single LFM pulse was developed in Section 4.2 and is given by

$$|\chi_1(\tau, \nu)| = \left|\left(1 - \frac{|\tau|}{T}\right) \text{sinc}\left[T(\nu + k\tau)\left(1 - \frac{|\tau|}{T}\right)\right]\right|,$$

$$|\tau| \leq T, \quad \text{zero elsewhere} \tag{9.3}$$

Next we create a uniform pulse train of N such LFM pulses separated by $T_r > 2T$. The complex envelope of the uniform pulse train, designated $u_N(t)$, is given by

$$u_N(t) = \frac{1}{\sqrt{N}} \sum_{n=0}^{N-1} u_1(t - nT_r) \tag{9.4}$$

Unit energy is maintained due to the division by $\sqrt{N}$.

For delay τ shorter than the pulse duration T, the AF of the coherent pulse train is related to the AF of a single pulse according to

$$|\chi_N(\tau, \nu)| = |\chi_1(\tau, \nu)| \left| \frac{\sin N\pi\nu T_r}{N \sin \pi\nu T_r} \right|, \qquad |\tau| \leq T \tag{9.5}$$

We will now add linear FM to the entire train of pulses, using an additional (but different) slope k_s. We designate the complex envelope of the LFM pulse train by $u_s(t)$ given by

$$u_s(t) = u_N(t) \exp(j\pi k_s t^2) = \frac{1}{\sqrt{N}} \exp(j\pi k_s t^2) \sum_{n=0}^{N-1} u_1(t - nT_r) \tag{9.6}$$

where

$$k_s = \pm \frac{\Delta f}{T_r}, \qquad \Delta f > 0 \tag{9.7}$$

where a plus sign stands for a positive frequency step and a minus sign stands for a negative frequency step. Henceforth we will assume a positive frequency step ($k_s > 0$), but the results apply to a negative step as well. Adding LFM modifies the signal's AF according to property 4 (see Section 3.1):

$$|\chi_s(\tau, \nu)| = |\chi_N(\tau, \nu + k_s\tau)| \tag{9.8}$$

Applying (9.8) to (9.5), we get

$$|\chi_s(\tau, \nu)| = |\chi_1(\tau, \nu + k_s\tau)| \left| \frac{\sin[N\pi(\nu + k_s\tau)T_r]}{N \sin[\pi(\nu + k_s\tau)T_r]} \right|, \qquad |\tau| \leq T \tag{9.9}$$

Using (9.3) in (9.9), we get the AF of the combined signal:

$$|\chi(\tau, \nu)| = \left| \left(1 - \frac{|\tau|}{T}\right) \mathrm{sinc}\left[T[\nu + (k + k_s)\tau]\left(1 - \frac{|\tau|}{T}\right) \right] \right|$$
$$\times \left| \frac{\sin[N\pi(\nu + k_s\tau)T_r]}{N\sin[\pi(\nu + k_s\tau)T_r]} \right|, \qquad |\tau| \leq T \tag{9.10}$$

Note that the LFM slope k_s, introduced to create the frequency step, adds to the original slope k of the single LFM pulse. Hence the ultimate bandwidth of each pulse in the train is

$$B = |k + k_s|T \qquad (9.11)$$

Using (9.7) and (9.11) in (9.10), the AF expression is simplified to

$$|\chi(\tau, \nu)| = \left|\left(1 - \frac{|\tau|}{T}\right) \text{sinc}\left[T\left(\nu + B\frac{\tau}{T}\right)\left(1 - \frac{|\tau|}{T}\right)\right]\right|$$

$$\times \left|\frac{\sin\{N\pi[\nu + \Delta f(\tau/T_r)]T_r\}}{N \sin\{\pi[\nu + \Delta f(\tau/T_r)]T_r\}}\right|, \quad |\tau| \leq T \qquad (9.12)$$

The expression in (9.12) is the AF for a coherent stepped-frequency train of LFM pulses, where Δf is the frequency step between pulses and B is the bandwidth of each individual pulse. For $|\tau| > T$ we have to use numerical methods. Appendix 9A creates a numerical stepped-frequency train of LFM pulses to be used with the AF plotting MATLAB code described in Appendix 3A.

9.2.2 Stepped-Frequency Train of Unmodulated Pulses

In a stepped-frequency train of unmodulated pulses, a large bandwidth is obtained from narrow bandwidth pulses by stepping the carrier frequency from pulse to pulse. Range resolution is determined by the overall bandwidth, while the extended duration of the coherent signal produces good Doppler resolution. An expected question is: What is the difference between this signal and a train of identical LFM pulses where the total bandwidth is found in each pulse?

One difference (in favor of stepped-frequency) is that in some radar systems, some components cannot operate properly over the entire bandwidth. An example is a phased-array antenna, where beam steering remains uniform only over a limited bandwidth. In a stepped-frequency pulse train the array can be adjusted to a new center frequency during the interval between pulses.

Another (unfavorable) difference is that if $T\Delta f > 1$, the autocorrelation function over the delay range $|\tau| < T$ exhibits additional lobes (one or more), referred to as *grating lobes*. The basis of grating lobes will become clear from the ambiguity function of this signal, obtained by setting $B = 0$ in (9.12):

$$|\chi(\tau, \nu)| = \left|\left(1 - \frac{|\tau|}{T}\right) \text{sinc}\left[T\nu\left(1 - \frac{|\tau|}{T}\right)\right]\right|$$

$$\times \left|\frac{\sin\{N\pi[\nu + \Delta f(\tau/T_r)]T_r\}}{N \sin\{\pi[\nu + \Delta f(\tau/T_r)]T_r\}}\right|, \quad |\tau| \leq T \qquad (9.13)$$

or from the magnitude of the autocorrelation function (ACF) obtained by setting $\nu = 0$ in (9.13):

$$|R(\tau)| = |\chi(\tau, 0)| = \left|\left(1 - \frac{|\tau|}{T}\right)\right|\left|\frac{\sin(N\pi\tau\Delta f)}{N\sin(\pi\tau\Delta f)}\right|, \quad |\tau| \leq T \qquad (9.14)$$

The first null of the ACF happens when the argument of the sine in (9.14) equals π:

$$\tau_{1'\text{st null}} = \frac{1}{N\Delta f} \approx \frac{1}{\text{bandwidth}} \quad (9.15)$$

Since N is the number of pulses and Δf is the frequency step between pulses, their product approximately equals the total bandwidth of the signal. Beyond the first null the ACF exhibits typical sidelobes, the first at -13 dB. In addition to the mainlobe (at $\tau = 0$) and the sidelobes, if $T\Delta f > 1$, the ACF will exhibit additional peaks (grating lobes) at

$$|\tau| = \frac{n}{\Delta f}, \quad n = 1, 2, \ldots, n_{\max} \quad \text{where} \quad n_{\max} = \lfloor T\Delta f \rfloor \quad (9.16)$$

This behavior of the ambiguity function will be demonstrated using two stepped-frequency signals, one with $T\Delta f = 0.8$ (no grating lobes) and one with $T\Delta f = 2.5$ (two grating lobes on each side). The other parameters are $N = 8$ pulses and $T_r/T = 10$. The amplitude, phase, and frequency characteristics of the first signal are shown in Fig. 9.5. The AF of that signal (limited in delay to

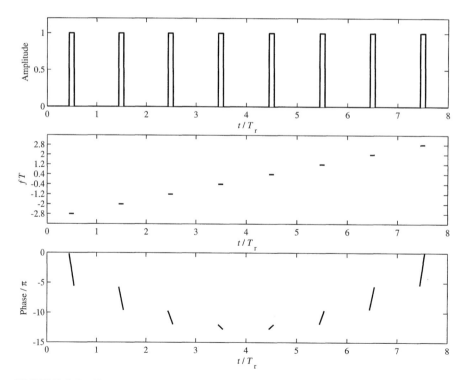

FIGURE 9.5 Characteristics of stepped-frequency pulse train with $T\Delta f = 0.8$, $N = 8$, $T_r/T = 10$.

DIVERSITY FOR BANDWIDTH INCREASE: STEPPED FREQUENCY

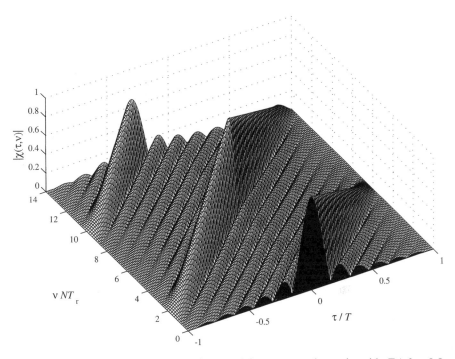

FIGURE 9.6 Ambiguity function of stepped-frequency pulse train with $T\Delta f = 0.8$, $N = 8$, $T_r/T = 10$.

$|\tau| \leq T$) is shown in Fig. 9.6. The AF of the second signal is given in Fig. 9.7, showing clearly the two grating lobes on each side of the mainlobe.

Clearly evident in the middle subplot in Fig. 9.5 are the normalized frequency steps between pulses ($T\Delta f = 0.8$). In the lower subplot, the phase of the signal is shown. Here the phase remained constant between pulses. Any other choice would not alter the AF over $|\tau| \leq T$ but will change the AF around the recurrent lobes: namely, at $|\tau - nT_r| \leq T$, $n = \pm 1, \pm 2, \ldots$.

An extension of the aperiodic AF plot, as far as the first recurrent lobe, is given in Fig. 9.8. The relatively low recurrent lobe, observed in Fig. 9.8, stem from the large diversity between pulses, caused by the frequency steps. The unique uniform pattern of the first recurrent lobe in the zero-Doppler cut of Fig. 9.8 is found whenever $T\Delta f$ is an integer multiple of 0.5. In such cases the number of peaks in the ACF's first recurrent lobe is $2(N-1)T\Delta f$, and their peak level is $1/(\pi NT\Delta f)$.

So far we learned how the various parameters of the stepped-frequency signal affect the AF and ACF. We noted that the first null of the ACF (i.e., the delay resolution) equals the inverse of the number of pulses N times the frequency step Δf. We also learned that the first grating lobe appears at a delay of $1/\Delta f$ (as long as it is shorter than the pulse duration T). The influence of the pulse interval T_r will be seen on the Doppler axis. The first null in Doppler will occur

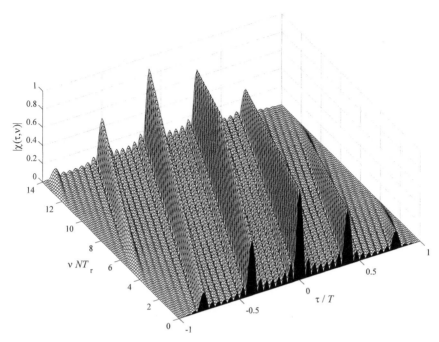

FIGURE 9.7 Ambiguity function of stepped-frequency pulse train with $T\Delta f = 2.5$, $N = 8$, $T_r/T = 10$.

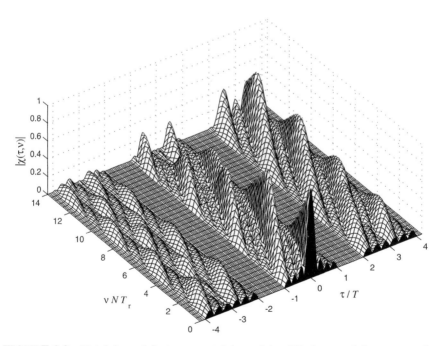

FIGURE 9.8 Mainlobe and first recurrent lobes of the AF of stepped-frequency pulse train. $T\Delta f = 1$, $N = 4$, $T_r/T = 3$.

DIVERSITY FOR BANDWIDTH INCREASE: STEPPED FREQUENCY

at $v_{\text{1'st null}} = 1/NT_r$ and the first recurrent Doppler ridge will cross the Doppler axis at $v_{\text{1'st peak}} = 1/T_r$. An overview of the AF ridges is summarized in Fig. 9.9.

One conclusion from Fig. 9.9 says that to avoid grating lobes on the zero-Doppler axis of the AF (i.e., on the ACF), it is necessary to keep $T\Delta f < 1$. However, because the grating lobe has some width, if $T\Delta f$ is close to 1, the tail of the grating lobe will still appear on the ACF. To prevent this it is necessary to keep $T\Delta f \leq 1 - 1/N$ when no weighting is used, and $T\Delta f \leq 1 - 2/N$ when weighting is used (because the grating lobe widens). Figure 9.9 shows that even if $T\Delta f = 1 - 1/N$, an off zero-Doppler cut of the AF will still exhibit two peaks (mainlobe and grating lobe). To avoid two peaks in delay, for any Doppler value the upper limit on $T\Delta f$ must be halved: namely, $T\Delta f < 0.5 - 1/N$. As shown later in this chapter, the problem of grating lobes can be mitigated by replacing the unmodulated pulses with linear FM pulses. It turns out that specific relationships between the interpulse frequency step Δf and the LFM pulse bandwidth B can nullify the grating lobes.

Figure 9.6 displays prominent sidelobes in both delay and Doppler. Adding interpulse weighting will affect both delay and Doppler sidelobes, because each pulse is at a different frequency. Figure 9.10 displays the amplitude of eight pulses using Chebyshev weighting (−50 dB). The frequency steps remain as in Fig. 9.5. The resulted ambiguity function is shown in Fig. 9.11. Comparing

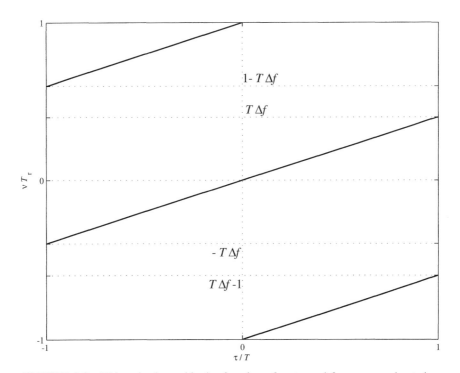

FIGURE 9.9 Ridges in the ambiguity function of a stepped-frequency pulse train.

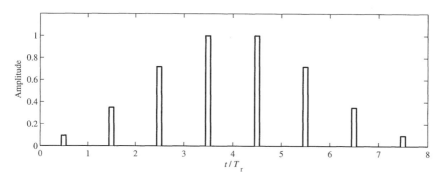

FIGURE 9.10 Amplitude of eight pulses with Chebyshev (−50 dB) weight window.

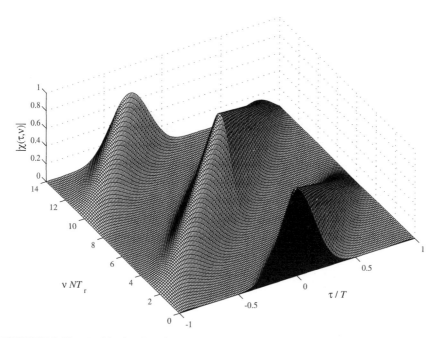

FIGURE 9.11 Ambiguity function of stepped-frequency pulse train with $T\Delta f = 0.8$, $N = 8$, $T_r/T = 10$. Interpulse Chebyshev weighting.

Fig. 9.11 with Fig. 9.6 shows significant reduction in sidelobes, in both Doppler and delay, but also significant widening (in both dimensions) of the mainlobe and recurrent Doppler lobes.

9.2.3 Stretch-Processing a Stepped-Frequency Train of Unmodulated Pulses

The stepped-frequency pulse-train signal is popular in radar systems that seek the narrow range resolution obtained with wide bandwidth signals, yet wish to avoid

DIVERSITY FOR BANDWIDTH INCREASE: STEPPED FREQUENCY

large instantaneous bandwidth. Reduced instantaneous bandwidth in the receiver processor is also important in such systems. Stretch processing offers such narrowband processing (Wehner, 1987). An important application where stretched processing is found is in automotive radars (Kajiwara, 1999; Rohling and Lissel, 1995; Rohling and Mende, 2000). The combination of stepped-frequency pulses and stretch processing is especially attractive for automotive radar applications because of the short range to the targets. The return delay is usually shorter than the pulse interval and may be even shorter than the pulse duration. Stretch processing can accommodate both situations. A simple design of automotive radar (Kajiwara, 1999; Levanon, 2002) will be used to demonstrate stretch processing of stepped-frequency pulse train.

A conceptual stepped-frequency radar (Kajiwara, 1999) is described in Fig. 9.12. The voltage staircase feeding the VCO (voltage-controlled oscillator) will generate the frequency coding. The stair period is T_r and the stair height will cause a frequency step of Δf. The switch on the left generates the two signals $x(t)$ and $s(t)$. Synchronously, the switch on the right connects the antenna to the VCO or to the mixer.

Figure 9.13 presents the transmitted $x(t)$ and reference $s(t)$ signals. The transmitted signal pulses (top) exhibit fixed amplitude A. The duty cycle T/T_r may be as high as 0.5. The frequency coding $f_n, n = 1, 2, \ldots, N$ follows a stepped-frequency law (linear or other). The reference signal (bottom) time-complements the transmitted one, indicating that the receiver is turned off during transmission. The variable pulse amplitudes represent the option to add a weight window at the receiver (the weights are inserted prior to the IFFT).

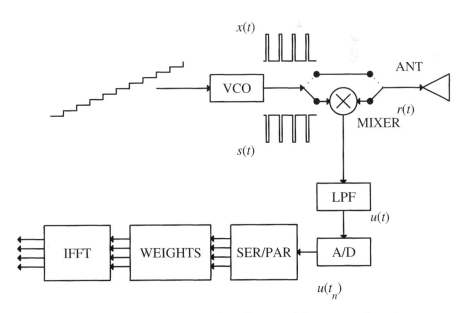

FIGURE 9.12 Stretch processing of a stepped-frequency pulse train.

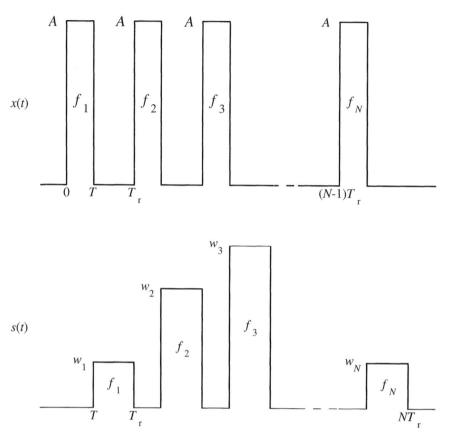

FIGURE 9.13 Stepped-frequency pulse-train: transmitted (top) and reference (bottom) signals.

Ignoring attenuation, the signal received from a stationary point target is a delayed version of the signal transmitted:

$$r(t) = x(t - \tau) \qquad (9.17)$$

The mixer, low-pass filter, and analog-to-digital converter create one complex sample of the low-pass filtered mixer output for each pulse. Those samples could be approximated by the expression

$$u(nT_r) = \int_{(n-1)T_r}^{nT_r} x(t - \tau) s^*(t)\, dt \qquad (9.18)$$

The N complex samples corresponding to the outcome of the N pulses in a batch are multiplied by the N weights representing the different amplitudes w_n

DIVERSITY FOR BANDWIDTH INCREASE: STEPPED FREQUENCY 239

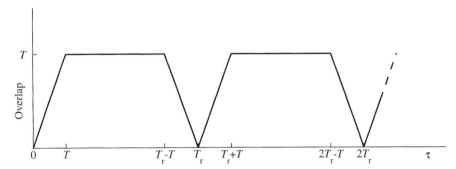

FIGURE 9.14 Overlap between a received signal delayed by τ and a reference signal.

of the reference signal. An inverse FFT yields the intensities of the returns at the effective range bins.

Some results from a simulated stretch processor are given in Figs. 9.14 to 9.19. To be able to see waveform details, the batch was constructed from only 16 pulses. A practical automotive radar may utilizes a 1000-pulse batch. The transmitter duty cycle is $\frac{1}{3}$: namely, $T_r = 3T$. Note that even if the delay τ is less than the pulse duration T, there is already some overlap between the received signal and the reference signal. Assuming infinitesimal rise times, the overlap as a function of delay is given in Fig. 9.14.

The phase of the received signal reflected from a point target with a delay of $\tau = 0.6T$ is shown in Fig. 9.15 (top) and compared to the reference signal (center). The phase is measured with respect to the phase of a virtual sine wave at the center of the frequency band, at $(f_8 + f_9)/2$. This choice of reference frequency explains why the phase ramps of the outer pulses are higher than of the inner pulses, and why the phase slope reverses between pulses 8 and 9. The phase of the mixer output is given at the bottom of Fig. 9.15. Note the constant phase during the entire overlap of a given pulse. Hence, the integral in (9.18) adds phasors that are all in the same direction. This constant phase during each pulse at the mixer's output results only for a *stationary* point target.

Thus, the $N = 16$ samples of the LPF (or integrator) output, one for each pulse, will have phases identical to those plotted at the bottom section of Fig. 9.15. The magnitude will depend on the mixer's output pulse duration (the overlap) and the intensity of the reflected signal. The overlap chart (Fig. 9.14) indicates a useful inherent gain control, which linearly suppresses close targets, up to a delay equal to the transmitted pulse width T. The phase value at the nth pulse is a function of the frequency and delay. For a delay τ and frequency f_n the phase of the nth pulse at the A/D output will be $2\pi f_n(T + \tau)$. For linear frequency stepping, $f_n = (n-1)\Delta f + f_1$, the delay τ translates directly to a pulse-to-pulse phase shift of $2\pi \Delta f(\tau + T)$, implying unambiguous delay measurement of $1/\Delta f$. The IFFT zero range bin corresponds to the delay in which all pulses have the same phase [i.e., when $\Delta f(\tau + T) = 0 \pmod 1$]. The range measurement resolution is a function of the number of pulses in the train (i.e., the number of IFFT bins)

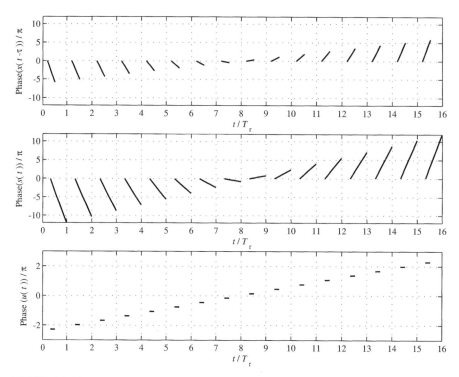

FIGURE 9.15 Phase of received (top), reference (middle), and mixer output (bottom) signals: $N = 16$, $T_r/T = 3$, $T\Delta f = 0.4$, $\tau/T = 0.6$.

and is given by $1/(N\Delta f)$, which is the inverse of the pulse train bandwidth. Note that we selected the initial phase of all the pulses to be zero. Selecting a different initial phase (e.g., a phase code similar to the ones suggested in Section 9.2.2) will have no effect on the receiver response to first-time-around echos (i.e., targets with delay lower than T_r) but will lower the receiver response to second-time-around echos (ambiguity function delay recurrent lobes).

To demonstrate the IFFT output for different scenes, the number of pulses (and frequencies) was raised to $M = 64$. Zero padding increased the length of the IFFT output vector by a factor of 4, to $M' = 256$. Figure 9.16 (top) displays the output assuming two *stationary* targets at normalized delays of $\tau/T = 0.5$ and 2 and $T\Delta f = 0.4$ ($\tau\Delta f = 0.2$ and 0.8). The relationship between the IFFT output bin number n_b and the normalized delay is given by

$$n_b = M'(\tau + T)\Delta f \quad \text{or} \quad \tau\Delta f = \frac{n_b}{M'} - T\Delta f \qquad (9.19)$$

where n_b is taken modulo M' and τ is unambiguous with period $1/\Delta f$.

The stepped-frequency (linear law) signal used to produce Fig. 9.16 had Chebyshev weighting (50-dB ripple) only on receive. The reflected signals exhibited

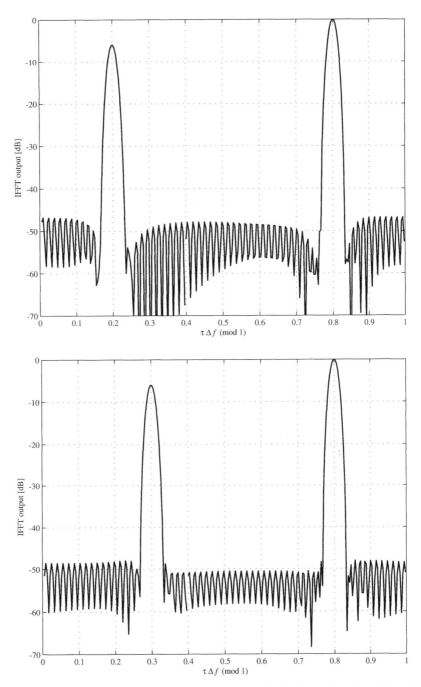

FIGURE 9.16 IFFT output with two targets at normalized delays of $\tau \Delta f = 0.2$ and 0.8. The targets exhibit identical received intensities. Top: both targets are stationary; bottom: the target at a normalized delay of $\tau \Delta f = 0.2$ is moving with a normalized Doppler of $\nu T_r = -0.1$.

equal intensities at the receiver input. The peak representing the closer target is lower because of the reduced overlap. The scene that produced Fig. 9.16 (bottom) is identical to the scene that produced Fig. 9.16 (top), except that the closer target is moving, causing a normalized Doppler shift of $\nu T_r = -0.1$ (i.e., receding). Note that the peak corresponding to this target has moved by $\Delta \tau = +0.1/\Delta f$ from its bin position without Doppler. This Doppler-shift-induced delay error follows the relationship

$$\Delta \tau \Delta f = -T_r \nu \qquad (9.20)$$

Doppler-induced delay errors pose a problem in radar scenarios where there are moving and stationary targets and both their distances and velocities have to be resolved. The simple stretch processor shown in Fig. 9.12 can be modified to produce cuts of the delay–Doppler response at nonzero Doppler.

9.2.3.1 Resolving Doppler in Stretch Processing A modified stretch processor that yields a two-dimensional delay–Doppler response is shown in Fig. 9.17. The vector U, at the output of the serial/parallel converter, contains M complex elements, one for each pulse, ordered chronologically. In the original processor

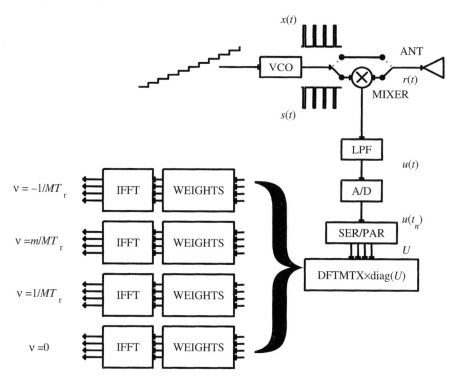

FIGURE 9.17 Stretch processing of a stepped-frequency pulse train, including a bank of Doppler filters.

DIVERSITY FOR BANDWIDTH INCREASE: STEPPED FREQUENCY

(Fig. 9.12) it was fed to a single IFFT block (after weighting) to yield a response matched to zero Doppler. To implement a bank of Doppler filters, we need to generate many versions of vector U, each one compensating a different Doppler shift. Such an operation is performed in the block marked DFTMTX*diag(U), which is exactly what this block performs in MATLAB terminology. DFTMTX(M) is the $M \times M$ complex matrix of values around the unit circle whose inner product with a column vector of length M yields the discrete Fourier transform of the vector. In our block we multiply this matrix by an $M \times M$ diagonal matrix, whose diagonal elements are the M elements of U. Each row in the resulted $M \times M$ complex array is a different (progressive) Doppler-compensated version of U. Each row is identically processed by adding weights and performing an IFFT.

The absolute value (linear scale) of the outputs of M such filters are presented in Fig. 9.18. The signal was a linear frequency-coded pulse train with $M = 144$ pulses, $T \Delta f = 0.4$ and $T_r = 3T$. Chebyshev weighting was used, with -50 dB ripple. There were two targets, with identical received intensities, and the following delays and Doppler shifts:

$$\text{Target 1}: \quad \tau \Delta f = 0.5, \nu T_r = -0.1$$

$$\text{Target 2}: \quad \tau \Delta f = 0.8, \nu T_r = 0$$

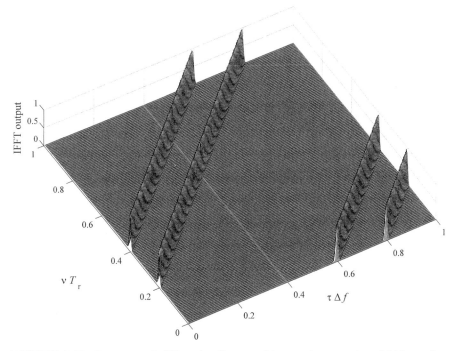

FIGURE 9.18 Response of all Doppler filters used in stretch processing of 144 step-frequency pulses, linearly coded with $T \Delta f = 0.4$, to two equal-intensity targets at $\tau \Delta f = 0.5$, $\nu T_r = -0.1$ (target 1) and $\tau \Delta f = 0.8$, $\nu T_r = 0$ (target 2).

The ridges typical of the ambiguity function of linear frequency stepped pulses are also seen in the delay–Doppler response of the stretch processor. From a delay–Doppler response as seen in Fig. 9.18, it is difficult to resolve delay from Doppler. Two more pulse trains are required, at opposite and at flat frequency slope, in order to separate delay from Doppler (Rohling and Mende, 2000).

Instead or in addition to using several batches of pulses at different frequency slopes, it is possible to use a randomlike coding law, such as that of Costas (1984). When using a Costas or other nonlinear law, the processor in Fig. 9.17 must be modified. It is necessary to reorder the inputs to the IFFT so that they will follow not the chronological order of the pulses, but will be sorted according to their carrier frequency. The response of this modified stretch processor to a Costas-coded train of 144 stepped-frequency pulses is shown in Fig. 9.19. The simulated targets are as in the previous (LFM) case. The peaks corresponding to the two targets are clearly visible at their expected locations, above a pedestal of approximately -20 dB. If the targets were not of equal intensity, the pedestal level would be approximately -20 dB below the peak of the stronger target. The relative height of the pedestal decreases as the number of pulses increases.

In Costas coding there is no good reason to add frequency weighting. Frequency weighting reduces sidelobes from interfering targets that happen to have

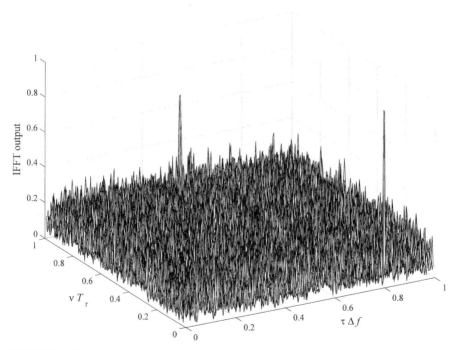

FIGURE 9.19 Response of all Doppler filters used in stretch processing of 144 step-frequency pulses, Costas coded with $T \Delta f = 0.4$, to two equal-intensity targets at $\tau \Delta f = 0.5$, $\nu T_r = -0.1$ (target 1) and $\tau \Delta f = 0.8$, $\nu T_r = 0$ (target 2).

the same Doppler shift as the processed target. Sidelobes from targets at different Doppler shifts remain high despite weighting. Hence, Fig. 9.19 was obtained without frequency weighting.

In all the examples used above, the product $T \Delta f$ was smaller than 1 in order to avoid grating lobes (and avoid range ambiguities smaller than the pulse width when using stretch processing). In the next section we show how, by adding LFM to the individual stepped-frequency pulses, we can reduce or even eliminate grating lobes, despite choosing $T \Delta f > 1$.

9.2.4 Stepped-Frequency Train of LFM Pulses

An effective way to increase the bandwidth of a coherent pulse train is to add a frequency step Δf between consecutive pulses. A large Δf value implies a large total bandwidth, hence improved range resolution. However, in Section 9.2.2 we showed that in the case of unmodulated pulses, when the product of the frequency step times the pulse duration T is larger than 1 ($T \Delta f > 1$), the autocorrelation function (ACF) of the stepped-frequency pulse train suffers from ambiguous peaks known as *grating lobes* (see Fig. 9.7). In Section 9.2.1 we derived the ambiguity function (AF) of a stepped-frequency train of LFM pulses, where in addition to the frequency step between pulses, each pulse is linearly frequency modulated with a frequency deviation B. Mitigating those grating lobes by combining stepped frequency and LFM was discussed by Rabideau (2002). In this section we follow Levanon and Mozeson (2003) and derive very simple relationships between $\Delta f, B$, and T that will place nulls exactly where the grating lobes are located and thus remove them completely. This will allow using larger-frequency steps: namely, using $T \Delta f > 1$.

9.2.4.1 Nullifying the Grating Lobes Setting $v = 0$ in the AF expression (9.12) derived for a stepped-frequency train of LFM pulses yields the magnitude of the ACF of the signal

$$|R(\tau)| = \left|\left(1 - \frac{|\tau|}{T}\right) \mathrm{sinc}\left[B\tau\left(1 - \frac{|\tau|}{T}\right)\right]\right| \left|\frac{\sin(N\pi\tau\,\Delta f)}{N\sin(\pi\tau\,\Delta f)}\right|, \quad |\tau| \leq T \quad (9.21)$$

Equation (9.21) is a product of two terms. The first, denoted $R_1(\tau)$, is due to a single LFM pulse:

$$|R_1(\tau)| = \left|\left(1 - \frac{|\tau|}{T}\right) \mathrm{sinc}\left[B\tau\left(1 - \frac{|\tau|}{T}\right)\right]\right|, \quad |\tau| \leq T \quad (9.22)$$

The second term, denoted $R_2(\tau)$, describes the grating lobes:

$$|R_2(\tau)| = \left|\frac{\sin(N\pi\tau\,\Delta f)}{N\sin(\pi\tau\,\Delta f)}\right|, \quad |\tau| \leq T \quad (9.23)$$

Clearly, $|R_2(\tau)|$ exhibits peaks (mainlobe and grating lobes) at

$$\tau_{\text{lobes}} = \frac{g}{\Delta f}, \qquad g = 0, \pm 1, \pm 2, \ldots, \lfloor T\Delta f \rfloor, \qquad |\tau| < T \qquad (9.24)$$

where $\lfloor x \rfloor$ implies the largest integer not exceeding x.

Nullifying the grating lobes of $|R_2(\tau)|$ is based on requiring that the nulls of $|R_1(\tau)|$ coincide with the grating lobes of $|R_2(\tau)|$. Finding such cases is based on requiring coincidence in two of the grating lobes. In some cases this requirement nullifies all the grating lobes. The general procedure is discussed in Section 9.2.4.3, but here we present a simple example. Requiring that $|R_1(\tau)|$ will exhibit its second and third nulls exactly at the first two gratings lobes, that is, at $\tau = 1/\Delta f$ and $\tau = 2/\Delta f$, respectively, yields the following two relationships:

$$T\Delta f = 5 \qquad (9.25)$$

and

$$TB = 12.5 \qquad (9.26)$$

The relationship in (9.25) implies that there must be exactly five grating lobes within the pulse duration. It turns out, however, that in this case, as in many others, not only the first two grating lobes are nullified when conditions (9.25) and (9.26) are met, but all five grating lobes are nullified. This property is demonstrated in Fig. 9.20 (top), where $|R_1(\tau)|$ (solid) and $|R_2(\tau)|$ (dotted) are plotted on top of each other (for $0 \leq \tau \leq T$). The resulting magnitude of the ACF on a logarithmic scale is plotted in Fig. 9.20 (bottom). No grating lobes can be seen. For comparison the ACF obtained with fixed-frequency pulses is presented in Fig. 9.21, where the grating lobes are prominent. Note that Figs. 9.20 and 9.21 were obtained using $N = 8$ pulses, which explains the six smaller lobes easily observed between every two grating lobes in Figs. 9.20 and 9.21. Other than that dependence on N, the figures are universal in the sense that they are functions only of TB and $T\Delta f$.

The ambiguity functions of a stepped-frequency train of LFM pulses [whose parameters are set according to (9.25) and (9.26): namely, $T\Delta f = 5$ and $TB = 12.5$] is presented in Fig. 9.22. The AF was plotted for a signal with $N = 8$ pulses. However, here we need to specify the duty cycle, which was $T/T_r = \frac{1}{10}$. Note that NT_r, which normalizes the Doppler axis, is the entire duration of the signal (all N pulses). Comparing Fig. 9.22 with Fig. 9.7 shows that nullifying the grating lobes is effective for extended Doppler. The grating lobes buildup with Doppler is relatively slow. However, it also shows that the height of the main diagonal ridge of the AF in Fig. 9.7 (around zero delay) decays very slowly in Doppler; while in Fig. 9.22 the ridge height decays faster.

When comparing Figs. 9.7 and 9.22, note the volume underneath the AF that was removed from the strip $|\tau| \leq T$. This volume must show up elsewhere. It will be found in the recurrent lobes (around multiples of T_r). As Fig. 9.4 shows, having $B > \Delta f$ creates some frequency overlap between pulses despite

DIVERSITY FOR BANDWIDTH INCREASE: STEPPED FREQUENCY

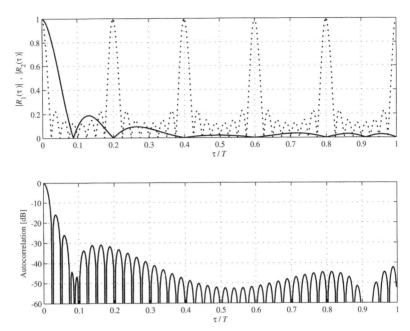

FIGURE 9.20 Stepped-frequency train of LFM pulses with grating lobe cancellation. Top: $|R_1(\tau)|$ (solid) and $|R_2(\tau)|$ (dotted); bottom: partial ACF (in dB). $N = 8$, $T\Delta f = 5$, $TB = 12.5$.

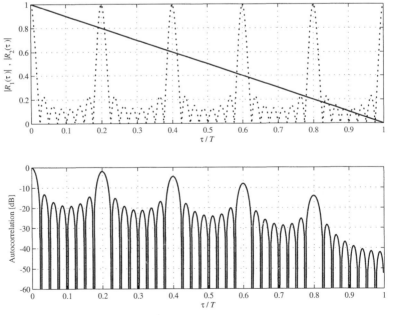

FIGURE 9.21 Stepped-frequency train of fixed-frequency pulses. Top: $|R_1(\tau)|$ (solid) and $|R_2(\tau)|$ (dotted); bottom: partial ACF (in dB). $N = 8$, $T\Delta f = 5$, $TB = 0$.

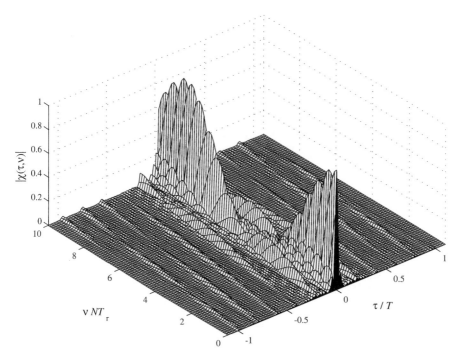

FIGURE 9.22 Ambiguity function of stepped-frequency train of LFM pulses with grating lobe cancellation. Delay axis limited to $|\tau| \leq T$. $N = 8$, $T\Delta f = 5$, $TB = 12.5$, $T/T_r = \frac{1}{10}$.

the frequency steps. Such an overlap does not exist with fixed-frequency pulses. The overlap is responsible for the relatively strong AF peaks at multiples of T_r. Figure 9.23 extends the delay axis of Fig. 9.22, to include the first recurrent lobes, and confirms the conclusion stated here. The AF volume distribution in the recurrent lobes depends on the order of pulses. If the pulse order is permuted, the volume in the recurrent lobes will be redistributed. Note that rearranging the order of pulses does not alter the ACF for $|\tau| \leq T$, hence does not affect the nullifying process, which is confined to this delay span.

We should compare the AF of a stepped-frequency train of LFM pulses, as seen in Fig. 9.22, to the AF of a train of identical LFM pulses (no interpulse frequency steps) as seen in Fig. 7.3. By adding the frequency step the signal ceased to be a coherent train of *identical* pulses, and lost some of its properties, among them the improved Doppler resolution. (Although exactly on the Doppler axis the AF in the area of the mainlobe is not affected by any frequency or phase modulation of the pulses.)

9.2.4.2 Special Cases When $T\Delta f \leq 3$ Before presenting a general search method for parameters that results in nullifying the grating lobes, we discuss the special case in which $T\Delta f \leq 3$, namely, when there are no more than three

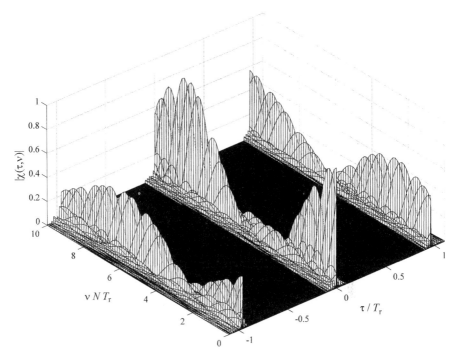

FIGURE 9.23 Ambiguity function of stepped-frequency train of LFM pulses with grating lobe cancellation. Delay axis extended to include the nearest recurrent lobes. $N = 8$, $T\Delta f = 5$, $TB = 12.5$, $T/T_r = \frac{1}{10}$.

grating lobes (positive delay). These cases are affected by the symmetry about $\tau = T/2$ of the sinc argument in (9.21), $B\tau(1 - |\tau|/T)$. Inserting the location of the first grating lobe $\tau_1 = 1/\Delta f$ in the expression of the sine argument and equating it to 1 will place the first null on the first grating lobe. The resulting relationship is

$$TB = \frac{(T\Delta f)^2}{T\Delta f - 1} \qquad (9.27)$$

When $1 < T\Delta f \le 2$, the first grating lobe is located in the second half of the pulse duration $T/2 < \tau_1 < T$. Placing a null on it will also place a symmetrical null in the first half, where there is no grating lobe. An example is shown in Fig. 9.24.

Clearly, in this case it is impossible to match the first grating lobe with the first null, but it is possible to match a null for any location of the single grating lobe; hence $T\Delta f$ can take any value within the limits $1 < T\Delta f \le 2$. For the specific value of $T\Delta f = 2$ and $TB = 4$, the first and second nulls coincide at $T/2$; hence the first grating lobe matches both the first and second nulls, and the second grating lobe, at $\tau_2 = T$, matches the "third" null. When $2 < T\Delta f \le 3$,

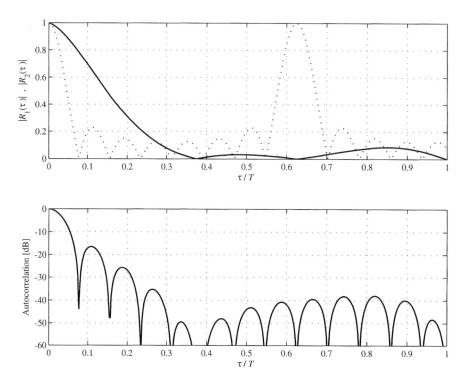

FIGURE 9.24 Stepped-frequency train of LFM pulses with grating lobe cancellation. Top: $|R_1(\tau)|$ (solid) and $|R_2(\tau)|$ (dotted); bottom: partial ACF (in dB). $N = 8$, $T\Delta f = 1.6$, $TB = 4.27$.

the first grating lobe is located in the first half of the pulse duration $0 < \tau_1 < T/2$; placing a null on it using (9.27) will also place a symmetrical null in the second half. For that second null to match the second grating lobe, located at $\tau_2 = 2/\Delta f$, it is necessary that (9.28) will also hold:

$$TB = \frac{(T\Delta f)^2}{2(T\Delta f - 2)} \qquad (9.28)$$

Both (9.27) and (9.28) will hold true only when $T\Delta f = 3$ and $TB = 4.5$. This case is shown in Fig. 9.25. It is the only case in which the first two grating lobes match the first two nulls. When $T\Delta f > 3$, there are three or more grating lobes, and matching nulls to the first two is more regular. A search procedure is described next.

9.2.4.3 General Search Procedure The search procedure tries to find coincidences between two grating lobes of $|R_2(\tau)|$ and two nulls of $|R_1(\tau)|$ when $T\Delta f > 3$. In some of the cases the coincidence will hold for all the grating

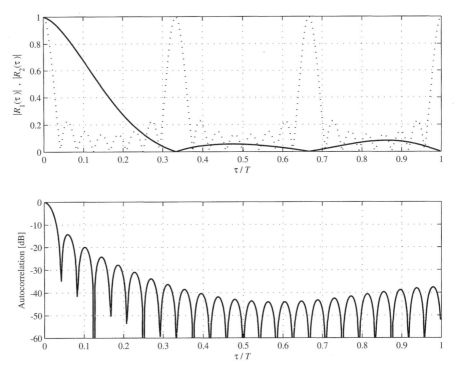

FIGURE 9.25 Stepped-frequency train of LFM pulses with grating lobe cancellation. Top: $|R_1(\tau)|$ (solid) and $|R_2(\tau)|$ (dotted); bottom: partial ACF (in dB). $N = 8$, $T\Delta f = 3$, $TB = 4.5$.

lobes of $|R_2(\tau)|$. A step in the search requires that $|R_1(\tau)|$ will exhibit its mth and nth $(n \geq m)$ nulls exactly at the qth and rth $(r > q)$ gratings lobes: namely, at $\tau = q/\Delta f$ and $\tau = r/\Delta f$, respectively. The entire search goes through all the possible choices of m, n, q, and r that meet the constraints above. Using a specific choice of m, n, q, and r in (9.22) and (9.23) yields the following two relationships:

$$\pi B \frac{q}{\Delta f} \left(1 - \frac{q}{T\Delta f}\right) = m\pi \qquad (9.29)$$

and

$$\pi B \frac{r}{\Delta f} \left(1 - \frac{r}{T\Delta f}\right) = n\pi \qquad (9.30)$$

Thus,

$$\frac{r}{n}\left(1 - \frac{r}{T\Delta f}\right) = \frac{q}{m}\left(1 - \frac{q}{T\Delta f}\right) \qquad (9.31)$$

Of special importance is the requirement to nullify the first two grating lobes, $q = 1$ and $r = 2$. Solving $T\Delta f$ yields

$$T\Delta f = \frac{mr^2 - nq^2}{mr - nq} \underset{q=1, r=2}{=} \frac{4m - n}{2m - n} \qquad (9.32)$$

Recall that n, m, q, and r should be all positive and that $m \leq n$ and $q < r$. Note that $\lfloor T\Delta f \rfloor$ is the number of grating lobes. To be a valid solution, it must be positive and larger than r. If a valid $T\Delta f$ was found for a specific selection of m, n, q, and r, using it in (9.30), we get that the pulse time–bandwidth product is

$$TB = \frac{n}{r(T\Delta f - r)}(T\Delta f)^2 = \frac{(mr^2 - nq^2)^2}{qr(r-q)(mr-nq)} \underset{q=1, r=2}{=} \frac{(4m-n)^2}{2(2m-n)} \qquad (9.33)$$

and the ratio $B/\Delta f$, which we will call *overlap ratio* (see Fig. 9.4), is

$$\frac{B}{\Delta f} = \frac{nT\Delta f}{r(T\Delta f - r)} = \frac{mr^2 - nq^2}{qr(r-q)} \underset{q=1, r=2}{=} \frac{4m - n}{2} \qquad (9.34)$$

To get a meaningful increase in bandwidth the number of pulses N should be much larger than the overlap ratio.

Once m, n, q, and r were selected (resulting in a given pulse time–bandwidth product and an overlap ratio) the autocorrelation function could be written as a function of TB and $T\Delta f$:

$$\left| R\left(\frac{\tau}{T}\right) \right| = \left| \left(1 - \left|\frac{\tau}{T}\right|\right) \operatorname{sinc}\left[TB \frac{\tau}{T} \left(1 - \left|\frac{\tau}{T}\right|\right) \right] \right|$$

$$\times \left| \frac{\sin[N\pi T\Delta f(\tau/T)]}{N \sin[\pi T\Delta f(\tau/T)]} \right|, \qquad \left|\frac{\tau}{T}\right| \leq 1 \qquad (9.35)$$

where the dimensionless parameters TB and $T\Delta f$ are functions of only m, n, q, and r. As pointed out before, the nulls of $|R_1(\tau)|$ are symmetric around $\tau = T/2$. Note also that the number of nulls is the minimal integer higher or equal to $TB/2$ (i.e., $\lceil TB/2 \rceil$). The number of peaks in $|R_2(\tau)|$ (grating lobes) is $\lfloor T\Delta f \rfloor$ (plus the mainlobe) and the peaks' positions are symmetric around $\tau = T/2$ when $T\Delta f$ is an integer.

Samples from a numerical search for possible candidates for m, n, q, and r produced the parameters listed in Table 9.1. (Only a few are listed; many more exist.) The table includes cases in which nullifying two grating lobes resulted in nullifying all the grating lobes and contains only a small fraction of the integer values of m and n over which the search was performed. Integer values of m and n were scanned such that TB is as high as 2000 (m and n are from 1 to 1500). It was found that for each value of TB, a solution is feasible only for relatively small values of $T\Delta f$ (e.g., for $TB = 1600$ it is impossible to find $T\Delta f > 60$ that still nullifies the first and second grating lobes). This result implies that when

DIVERSITY FOR BANDWIDTH INCREASE: STEPPED FREQUENCY   253

TABLE 9.1 Examples of Valid Cases

m	n	q	r	$T\Delta f$	TB	$B/\Delta f$
—	—	1	2	2	4	2
1	1	1	2	3	4.5	1.5
2	2	1	2	3	9	3
2	3	1	2	5	12.5	2.5
3	3	1	2	3	13.5	4.5
3	4	1	2	4	16	4
4	4	1	2	3	18	6
4	5	1	2	3.667	20.1667	5.5
5	6	1	2	3.5	24.5	7
4	7	1	2	9	40.5	4.5

TB is high, a valid frequency step Δf must be small and the number of pulses N should be very high in order to get a noticeable increase in bandwidth.

In addition to the fact that the grating lobes were nullified, there is interest in the mainlobe width and the height of the near sidelobes. Both are also affected by the number of pulses, N. Regarding the mainlobe width, the location of the first overall null is determined by which of the two expressions, $|R_1(\tau)|$ or $|R_2(\tau)|$, exhibits an earlier first null [see (9.22) and (9.23)]. When $TB \gg 1$, the location of the first null of $|R_1(\tau)|$ is approximately the inverse of TB. This allows a simple expression for the location of the first overall null:

$$\frac{\tau_{1'\text{st null}}}{T} = \min\left(\frac{1}{TB}, \frac{1}{NT\Delta f}\right) \qquad (9.36)$$

The height of the near sidelobes cannot be described by a simple expression. However, when $TB \approx 0.67 NT\Delta f$, the first null of $|R_1(\tau)|$ will coincide approximately with the peak of the first (and highest) sidelobe of $|R_2(\tau)|$, splitting it into two lower sidelobes. An example is given in Fig. 9.26. Thus, in addition to nullifying the first grating lobes, we have used the additional parameter, N, to place a null on the first sidelobe peak.

Figure 9.26 was obtained using the case $m = 4, n = 4, q = 1, r = 2, T\Delta f = 3, TB = 18$. This results in $B/\Delta f = 6 \gg 1$ and a relatively large m. As $B/\Delta f$ and hence m are increased, the mainlobe of the ACF will be determined more by $|R_1(\tau)|$ than by $|R_2(\tau)|$. In the limit, when $\Delta f = 0$, the signal reduces to a train of identical LFM pulses. Figure 9.27 presents a comparison of the partial $(0 \leq \tau \leq T)$ ACFs of the signal above, with the ACF of a signal in which $T\Delta f = 0$ and $TB = 25$, that exhibits the same ACF mainlobe width.

The different partial ACFs seen in Fig. 9.27 are related to differences in spectral shape. The reduced frequency overlap between pulses (see Fig. 9.28) in the first case $(T\Delta f = 3, TB = 18)$ shapes the spectrum in a manner that emphasizes the central frequencies, thus creating a similar effect to that of a spectral window, designed to reduce autocorrelation sidelobes. An intuitive feeling for the expected

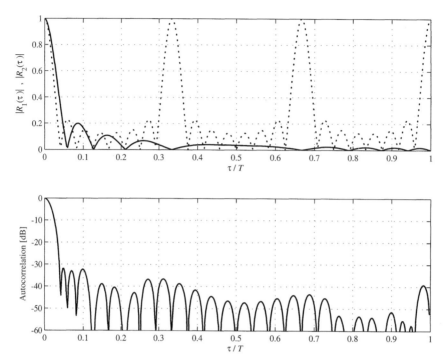

FIGURE 9.26 Stepped-frequency train of LFM pulses with grating lobe cancellation. Top: $|R_1(\tau)|$ (solid) and $|R_2(\tau)|$ (dotted); bottom: partial ACF (in dB). $N = 8$, $T\Delta f = 3$, $TB = 18$.

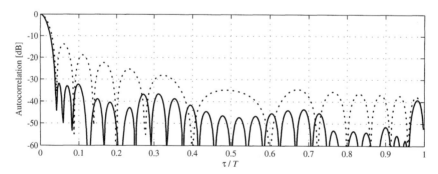

FIGURE 9.27 Comparing the ACF (in dB) of (1) $N = 8$, $T\Delta f = 3$, $TB = 18$ (solid); (2) $N = 8$, $T\Delta f = 0$, $TB = 25$ (dotted).

spectrum can be obtained from the histogram of the frequencies used during the eight pulses, as shown schematically in Fig. 9.29.

The exact power spectral density of the train of LFM pulses (with and without interpulse frequency steps) can be derived numerically by performing Fourier transform on the entire signal (and squaring the result) or by performing Fourier

DIVERSITY FOR BANDWIDTH INCREASE: STEPPED FREQUENCY 255

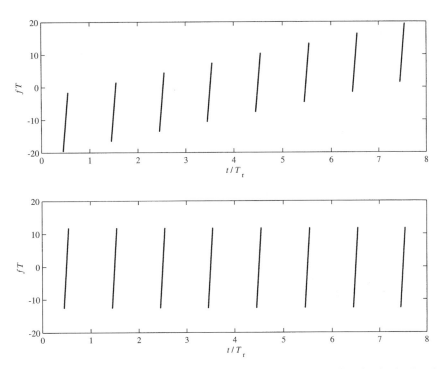

FIGURE 9.28 Comparing the frequency evolution of the two signals: (top) $N = 8$, $T\Delta f = 3$, $TB = 18$, $T_r/T = 10$; (bottom) $N = 8$, $T\Delta f = 0$, $TB = 25$, $T_r/T = 10$.

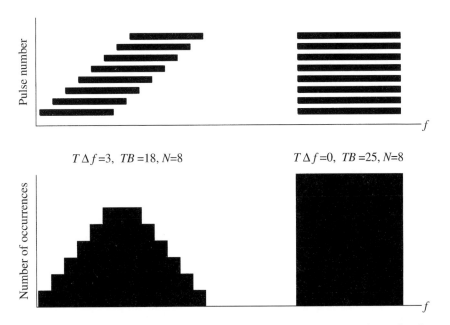

FIGURE 9.29 Qualitative construction of the frequency histogram of two signals.

transform on the entire autocorrelation function (ACF). Because of the periodicity, the detailed spectrum will not clearly reveal the difference between the two spectrums. On the other hand, a Fourier transform of the analytic expression of the partial ACF, $R(\tau)$, $|\tau| \leq T$, as given in (9.12), will display a smoothed and more comprehensible spectrum. The resulting smoothed spectrum, for the two signals whose ACFs are given in Fig. 9.27, is presented in Fig. 9.30. The dotted curve in Fig. 9.30 is of a train of identical LFM pulses (without interpulse steps), and this smoothed spectrum resembles the spectrum of a single LFM pulse. The solid curve (of the stepped LFM pulses) shows a typical windowed spectrum, in which the center frequencies are favorably weighted. This result explains the reduced ACF sidelobes of stepped-frequency LFM pulses compared to a train of identical LFM pulses.

The corresponding ambiguity functions, extended as far as the first recurrent lobe, are shown in Fig. 9.31. The top part is the AF of the stepped-frequency train of LFM pulses, and the bottom part is the AF of a train of identical LFM pulses. Recall (see Fig. 9.28) that the intrapulse frequency deviation of the latter signal was increased to achieve the same ACF mainlobe width as the first signal. At and near zero Doppler, the top AF shows reduced near-sidelobes (due to spectral shaping) and lower recurrent lobes (due to less overlap between pulses). However,

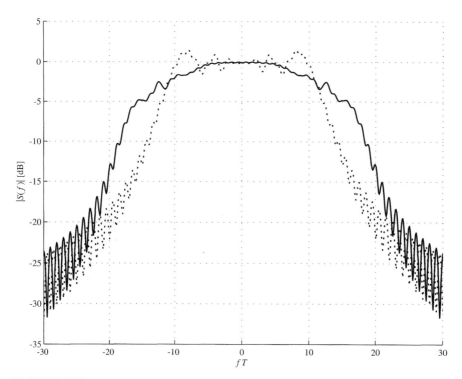

FIGURE 9.30 Comparing the power spectral density (in dB) of (1) $N = 8$, $T\Delta f = 3$, $TB = 18$ (solid); (2) $N = 8$, $T\Delta f = 0$, $TB = 25$ (dotted).

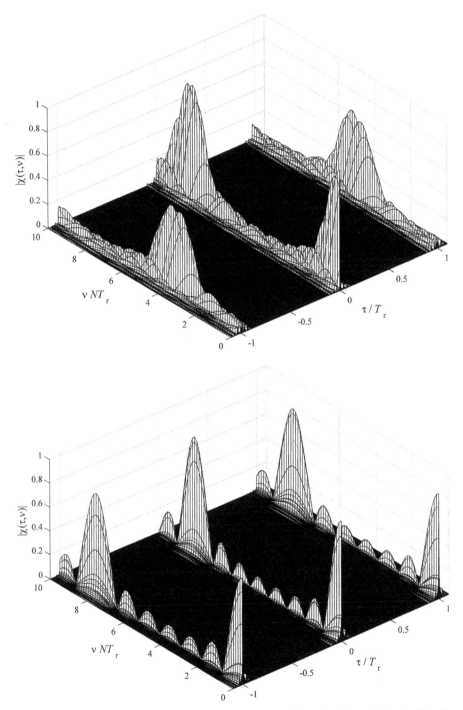

FIGURE 9.31 Ambiguity function of two signals for $|\tau| \leq T_r + T$: (top) $N = 8$, $T\Delta f = 3$, $TB = 18$, $T_r/T = 10$; (bottom) $N = 8$, $T\Delta f = 0$, $TB = 25$, $T_r/T = 10$.

it does not exhibit the ordered "bed of nails" mesh that the lower AF exhibits (due to the periodicity of identical pulses), hence loses some Doppler resolution.

We can conclude that combining interpulse frequency steps with intrapulse LFM allows increased bandwidth (not instantaneously) without causing ACF grating lobes, and can also reduce ACF near-sidelobes. What was demonstrated in Fig. 9.26 is a different approach to shaping the spectrum, not through a predetermined frequency weight window, but through manipulating the locations of nulls and peaks of the two components of the ACF. On the other hand, the signal is not periodic anymore, hence loses some Doppler resolution.

9.2.4.4 Costas-Ordered Stepped-Frequency Train of LFM Pulses The interpulse frequency steps can follow a different-than-linear order. In Section 9.2.3.1 we demonstrated Costas ordering of the frequency steps when the pulses were unmodulated. Costas ordering can also be applied when the pulses are LFM pulses. Figure 9.32 presents the frequency evolution of a train of 16 LFM pulses with Costas-ordered steps. The parameters ($T \Delta f = 5, TB = 12.5$) still result in nullifying of the ACF grating lobes. Recall that the ACF over $|\tau| \leq T$ is not affected by the order of the interpulse frequency steps. The order affects the AF at higher Doppler shifts and the recurrent lobes of both the ACF and the AF.

The AF of the Costas-ordered signal is presented in Fig. 9.33. The top part zooms on $|\tau| \leq T$, while the bottom part extends in delay as far as the first recurrent lobe. Comparing Fig. 9.33 (top) with Fig. 9.22, note the more random distribution of the AF volume at the vicinity of the Doppler axis. Comparing Fig. 9.33 (bottom) with Fig. 9.23, note a drastic reduction of the first recurrent lobe. Due to the Costas ordering, higher recurrent lobes will be similar to the first recurrent lobe. Figure 9.34 displays the ACF in decibels as far as the second recurrent lobe, demonstrating peak ACF recurrent lobes of -28 dB.

Further reduction and reshaping of the ACF recurrent lobes can be obtained by inverting the LFM slope of some of the pulses. Such a modification will not

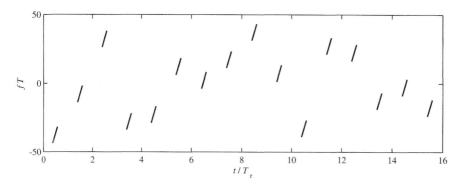

FIGURE 9.32 Frequency evolution of a Costas-ordered stepped-frequency train of LFM pulses: $N = 16, T\Delta f = 5, TB = 12.5, T_r/T = 5$. Frequency order is 0, 6, 14, 2, 3, 10, 8, 11, 15, 9, 1, 13, 12, 5, 7, 4.

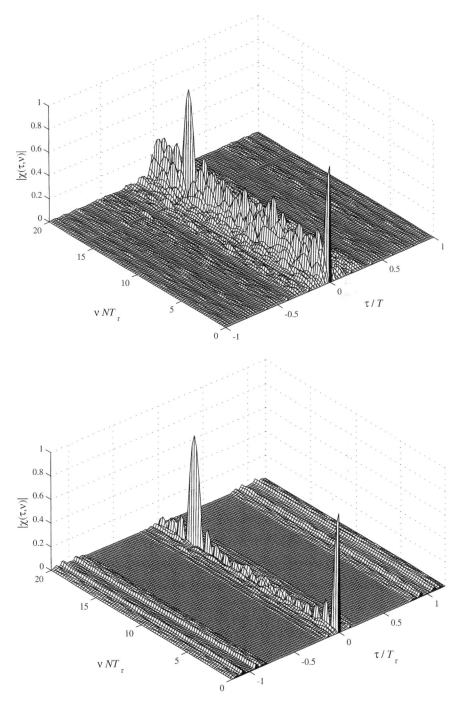

FIGURE 9.33 AF of the Costas-ordered signal, $N = 16$, $T\Delta f = 5$, $TB = 12.5$, $T_r/T = 5$: (top) $|\tau| \leq T$; (bottom) $|\tau| \leq T_r + T$.

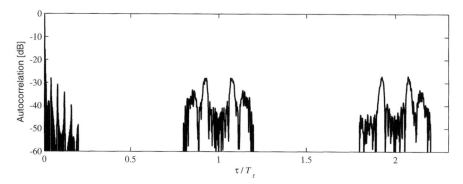

FIGURE 9.34 ACF of the Costas-ordered signal, $N = 16$, $T\Delta f = 5$, $TB = 12.5$, $T_r/T = 5$, $0 \leq \tau \leq 2T_r + T$.

alter the grating lobe nullifying. An example of the frequency evolution of such a signal is presented in Fig. 9.35. The inverting rule caused any two pulses, at neighboring frequency steps, to be of opposite frequency slopes (e.g., pulses 1 and 11). The resulting ACF is shown in Fig. 9.36, which should be compared to Fig. 9.34 (no slope inversion).

9.2.4.5 Modified Single Costas Pulse The signal described in Section 9.2.4.4, whose frequency evolution appears in Fig. 9.35, can be squeezed into a single pulse by removing the interval between pulses (i.e., setting $T_r = T$). This will result in a modified single Costas pulse, which differs from the original Costas pulse (see Section 5.1) in two aspects: (1) The interval between subcarriers is not equal but is larger than the inverse of the bit duration; and (2) each bit is LFM rather than fixed-frequency. If the relationships between $t_b \Delta f$ and $t_b B$ follow the rules described in Section 9.2.4.3, grating lobes will be avoided. The frequency evolution of such a modified Costas pulse is shown in

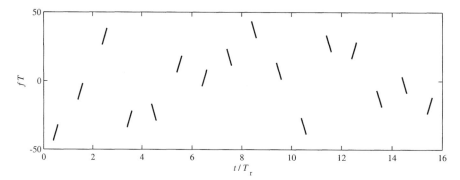

FIGURE 9.35 Frequency evolution of a Costas-ordered stepped-frequency train of LFM pulses (some inverted slopes), $N = 16$, $T\Delta f = 5$, $TB = 12.5$, $T_r/T = 5$.

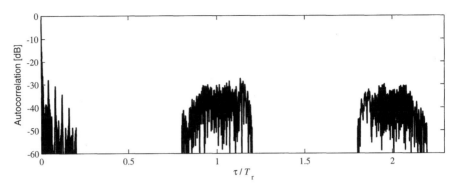

FIGURE 9.36 ACF of the Costas-ordered signal (some inverted slopes), $N = 16$, $T\Delta f = 5$, $TB = 12.5$, $T_r/T = 5$, $0 \leq \tau \leq 2T_r + T$.

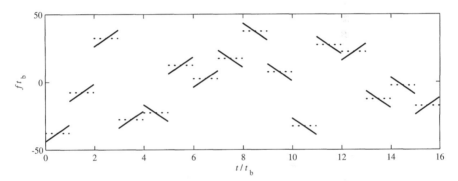

FIGURE 9.37 Frequency evolution of a modified Costas pulse, $N = 16$, $t_b \Delta f = 5$, $t_b B = 12.5$.

Fig. 9.37 (solid line). The dotted line is the frequency without the added LFM. The peak-to-peak frequency deviation, multiplied by the total pulse duration T, is given by

$$T(f_{\max} - f_{\min}) = \frac{T}{t_b}[(N-1)t_b\Delta f + t_b B] = N[(N-1)t_b\Delta f + t_b B] \quad (9.37)$$

This product is approximately the time–bandwidth product of the signal, and also its time compression ratio. For the example given in Fig. 9.37 (in which $N = 16$), equation (9.37) yields $T(f_{\max} - f_{\min}) = 1400$. A conventional Costas signal needs 37 or 38 ($\approx \sqrt{1400}$) elements to get a similar compression ratio. Partial plots of the ambiguity function and autocorrelation function of this signal are given in Figs. 9.38 and 9.39, respectively.

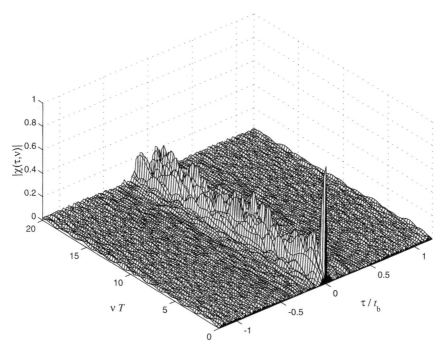

FIGURE 9.38 AF of a modified Costas pulse, $N = 16$, $t_b \Delta f = 5$, $t_b B = 12.5$, $|\tau| \leq 1.2 t_b = 1.2 T/16$.

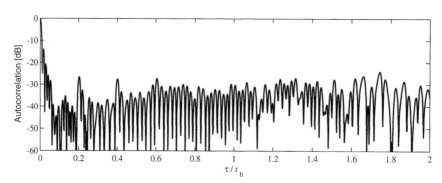

FIGURE 9.39 ACF of a modified Costas pulse, $N = 16$, $t_b \Delta f = 5$, $t_b B = 12.5$, $0 \leq \tau \leq 2 t_b = T/8$.

9.3 TRAIN OF COMPLEMENTARY PULSES

In previous sections we demonstrated how phase, frequency, and LFM slope diversity between pulses can reduce the recurrent lobe and sidelobe level of a coherent train of modulated pulses. We also showed that careful selection of the frequency stepping and LFM slopes yield nullifying the sidelobes and eliminating

TRAIN OF COMPLEMENTARY PULSES

grating lobes. In this section we present a method that by using specified phase coding for the different pulses in the train yields zero ACF sidelobes around the mainlobe area. Phase codes that have this property are termed *complementary*.

The origin of *complementary sets* traces back to the 1960s, when Golay (1961) first studied pairs of *binary complementary sequences* for radar astronomy. When the complementary sequences (or codes) are used to phase-modulate a radar pulse, the resulting are *complementary pulses*. A pair of finite sequences are said to be complementary if the sum of their autocorrelation functions is zero except for the zero shift, and each member of the pair is called a complementary sequence. For example, take the set of two sequences with four elements each, given by {1 −1 1 1} and {1 1 1 −1}. The autocorrelations of the sequences are given by {1 0 −1 4 −1 0 1} and {−1 0 1 4 1 0 −1}. It is easy to see that the two sequences form a complementary pair since the sum of their autocorrelations is {0 0 0 8 0 0 0}.

Using a complementary pair to phase-modulate a pair of phase-coded pulses with four chips in each pulse is straightforward. The resulting two diverse pulse trains and the autocorrelation function are shown in Fig. 9.40. In the area surrounding the zero delay response, note how the autocorrelation function is zero excluding $|\tau| \leq t_b$. The partial ambiguity function for $0 \leq \nu \leq 12/T_r$ of a pulse

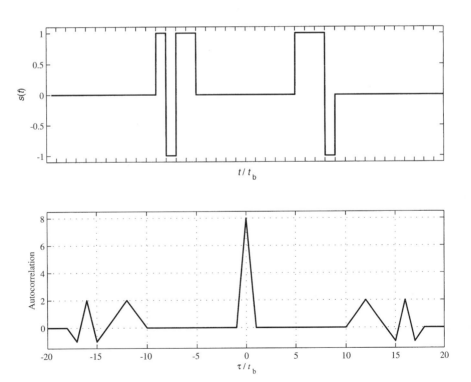

FIGURE 9.40 Complementary phase-coded pulse pair and the resulting autocorrelation.

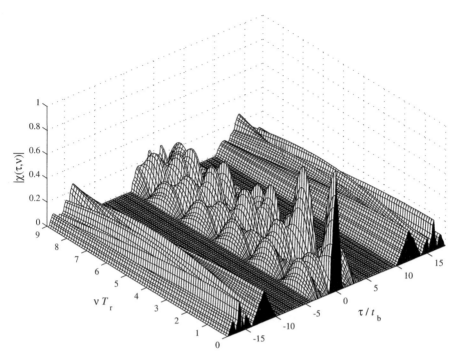

FIGURE 9.41 Ambiguity function of a pulse train based on the binary complementary pair {1 −1 1 1} and {1 1 1 −1}.

train phase-coded using the complementary pair used for Fig 9.40 is shown in Fig. 9.41 using a linear scale. Note that although the autocorrelation sidelobe level is zero, the ambiguity function exhibits relatively high sidelobes for nonzero Doppler.

Generalization from a complementary *pair* to a complementary *set* is straightforward. A set of finite sequences are said to be complementary if the sum of their autocorrelation functions is zero for all nonzero shifts. Each member of the set is called a complementary sequence. Complementary sets can be generated directly or by recursive steps based on complementary sets with fewer elements in each sequence or fewer members in the set. Complementary sets that are generated directly and not by recursion are called *kernels*. Golay (1961) gave kernels of binary complementary pairs of length 2, 10, and 26. Sivaswamy (1978a) extended the work to complementary pairs of polyphase sequences and gave kernels of a complementary set of three sequences with length 3 and a complementary triplet of sequences with length 2. Frank (1980) gives some additional kernels and lengths of polyphase complementary sets. Kretschmer and Gerlach (1991) give methods for constructing complementary sets of L sequences with length L by using orthogonal matrices. An important method first proposed by Kretschmer and Gerlach (1991) and independently by Popovic (1990) is to use all cyclic shifts of a perfect sequence (a sequence having an ideal periodic autocorrelation;

TABLE 9.2 Some Kernels of Known Polyphase Complementary Sets

S	L	Phase Sequence
2	2	[0 0], [0 π]
2	10	[0 0 π π π π π 0 π π], [0 0 π 0 π 0 π π 0 0]
2	26	[0 0 0 π π 0 0 0 π 0 π π 0 π 0 π π 0 0 π 0 0 0 0],
		[0 0 0 0 π 0 0 π π 0 π 0 0 0 0 0 π 0 π π π 0 0 π π π]
2	3	[0 0 π], [0 π/2 0]
2	4	[0 3π/2 0 π/2], [0 π/2 0 3π/2]
3	3	[0 π π], [0 2π/3 7π/3], [0 π/3 5π/3]
3	2	[0 0], [0 2π/3], [0 4π/3]
S	L < S	Rows of an S-by-L matrix with orthogonal columns
S	S	All cyclic shifts of any ideal sequence with length S
ab	a^2b	All a-position cyclically shifted ideal sequence with length a^2b

see Chapter 6 for some examples). Popovic (1992) also gives conditions where all cyclic shifts of an ideal chirplike sequence (also described in Chapter 6) form a *supercomplementary set* (formed by the union of some complementary sets). Note that the complementary pair used for Fig. 9.40 is such a set (formed by two cyclic shifts of an ideal sequence of length 4, the Barker sequence). Table 9.2 gives some kernels of known complementary sets of S polyphase sequences with length L.

Note that adding a constant phase to either of the sequences does not change the sequence autocorrelation and thus has no effect on the complementary property. Other, less trivial operations that preserve the complementary property are described in Box 9A. We will refer to all complementary sets connected by the operations in Box 9A as *equivalent*.

BOX 9A: Operations That Yield Equivalent Complementary Sets

Property 9A.1: *Reversal transformation.* The cross-correlation of reversed-order sequences is the reversed-order cross-correlation of the original sequences. Due to the symmetry property of the autocorrelation, the autocorrelation of the reversed-order sequence equals the conjugate of the original sequence autocorrelation. When the original sequence is binary, the autocorrelation is also binary and the conjugate operation has no effect. Thus, reversing *any* number of the sequences in a *binary* complementary set or reversing *all* sequences in a *polyphase* complementary set yields an equivalent complementary set.

Property 9A.2: *Conjugation transformation.* The autocorrelation of a conjugated sequence is the conjugated autocorrelation of the original sequences.

Thus, conjugating *all* sequences in a *polyphase* complementary set yields an equivalent complementary set.

Property 9A.3: *Constant multiplication transformation.* The autocorrelation of a sequence multiplied by a constant is the original autocorrelation multiplied by the absolute value (squared) of the constant used. Thus, multiplying all sequences in a complementary set by different constants having the same absolute value yields an equivalent complementary set.

Property 9A.4: *Progressive multiplication transformation.* The autocorrelation of progressively multiplied sequences is the progressively multiplied autocorrelation of the original sequences. Thus, progressively multiplying all sequences in a complementary set by the same constant yields an equivalent complementary set.

Property 9A.5: *Even–odd position transformation.* Tseng and Liu show that for a binary complementary set $\{A_i, 1 \leq i \leq S\}$ of S sequences, denote by A^* the sequence consisting of the elements of odd subscripts in A and by A^{**} the sequence consisting of the even subscripts of A. Then $\{A_i^*, A_i^{**} : 1 \leq i \leq S\}$ is also a complementary set.

9.3.1 Generating Complementary Sets Using Recursion

The operations described in Box 9A provide the possibility of finding equivalent complementary sets with the same sequence length. Finding complementary sets of different sequence length is not trivial. Golay (1949), Shapiro (1951), and later Tseng and Liu (1972) proposed a simple recursive procedure for extending complementary pairs of sequences with length M to length $2M$ (and recursively to any length $2^k M$). The procedure assumes knowledge of a complementary pair S_1 and S_2. Let $\acute{S}_2$ represent the 180° phase-shifted sequence of S_2. Let $C_1 = \{S_1 S_2\}$ and $C_2 = \{S_1 \acute{S}_2\}$ represent the concatenated sequences with double the original length. Tseng and Liu (1972) gave a simple proof showing that the new sequence C_1 and C_2 are also a complementary pair. Sivaswamy (1978a) extended the proof to complementary pairs of polyphase sequences and also showed that it holds for the case of a complementary triplet. Budisin (1990) generalized the generation of complementary pairs such that multilevel and complex sequences of magnitude different from unity can also be generated.

The procedure for recursively generating complementary pairs of sequences is exemplified using the complementary pair given by $\{j \quad -1\}$ and $\{j \quad 1\}$. The autocorrelation of the two sequences is calculated below and is given by $\{j \quad 2 \quad -j\}$ and $\{-j \quad 2 \quad j\}$. The sum of the autocorrelations equals $\{0 \quad 4 \quad 0\}$.

$\{j \quad -1\}^*$	$-j \quad -1$	$\{j \quad 1\}^*$	$-j \quad 1$
$\{j \quad -1\}$		$\{j \quad 1\}$	
j -1	$\begin{array}{cc} & 1 & -j \\ j & 1 & \end{array}$	j 1	$\begin{array}{cc} & 1 & j \\ -j & 1 & \end{array}$
	$j \quad 2 \quad -j$		$-j \quad 2 \quad j$

(9.38)

Note that the complementary pair can also be written in matrix form as

$$\begin{bmatrix} j & -1 \\ j & 1 \end{bmatrix} \tag{9.39}$$

where the two rows are the two sequences used. Note that the two matrix columns are orthogonal since $j \cdot (-1) + j \cdot 1 = 0$. This is no surprise, recalling that the rows of a matrix with orthogonal columns form a complementary set (Kretschmer and Gerlach, 1991).

Generating a new pair of four-element complementary sequences is demonstrated in Fig. 9.42. The new sequences are generated by concatenating the first two-element sequence with the second two-element sequence (left-hand side of Fig. 9.42) and by concatenating the first two-element sequence with the 180° phase-shifted version of the second two-element sequence (right-hand side of Fig. 9.42). The autocorrelations of the two four-element sequences are $\{-j \quad 0 \quad -j \quad 4 \quad j \quad 0 \quad j\}$ and $\{j \quad 0 \quad j \quad 4 \quad -j \quad 0 \quad -j\}$, adding up to $\{0 \quad 0 \quad 0 \quad 8 \quad 0 \quad 0 \quad 0\}$. Thus, the pair $\{j \quad -1 \quad j \quad 1\}\{j \quad -1 \quad -j \quad -1\}$ is complementary. Switching the roles of the two sequences gives a different pair of complementary sequences with length 4, given by $\{j \quad 1 \quad j \quad -1\}$ and $\{j \quad 1 \quad -j \quad 1\}$, which is also complementary (see Fig. 9.43). Combining the two pairs gives a set of four sequences of length 4 which are supercomplementary. Applying the method to the four-element sequences leads to complementary pairs of sequences with eight elements which can be combined to a supercomplementary set of eight sequences with eight elements each.

9.3.2 Complementary Sets Generated Using the PONS Construction

The procedure for generating 2^N sequences of length 2^N just demonstrated for $N = 2$ and $N = 3$ can be generalized for any N using matrix notations and is referred to as the *PONS construction* (Zulch et al., 2002). The PONS construction starts with any pair of complementary sequences s_1 and s_2 of length M and obtains four sequences of length $2M$:

$$\text{PONS construction}: \begin{bmatrix} s_1 & s_2 \\ s_1 & -s_2 \\ s_2 & s_1 \\ -s_2 & s_1 \end{bmatrix} \tag{9.40}$$

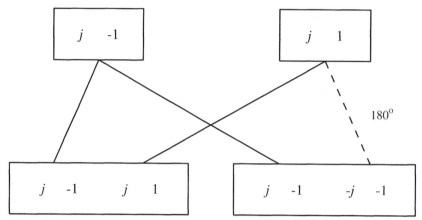

FIGURE 9.42 Generating a complementary pair of four-element sequences using a complementary pair of two-element sequences.

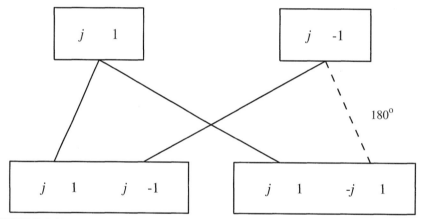

FIGURE 9.43 Generating a complementary pair of four-element sequences using a complementary pair of two-element sequences (switched roles).

To produce a PONS matrix of size 2^N, we start with the matrix

$$P_1 : \begin{bmatrix} 1 & 1 \\ 1 & -1 \end{bmatrix} \qquad (9.41)$$

and apply the PONS construction (9.40) recursively to each pair of adjacent rows. Thus the 4×4 PONS matrix is formed by substituting the first row of P_1 for s_1 and the second row for s_2 in the PONS construction defined in (9.40), to obtain

$$P_2 : \begin{bmatrix} 1 & 1 & 1 & -1 \\ 1 & 1 & -1 & 1 \\ 1 & -1 & 1 & 1 \\ -1 & 1 & 1 & 1 \end{bmatrix} \qquad (9.42)$$

The next PONS matrix P_3 of size 8×8 [see (9.43)] is formed by applying the PONS construction [defined in (9.40)] using the first two rows of P_2 (9.42) to generate the first four rows of P_3 and the last two rows of P_2 to generate the last four rows of P_3. The PONS matrix consists of rows where each row has a matching row forming a complementary set. In the case of the 8×8 PONS matrix shown in (9.43), it is possible to verify that rows 1 and 2, rows 3 and 4, 5 and 6, 7 and 8, 1 and 6, 2 and 5, 3 and 8, and 4 and 7 form complementary pairs. Furthermore, the quartets formed by rows 1, 3, 6, and 8, and rows 2, 4, 5, and 7 form complementary quartets. Systematic methods for finding complementary quartets using a different PONS construction were also described by Zulch et al. (2002).

$$P_3: \begin{bmatrix} 1 & 1 & 1 & -1 & 1 & 1 & -1 & 1 \\ 1 & 1 & 1 & -1 & -1 & -1 & 1 & -1 \\ 1 & 1 & -1 & 1 & 1 & 1 & 1 & -1 \\ -1 & -1 & 1 & -1 & 1 & 1 & 1 & -1 \\ 1 & -1 & 1 & 1 & -1 & 1 & 1 & 1 \\ 1 & -1 & 1 & 1 & 1 & -1 & -1 & -1 \\ -1 & 1 & 1 & 1 & 1 & -1 & 1 & 1 \\ 1 & -1 & -1 & -1 & 1 & -1 & 1 & 1 \end{bmatrix} \quad (9.43)$$

Using the PONS construction makes it possible to choose from many sets of complementary pairs such that the recurrent lobe sidelobes and the off-zero Doppler behavior are optimized.

9.3.3 Complementary Sets Based on an Orthogonal Matrix

An alternative method of finding large sets of complementary sequences is based on orthogonal matrices. Hadamard matrices have the orthogonality property and thus can be used for the construction process. For example, reading the rows of an 8×8 Hadamard matrix in (9.44) gives a set of eight complementary sequences with eight elements each:

$$8 \times 8 \text{ Hadamard matrix}: \begin{bmatrix} 1 & 1 & 1 & 1 & 1 & 1 & 1 & 1 \\ 1 & -1 & 1 & -1 & 1 & -1 & 1 & -1 \\ 1 & 1 & -1 & -1 & 1 & 1 & -1 & -1 \\ 1 & -1 & -1 & 1 & 1 & -1 & -1 & 1 \\ 1 & 1 & 1 & 1 & -1 & -1 & -1 & -1 \\ 1 & -1 & 1 & -1 & -1 & 1 & -1 & 1 \\ 1 & 1 & -1 & -1 & -1 & -1 & 1 & 1 \\ 1 & -1 & -1 & 1 & -1 & 1 & 1 & -1 \end{bmatrix}$$
(9.44)

Note that the Hadamard-based construction only implies complementarily of the eight sequences, whereas the PONS construction gives pairs of complementary sequences where the set of all eight sequences is supercomplementary.

We can use only part of the Hadamard matrix columns (the subset of columns is still orthogonal) and get a set of eight complementary sequences (rows) with fewer elements in each sequence. For example, using only the five leftmost columns of the 8×8 Hadamard matrix gives eight complementary sequences with five elements $\{+++++\}$ $\{+-+-+\}$ $\{++---+\}$ $\{+---++\}$ $\{+++-\}$ $\{+-+--\}$ $\{++---\}$ $\{+---+-\}$. In the same way, only part of the columns of the PONS matrix can be used to form a complementary set of sequences with fewer than 2^N elements. (The PONS matrix also has orthogonal columns since the rows form a complementary set and the matrix is square.) Note that in this case the subsets of the rows will no longer remain complementary.

9.4 TRAIN OF SUBCOMPLEMENTARY PULSES

The complementary coded waveform sets (also known as *Welti codes* or *δ-codes*) described in Section 9.3 have an autocorrelation function that gives zero sidelobes for $\tau \geq t_b$ and low recurrent lobes (t_b is the coded signal bit length). Sivaswamy (1978b) introduced the concept of digital and analog *subcomplementary* sequences for pulse compression. The subcomplementary signals exhibit a zero correlation zone for $\tau \geq \tau_0$ and low recurrent lobes (τ_0 being the duration of the subcomplementary signal kernel). Extending any analog or digital signal with duration τ_0 using simple transformations generates subcomplementary signals. The transformations used are the same as those given by Shapiro (1951), Golay (1961), and Tseng and Liu (1972) to transform a short complementary set to a longer complementary set (see Section 9.3.2). Kretschmer and Gerlach (1991) give the same result by stating that for any coded sequence S_0, a subcomplementary set of sequences results from the Kronecker product of S_0 and a matrix consisting of a set of complementary sequences. That result, although stated by Kretschmer and Gerlach for a coded sequence, also applies to analog signals.

An example of a subcomplementary train of eight pulses is given in Fig. 9.44. The train of eight pulses is based on an analog kernel (LFM bit) and a complementary set of eight sequences of length 8 based on the rows of the Hadamard matrix and given by $\{++++++++\}$ $\{++-----++\}$ $\{+---++--+\}$ $\{++++----\}$ $\{+-+-+-+-\}$ $\{++--++--\}$ $\{+-+--+-+\}$ $\{+---+-++-\}$. The LFM slope was selected such that the ACF mainlobe extends as far as $\tau = t_b/5 = T/40$ (T being the pulse width). The train duty cycle was selected to be $\frac{1}{3}$, ($T_r = 3T$). Figure 9.44 shows the frequency nature of the basic pulse (top) and the phase history of each pulse in the subcomplementary pulse train (after introducing diversity by phase coding). Note the $180°$ phase jumps in locations where the modulating sequences change polarity.

The partial ambiguity function of the 8×8 LFM-based subcomplementary pulse train is shown in Fig. 9.45. Note that using the complementary set as an outer code leads to an autocorrelation function with zero sidelobes for $|\tau| > t_b$ (excluding recurrent lobes not shown in the figure). Note also that using a high-bandwidth kernel leads to a mainlobe width much smaller than t_b (but with

TRAIN OF SUBCOMPLEMENTARY PULSES

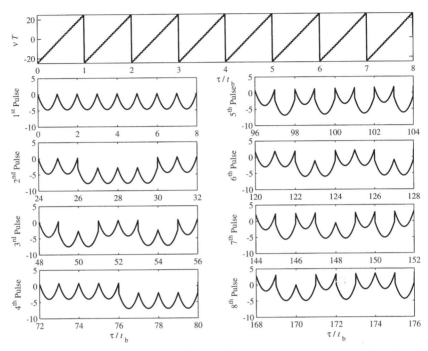

FIGURE 9.44 Structure and phase history of a subcomplementary train of eight pulses with 8 bits based on a LFM kernel and a binary 8×8 Hadamard-based complementary set.

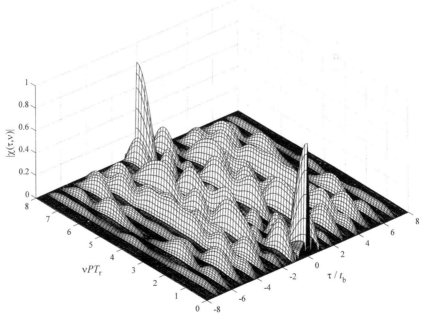

FIGURE 9.45 Partial ambiguity function of a subcomplementary train of eight pulses with 8 bits (LFM bit, Hadamard-based complementary set).

substantial sidelobes for $|\tau| < t_b$). To get the same mainlobe width with zero sidelobes for delays smaller than $|\tau| = t_b$ (a complementary code), one needs to increase the number of bits in each pulse (to 40 instead of 8) and thus also increase the number of pulses in the train.

A second example uses the 8×8 binary complementary set based on the P_3 PONS matrix given in (9.43). The AF of the resulting pulse train is shown in Fig. 9.46. Note that the ambiguity sidelobes are much lower than in Fig. 9.45. The ambiguity volume does not disappear; it is just spread differently in the delay–Doppler plane. Since the difference between the waveforms is only in the various phase codes used, the volume only moves to a different position in Doppler (the ambiguity volume distribution for different delays remains the same as for the signal prior to coding).

A third example using an 8×8 polyphase complementary set based on all eight cyclic shifts of a P4 sequence of length 8 is shown in Fig. 9.47. The ordered nature of the polyphase complementary set due to using cyclic shifts and a P4 sequence (derived from the phase history of a LFM pulse) yields the diagonal ridge structure of the AF shown clearly in Fig. 9.47. The code used is written in matrix form in (9.45), showing the phase values of the various code elements. Reversing the cyclic shift direction or changing the P4 sequence to a generalized P4 (or Zadoff–Chu code; see Chapter 6) gives additional parameters, allowing

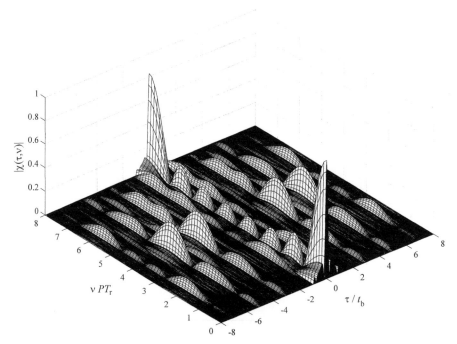

FIGURE 9.46 Partial ambiguity function of a subcomplementary train of eight pulses with 8 bits (LFM bit, PONS-based complementary set).

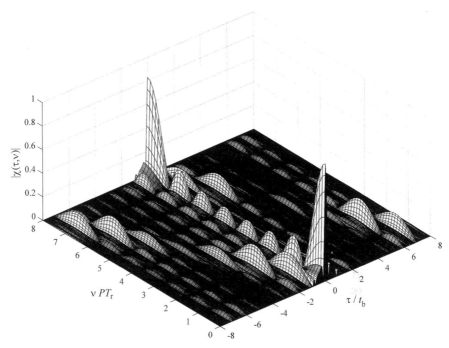

FIGURE 9.47 Partial ambiguity function of a subcomplementary train of eight pulses with 8 bits (LFM bit, cyclic shifts of P4 complementary set).

control of the AF diagonal ridge slope and direction.

8×8 cyclic shifts of P4 complementary set:

$$\frac{\pi}{8} \begin{bmatrix} -1 & -4 & 7 & 0 & 7 & -4 & -1 & 0 \\ -4 & 7 & 0 & 7 & -4 & -1 & 0 & -1 \\ 7 & 0 & 7 & -4 & -1 & 0 & -1 & -4 \\ 0 & 7 & -4 & -1 & 0 & -1 & -4 & 7 \\ 7 & -4 & -1 & 0 & -1 & -4 & 7 & 0 \\ -4 & -1 & 0 & -1 & -4 & 7 & 0 & 7 \\ -1 & 0 & -1 & -4 & 7 & 0 & 7 & -4 \\ 0 & -1 & -4 & 7 & 0 & 7 & -4 & -1 \end{bmatrix} \qquad (9.45)$$

9.5 TRAIN OF ORTHOGONAL PULSES

A different method, which like the subcomplementary signals also results in signals with a zero correlation zone for $|\tau| \geq t_b$ and low recurrent lobes, was introduced by Mozeson and Levanon (2003). Unlike the subcomplementary method proposed by Sivaswamy, the new approach is based on overlaying an orthogonal coding over any analog or digital signal. Figure 9.48 provides a graphical

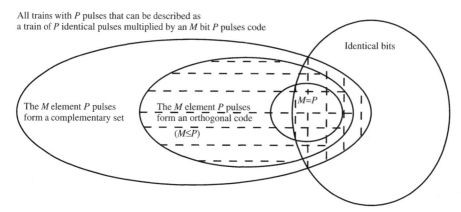

FIGURE 9.48 Connection between Sivaswamy subcomplementary signals, orthogonal-coded pulse trains, and complementary sets.

interpretation of the relation between the construction of Sivaswamy subcomplementary signals, orthogonal coded pulse trains, and complementary sets.

We limit the discussion to all pulse trains that can be described as a train of P identical pulses, multiplied by an M-bit P pulses code for any M and P. The area in Fig. 9.48 covered with vertical dashed lines represents Sivaswamy's subcomplementary signals. These signals are designed such that (1) the bits are identical, and (2) the M-element P-pulse code is a complementary set. The area in Fig. 9.48 covered with horizontal dashed lines represents the orthogonal-coded pulse trains. The orthogonal-coded pulse train design requires orthogonality of the code (a demand stronger than the demand for a complementary set) but does not require the identity of bits.

A simple analytic expression for the autocorrelation function of an orthogonal-coded pulse train is derived in Box 9B. Note that adding a constant phase shift to any of the bits of the original pulse does not change the partial ACF in the area of the mainlobe ($0 \leq \tau \leq T_r - T$) but can help reduce recurrent lobes or shape the AF.

BOX 9B: Autocorrelation Function of Orthogonal-Coded Pulse Trains

We write the orthogonal-coded pulse train complex envelope as

$$g(t) = \sum_{p=1}^{P} \sum_{m=1}^{M} a_{p,m} s_m[t - (p-1)T_r] \qquad (9B.1)$$

where $a_{p,m}$ is the coding element used for the mth bit in the pth pulse in the train ($1 \leq m \leq M$, $1 \leq p \leq P$), T_r is the pulse repetition interval, and $s_m(t)$ is the complex envelope of the mth bit of the original pulse during $(m-1)t_b \leq t < mt_b$ and zero elsewhere. The ACF is calculated directly using (9B.1) in

$$R(\tau) = \int_0^{PT_r} g(t)g^*(t-\tau)\,dt, \qquad -\infty < \tau < \infty \qquad (9B.2)$$

Rearranging the order of summation and integration and taking out of the integral the parts that are not a function of t gives

$$R(\tau) = \sum_{m=1}^{M}\sum_{k=1}^{M}\sum_{p=1}^{P}\sum_{q=1}^{P} a_{p,m}a_{q,k}^*$$

$$\times \int_0^{PT_r} s_m[t-(p-1)T_r]s_k[t-\tau-(q-1)T_r]\,dt \qquad (9B.3)$$

To get a simple expression for the autocorrelation function we assume that $\tau \geq 0$ (the autocorrelation function is symmetric around zero delay) and define $\tau = rT_r + i_r t_b + \eta$, where r is the integer part of T_r in τ ($r \geq 0$), i_r is the integer part of t_b in $\tau - rT_r$ ($0 \leq i_r < T_r/t_b$), and η is the reminder of τ after division by t_b ($0 \leq \eta < t_b$). We can write the autocorrelation function as

$$R(rT_r + i_r t_b + \eta) = \sum_{m=1}^{M}\sum_{k=1}^{M}\sum_{p=1}^{P}\sum_{q=1}^{P} a_{p,m}a_{q,k}^* \int_0^{PT_r} s_m[t-(p-1)T_r]$$

$$\times s_k[t - \eta - i_r t_b - (q+r-1)T_r]\,dt \qquad (9B.4)$$

Assuming that $T_r > 2T$ (duty cycle is below $\frac{1}{2}$) the multiplication of $s_m[t-(p-1)T_r]$ by $s_k[t - \eta - i_r t_b - (q+r-1)T_r]$ is zero for all m, k, t, η, i_r, and $p \neq q+r$. Thus, we can write the ACF as

$$R(rT_r + i_r t_b + \eta) = \sum_{m=1}^{M}\sum_{k=1}^{M}\sum_{p=r+1}^{P} a_{p,m}a_{p-r,k}^* \int_0^{PT_r} s_m[t-(p-1)T_r]$$

$$\times s_k[t - \eta - i_r t_b - (p-1)T_r]\,dt \qquad (9B.5)$$

Changing the variable in the integral, we can write

$$R(rT_r + i_r t_b + \eta) = \sum_{m=1}^{M}\sum_{k=1}^{M}\sum_{p=r+1}^{P} a_{p,m}a_{p-r,k}^*$$

$$\times \int_{-(p-1)T_r}^{PT_r-(p-1)T_r} s_m(t)s_k(t - \eta - i_r t_b)\,dt \qquad (9B.6)$$

For the integrand not to be zero we must have that $s_m(t)$ is not zero or that $0 \leq (m-1)t_b \leq t < mt_b \leq Mt_b$. Since the integral's lower limit is negative and the upper limit is at least $T_r > Mt_b$ for all p, we can replace the integral lower limit with $(m-1)t_b$ and the upper limit with mt_b without changing the result. The ACF can thus be written as

$$R(rT_r + i_r t_b + \eta) = \sum_{m=1}^{M} \sum_{k=1}^{M} \sum_{p=r+1}^{P} a_{p,m} a_{p-r,k}^* \int_{(m-1)t_s}^{mt_s} s_m(t) s_k(t - \eta - i_r t_b) \, dt$$
(9B.7)

where the integral is not a function of p or r. When summing over all possible values of m and k, the integrand will not be zero only for $(k-1)t_b \leq t - \eta - i_r t_b < k t_b$. Using the limits of the integral, we can also write that $m t_b > t \geq \eta + i_r t_b + (k-1)t_b > t - t_b \geq (m-2)t_b$. For a given delay ($i_r$ and η are fixed) and m, we get that the integral will be nonzero only for at most two values of k that solve $m - \eta/t_b - i_r + 1 > k > m - i_r - \eta/t_b - 1$. Denote these two values of k by $k_1 = m - i_r - 1$ and $k_2 = m - i_r$. Note that in the case where $\eta = 0$ we get only one value of $k = k_2 = m - i_r$, and that k must also be between 1 and M.

We can solve the integral explicitly for $k = k_1 = m - i_r - 1$ (denote this by I_1) and $k = k_2 = m - i_r$ (denoted I_2). For mathematical simplicity we also define I_1 and I_2 to be zero when the value of k is not between 1 and M. This gives I_1 and I_2 as

$$I_1 = \begin{cases} \int_{(m-1)t_b}^{(m-1)t_b+\eta} s_m(t) s_{m-i_r-1}(t - \eta - i_r t_b) \, dt, & 1 \leq m - i_r - 1 \leq M \\ 0, & \text{otherwise} \end{cases}$$
(9B.8)

$$I_2 = \begin{cases} \int_{(m-1)t_b+\eta}^{mt_b} s_m(t) s_{m-i_r}(t - \eta - i_r t_b) \, dt, & 1 \leq m - i_r \leq M \\ 0, & \text{otherwise} \end{cases}$$
(9B.9)

Note that when $\eta = 0$, $I_1 = 0$, which coincides with our previous observation on the case of integer τ/t_b.

Since I_1 and I_2 are not a function of p or r, we can write the autocorrelation function as

$$R(rT_r + i_r t_b + \eta) = \sum_{m=1}^{M} \left(I_1 \sum_{p=r+1}^{P} a_{p,m} a_{p-r,m-i_r-1}^* + I_2 \sum_{p=r+1}^{P} a_{p,m} a_{p-r,m-i_r}^* \right)$$
(9B.10)

and the partial ACF in the area of the mainlobe ($0 \leq \tau \leq T_r - T$) as

$$R(\tau) = \begin{cases} P \sum_{m=1}^{M} R_m(\tau), & 0 \leq \tau < t_b \\ 0, & t_b \leq \tau < T_r - T \end{cases}$$
(9B.11)

where $R_m(\tau)$ is the ACF of $s_m(t)$.

9.5.1 Orthogonal-Coded LFM Pulse Train

An example of an orthogonal-coded pulse train is shown in Fig. 9.49. The original pulse is linearly frequency modulated (LFM) to yield the desired mainlobe width (the same mainlobe width as that obtained with the subcomplementary pulse trains of Figs. 9.44 to 9.47). A train of eight identical LFM pulses is overlaid with the 8×8 Hadamard-based complementary set used for the subcomplementary pulse train design in Fig. 9.44, yielding the phase history shown in the lower part of Fig. 9.49.

The partial ambiguity function of the orthonormal-phase-coded LFM pulse train is shown in Fig. 9.50. Comparing Fig. 9.50 to Fig. 9.45, we note that the orthogonal-coded LFM pulse train partial ambiguity function has much lower sidelobes (peaks are reduced exponentially as the delay increases) than for the subcomplementary pulse train (peaks reduce linearly when delay increases). Figure 9.51 gives the AF of an orthogonal-coded LFM pulse train when using an orthogonal code based on the PONS complementary set. Note that the ambiguity sidelobe is lower than for the Hadamard code.

Finally, Fig. 9.52 shows the AF of the orthogonal-coded LFM pulse train using all cyclic shifts of a P4 sequence as the orthogonal code. Note that the orthogonal-coded LFM pulse train can be looked at as an identical slice signal with linear

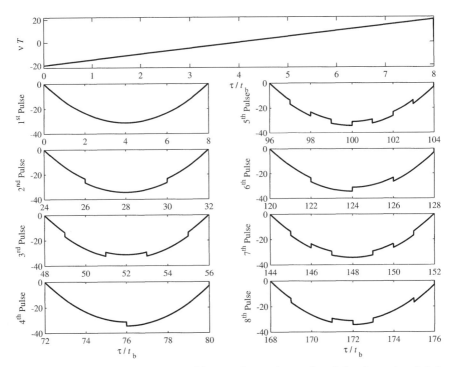

FIGURE 9.49 Structure and phase history of an orthogonal-coded pulse train of eight pulses with 8 bits based on a LFM kernel and an 8×8 complementary set.

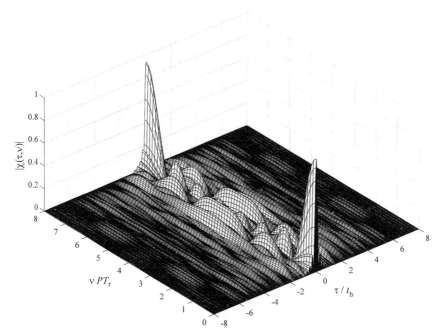

FIGURE 9.50 Partial ambiguity function of an 8×8 orthogonal-coded pulse train based on an LFM pulse (Hadamard-based orthogonal code).

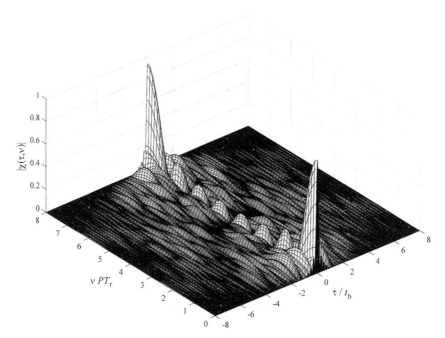

FIGURE 9.51 Partial ambiguity function of an 8×8 orthogonal-coded pulse train based on an LFM pulse (PONS-based orthogonal code).

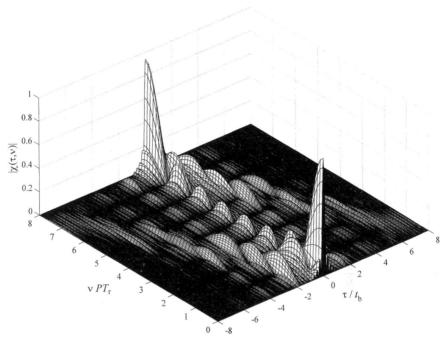

FIGURE 9.52 Partial ambiguity function of an 8 × 8 orthogonal-coded pulse train based on an LFM pulse (cyclic shifts of P4 orthogonal code).

frequency stepping between slices. In this case, based on the analytic expression in Box 9B, it can easily be shown that the partial ACF can be written as

$$R(\tau) = \rho(\tau)\frac{\sin(\pi M \Delta f \tau)}{\sin(\pi \Delta f \tau)} \qquad (9.46)$$

where $\rho(\tau)/M$ is the autocorrelation function of a single LFM bit with $1/M$ of the full LFM pulse frequency deviation, and Δf is the frequency shift between slices (equal to the slice bandwidth to get a continuous LFM pulse). The first autocorrelation sidelobe level of the orthogonal-coded LFM pulse train is -15.9 dB; for a train of identical LFM pulses without orthonormal coding the sidelobe level is -13.6 dB. Further lowering the ACF sidelobe through shaping $\rho(\tau)$ (and increasing the time–bandwidth product within the slice) is addressed next.

9.5.2 Orthogonal-Coded LFM–LFM Pulse Train

Consider an orthonormal phase-coded pulse train where each bit is a linear frequency-modulated pulse and the slices are also linearly frequency shifted (see Fig. 9.53, top). This signal was named by Mozeson and Levanon (2003) as an LFM–LFM orthonormal-phase-coded pulse train. Denote the slice frequency deviation B and the frequency shift between slices Δf. The total pulse

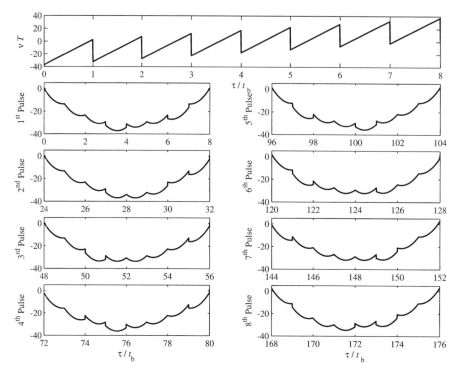

FIGURE 9.53 Structure and phase history of an orthogonal-coded LFM–LFM pulse train. $B = 4.96/t_b$. $\Delta f = 5/T$. PONS orthonormal phase code.

frequency deviation is given by $(M − 1)\Delta f + B$. Note that in the continuous orthogonal-phase-coded LFM signal described in Section 9.5.1, $\Delta f = B$, and for the continuous LFM Sivaswamy subcomplementary signal all slices are identical and $\Delta f = 0$. Note also that the LFM–LFM construction effectively shapes the signal spectrum because the instantaneous frequency histogram has higher values near the center frequency. Using the analytic expression in 9B, it is easy to show that for the LFM–LFM signal the normalized partial ACF for $0 \leq \tau \leq t_b$ is given by

$$R(\tau) = \frac{1}{M} \frac{\sin[\pi B \tau (1 - \tau/t_b)]}{\pi B \tau} \frac{\sin(\pi M \Delta f \tau)}{\sin(\pi \Delta f \tau)} \tag{9.47}$$

Using a relatively simple expression for the autocorrelation function, it is easy to minimize the sidelobe in the mainlobe area while holding the mainlobe width fixed. We hold the first ACF null fixed at $t_b/5 = T/40$ and use B and Δf as the free parameters. Using numerical search methods the minimum peak sidelobe level found for $M = 8$ was -37.39 dB (obtained with $B = 4.96/t_b$ and $\Delta f = 5/T$). The total frequency deviation normalized by the pulse duration in this case is 74.68. Figure 9.53 shows the frequency and phase history of the minimum peak sidelobe LFM–LFM pulse with $B = 4.96/t_b$ and $\Delta f = 5/T$ overlaid

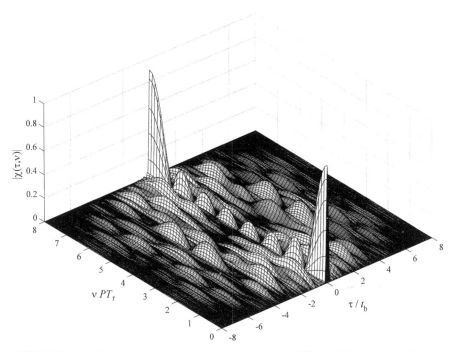

FIGURE 9.54 Partial ambiguity function of an 8×8 LFM–LFM orthogonal-coded pulse train. $B = 4.96/t_b$. $\Delta f = 5/T$. PONS-based orthogonal code.

with the PONS binary orthonormal phase code. The partial ambiguity function for $-Mt_b \leq \tau \leq Mt_b$ and $0 \leq \nu \leq 1/T_r$ are given in Fig. 9.54. The full autocorrelation and autocorrelation zoom for the first slice are shown in Fig. 9.55. The repetition interval (T_r) used for the plots was $3T$. Note that the ACF nulls can be moved by changing B or Δf from batch to batch with a small change in the peak sidelobe level.

When using 16 slices per pulse ($P = M = 16$), we limit the mainlobe to $t_b/2.5$ and get a minimum peak sidelobe of -41.7 dB (obtained for $B = 4.16/t_s$ and $\Delta f = 5/T$) with total frequency deviation (normalized by the pulse duration) of 104.1. When we keep increasing $M (= P)$, the minimum peak sidelobe is reduced but with an increase in the overall frequency deviation. When $M = P = 40$, the slice width becomes the mainlobe width and the remaining sidelobes vanish (the resulting signal is then a 40×40 complementary pulse train).

9.5.3 Orthogonal-Coded LFM–NLFM Pulse Train

The LFM–LFM construction effectively shapes the signal spectrum because the instantaneous frequency histogram has higher values near the center frequency. This can be further generalized by allowing B, Δf, or both to change from slice to slice. We refer to this generalized case as LFM–NLFM signal (each slice is

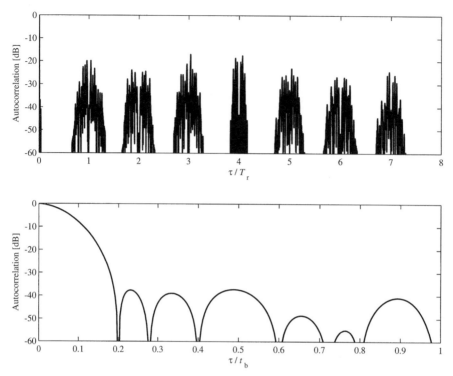

FIGURE 9.55 ACF of an 8 × 8 LFM–LFM orthogonal-coded pulse train. $B = 4.96/t_b$. $\Delta f = 5/T$. PONS-based orthogonal code.

LFM, but the slopes and bandwidths are changed nonlinearly between slices). It can be shown, based on the results in Box 9B, that for the LFM–NLFM signal the normalized partial autocorrelation function for $0 \leq \tau \leq t_b$ is given by

$$R(\tau) = \sum_{m=1}^{M} \exp(j2\pi f_m \tau) \frac{\sin[\pi B_m \tau (1 - \tau/t_b)]}{\pi B_m \tau} \quad (9.48)$$

where B_m and f_m are, respectively, the frequency deviation and center frequency of the mth slice. When the mainlobe width is maintained as in previous examples and the pulse time–bandwidth product is limited to one of the minimal peak sidelobe LFM–LFM pulses, numerical optimization lowers the autocorrelation peak sidelobe to -41.87 dB. Allowing the overall signal frequency deviation to take higher values will further decrease the minimum peak sidelobe level obtainable. For maximal time–frequency span products of 80 and 90, minimum peak sidelobes of -43.6 and -47.33 dB, respectively, are found. Note that changing the order of bits or the slope direction will not change the ACF in the mainlobe area but can help control the recurrent lobes, AF, and spectrum (there are $2^M M!$ options for permuting the bits and changing slope direction).

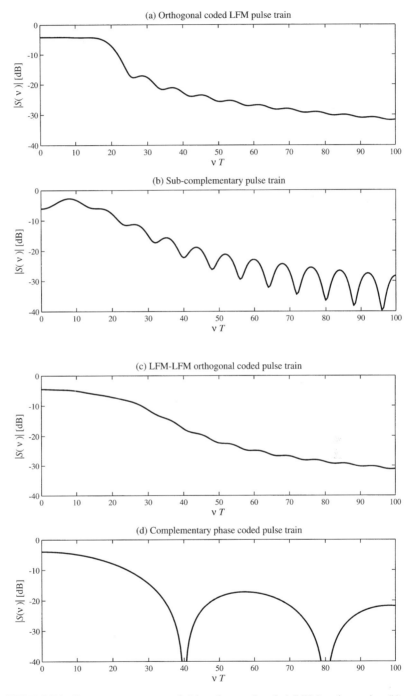

FIGURE 9.56 Frequency spectra of (a) orthogonal-coded LFM pulse train, (b) sub-complementary pulse train, (c) LFM–LFM orthogonal-coded pulse train, and (d) complementary phase-coded pulse train.

9.5.4 Frequency Spectra of Orthogonal-Coded Pulse Trains

Figure 9.56 compares the frequency spectra of (a) an orthogonal-coded LFM pulse train, (b) a subcomplementary pulse train based on an LFM kernel, (c) an orthogonal-coded LFM–LFM pulse train, and (d) a complementary phase-coded pulse train. Note that the orthogonal-coded LFM pulse train has the lowest effective bandwidth (although not shown, the spectrum of a train with identical LFM pulses has a slightly lower effective bandwidth). The subcomplementary pulse train has a higher effective bandwidth. The complementary phase-coded pulse train has a sinc-squared spectrum with the largest effective bandwidth and frequency spectrum sidelobe level.

APPENDIX 9A: GENERATING A NUMERICAL STEPPED-FREQUENCY TRAIN OF LFM PULSES

The following MATLAB code `steplfm.m` prepares the signal element vector and frequency coding vector needed by the ambiguity function plotting code `ambfn1.m` described in Appendix A. The signal generated is defined by:

- N number of pulses
- TB time–bandwidth product of a single LFM pulse
- $T \Delta f$ normalized interpulse frequency step
- T_r/T inverse of duty cycle

Many other signals, which can be described as special cases of a stepped-frequency train of LFM pulses, can also be generated by `steplfm.m`, among them:

- Stepped-frequency train of unmodulated pulses (set $TB = 0$)
- Train of LFM pulses (set $T \Delta f = 0$)
- Train of unmodulated pulses (set $TB = 0$ and $T \Delta f = 0$)
- Single LFM pulse (set $N = 1$, $T \Delta f = 0$, and $T_r/T = 1$)

In `steplfm.m` each pulse is constructed from 50 chips (bits). This choice is usually sufficient to approximate a continuous LFM pulse, yet makes it possible to create a train with a relatively large number of pulses (8, 16, ...) at a reasonable duty cycle (e.g., 1/10). `steplfm.m` creates monotonically increasing interpulse frequency steps. However, for some lengths a Costas order can be chosen. Furthermore, it offers an option to create alternating frequency slopes.

The advanced MATLAB program appendix at the end of the book includes a graphic user interface (GUI) implementation of the `steplfm.m` code through the program pair `steplfm_gui.m` and `steplfm_cal.m`.

The Code

```
% steplfm.m - stepped train of LFM pulses
% creates stepped-LFM signal for use with ambiguity function plotting programs

tpdf=input(' T*Df (typ. = 0 or 5) = ? ');
tpb=input(' T*B (typ. = 0 or 12.5) = ? ');
trovertp=input('Tr/T (typ. = 4) (Using 1 implies contiguous pulses) = ? ');
nn=input(' Number of pulses (integer, typ. = 3, 4, 8 or 16) = ? ');

mm=50;
mm2=mm/2-1;
ufm=ones(1,mm);
ddf=tpb/mm^2;
ffm=ddf*(-mm2:mm2+1)-ddf/2;
mtr=(trovertp-1)*mm;
space1=zeros(1,mtr);
u_step1=[ufm space1];
f_step1=[ffm space1];
f1=-(nn-1)/2;
u_step=u_step1;
f_step=f_step1;
f_add=[f1*u_step1];

reverse=input(' Alternating slopes (yes=1, no=0) = ? ');
if reverse==1
   rv=-1;
else
   rv=1;
end
cost=input('Linear (=0) or Costas (=1)steps = ? ');
if cost ==1;
   if nn==16
      qq=[6 14 2 3 10 8 11 15 9 1 13 12 5 7 4] ;
   elseif nn==8
         qq=[7 2 5 1 6 4 3];
      elseif nn==4
            qq=[2 3 1];
         elseif nn==3
               qq=[2 1];
   end
end
if cost==0 % linear steps
   qq=[1:nn-1];
end
for q= qq
      u_step=[u_step u_step1];
      f_step=[f_step (rv)^q*f_step1];
      f_add=[f_add (f1+q)*u_step1];
end
f_total=f_step+(tpdf/mm)*f_add;

disp('')
disp(' In ambfn1 ')
disp('  use "u_step" as Signal elements and  "f_total" as Frequency coding')
disp(' ')
disp(' In ambfn7 ')
disp('  use  "u_amp = u_step;"    and   "f_basic = f_total;" ')
disp(' ')
```

PROBLEMS

9.1 Phase-coded pulse train

(a) Plot the autocorrelation function (ACF) and ambiguity function (AF) of a train of eight constant-frequency pulses, with $T_r = 4T$ and polarity reversal as follows: $+-++ +-+$.

(b) What is the difference between the ACF of a phase-coded pulse train and a phase-coded pulse using the same phase code?

(c) Describe a method for drawing the continuous ACF of a phase-coded pulse train based on the discrete ACF of the phase code. What is the difference between this method and the method described in Chapter 6 for calculating the ACF of a phase-coded pulse?

9.2 Stepped-frequency pulse train

(a) Plot the AF of a train of $N = 2$ and 3 stepped-frequency unmodulated pulses with $T_r = 3T$ and $T\Delta f = 0.5$.

(b) Repeat the calculations for $T\Delta f = 1$.

9.3 Stepped-frequency pulse train ACF first recurrent lobe

(a) Prove analytically or show numerically (using enough examples) that in a stepped-frequency train of unmodulated N pulses, whenever $T\Delta f$ is an integer multiple of 0.5, the number of uniform peaks in the ACF's first recurrent lobe is $2(N-1)T\Delta f$.

(b) Show that the ACF first recurrent lobe peak level is $1/(\pi N T \Delta f)$.

9.4 Amplitude weighting of stepped-frequency pulse train

(a) Plot the AF of a stepped-frequency train of unmodulated $N = 16$ pulses with $T\Delta f = 0.5$ and $T_r = 3T$.

(b) Assume that the flat-top pulses are amplitude weighted according to the square root of a Hamming window. Recalculate the AF of the weighted pulse train. Extend the delay axis to cover the first recurrent lobe, and the Doppler axis as far as $\nu N T_r = 32$.

(c) Comment on sidelobes and recurrent lobes in both dimensions.

(d) Compare the results to the case of using a full Hamming window at the receiving end only.

9.5 Costas-ordered stepped-frequency pulse train

(a) Plot the AF of a stepped-frequency train of unmodulated and unweighted $N = 16$ pulses, with $T\Delta f = 0.8$ and $T_r = 3T$.

(b) Replace the linear stepping order with Costas order { 1 7 15 3 4 11 9 12 16 10 2 14 13 6 8 5 }. Extend the delay axis to cover the first recurrent lobe and the Doppler axis as far as $\nu N T_r = 17$.

(c) Comment on sidelobes and recurrent lobes in both dimensions.

PROBLEMS

9.6 Amplitude-weighted stepped-frequency LFM pulse train
Consider a train of $N = 8$ stepped-frequency LFM pulses, with $TB = 4$, $T\Delta f = 2$, and $T_r = 3T$.

(a) Plot the ACF and AF ($|\tau| \leq T, 0 \leq \nu \leq 12/NT_r$).

(b) Change the pulse amplitudes according to an eight-element square-root of a Hamming weight window and repeat the plots requested in part (a). Explain the changes in the ACF.

(c) Repeat parts (a) and (b) when the even pulses exhibit opposite FM slope.

9.7 ACF grating lobes of a stepped-frequency LFM pulse train
Consider a coherent train of stepped-frequency LFM pulses with $T\Delta f = 7$.

(a) What is the lowest TB that will nullify the grating lobes?

(b) For the given $T\Delta f$ and the resulting TB, what is the number of pulses N that will cause a null of $|R_1(\tau)|$ [equation (9.22)] to coincide approximately with the first sidelobe of $|R_2(\tau)|$ [equation (9.23)]? Plot the ACF of the resulting signal.

9.8 Changing the LFM slope to get pulse diversity
Consider a coherent train of two LFM pulses with $T\Delta f = 0$ (no frequency step), $TB = 40$, and $T_r = 3T$.

(a) Plot the ACF and AF ($0 \leq \nu \leq 9/NT_r$) when the LFM slopes in the two pulses are identical. Comment on the behavior of the mainlobe and the recurrent lobe.

(b) Plot the ACF and AF ($0 \leq \nu \leq 9/NT_r$) when the LFM slope in the second pulse is reversed. Comment on the behavior of the mainlobe and the recurrent lobe. Explain the uniformity of the ACF's recurrent lobe.

(c) Find the recurrent lobe level in the case of part (b) for $TB = 0, 5, 10, 20, 40$, and 80.

9.9 Changing the LFM slope to get diversity within the pulse

(a) Repeat Problem 9.8(b) when the two pulses are contiguous: namely, when $T_r = T$. Plot the AF and the ACF of the combined single pulse.

(b) Comment on the mainlobe behavior with increasing Doppler and on the sidelobe pedestal.

9.10 Binary complementary pairs

(a) Calculate the aperiodic autocorrelation functions of the Golay 26-element binary pair {00011000101101010110010000} and {00001001101000001011100111}.

(b) Add the two autocorrelation functions to show that the two codes are complementary.

(c) Show by direct calculation of the ACF that the Golay 10-element binary pair {0011111011} and {0010101100} is also a complementary pair.

9.11 AF of a binary complementary pair
Plot the AF, ACF, and spectrum of the Golay 10-element binary pair {0011111011} and {0010101100}. The length of the space between the two pulses should be 20 bits (resulting in a total signal length of 40 bits).

9.12 Spectral shaping of a binary complementary pair
(a) Repeat Problem 9.11 when the rectangular bit is replaced by a Gaussian-windowed sinc ($\sigma = 0.7$), as described in Fig. 6.38. Decrease the spacing between pulses to 14 bits, to maintain a total signal length of 40 bits.
(b) Compare the ACF near-sidelobes and the spectrum to the corresponding sidelobe and spectrum in Problem 9.11.

9.13 Recurrent lobes of a complementary phase-coded pulse train
Four complementary phase-coded pulses are used in a coherent train. The pulses are phase coded as follows: (A) $+++-$ (B) $++-+$ (C) $+-++$ (D) $-+++$. The spacing after each pulse (including the last one) is 8 bits, implying that: $T = 4t_b$, $T_r = 3T$, and $N = 4$.
(a) Plot the ambiguity function cut at $|\chi(-4t_b \leq \tau \leq 4t_b, \nu 4T_r = 1)|$ when the pulses in the train are ordered as follows: (sequence 1) ABCD, (sequence 2) BDCA.
(b) For the two sequences, plot the first recurrent lobe of the autocorrelation function.
(c) Using the results of parts (a) and (b), point out the performance differences between the two sequences and find a qualitative explanation.

9.14 Operations that yield equivalent complementary sets
Prove by using any of the operations of Box 9A on the 10-element binary Golay pair that the following sets are also complementary.
(a) {0011010100} {0011111011}
(b) {1100000100} {0011010100}
(c) {0111111001} {0110101110}
(d) {0011010100} {0011111011} {0010101100} {0011111011}
(e) {1100000100} {0011010100} {0011010100} {0011111011}

9.15 Complementary sets formed using cyclic shifts of a perfect sequence
(a) Calculate the eight-element P4 code described in Chapter 6 and verify that it is a perfect code (i.e., the periodic ACF is ideal).
(b) Write an 8×8 matrix where the rows are given by all (eight) cyclic shifts of the code just calculated. Verify that the columns of the matrix are orthonormal.

PROBLEMS 289

   (c) Calculate the ACF of all (eight) codes formed by reading the rows of the 8 × 8 matrix and show that they are complementary.
   (d) Truncate the matrix to contain eight rows but only four columns by erasing four of the original columns. Verify that all (eight) codes formed by reading the rows of the 8 × 4 matrix are complementary. Explain why.

9.16 **Supercomplementary sets**
   (a) Find subsets of the eight codes calculated in Problem 9.15 which are also complementary (note that there are many possible subsets of size 2 and 4).
   (b) Are the eight codes derived from the 8 × 8 Hadamard matrix also supercomplementary? Explain the difference.

9.17 **Generating complementary sets of high order using nested operations**
   (a) Verify that $\{0 \ 0 \ \pi\} \{0 \ \pi/2 \ 0\}$ and $\{0 \ 0\} \{0 \ \pi\}$ are two complementary pairs.
   (b) Use nested operations with the two-element code as the outer code and the three-element code as the inner code to derive a six-element complementary pair.
   (c) Repeat the operation to derive a 12-element complementary pair.

9.18 **Asymptotically complementary sets**
   (a) Calculate the aperiodic autocorrelation of the eight codes formed by cyclically shifting an eight-element m-sequence.
   (b) Is the set of eight codes complementary?
   (c) When the code length increases, the sidelobe level of the m-sequence PACF remains constant (i.e., the m-sequence is asymptotically perfect). What happens to the sidelobe level of the train of all cyclically shifted versions of m-sequences when the code length increases?
   (d) Is the set of codes asymptotically complementary? Show that the peak sidelobe level decreases by 6 dB when the code length is doubled.
   (e) Is the set also asymptotically supercomplementary? What are the subsets that are asymptotically complementary?
   (f) Write an expression for the sidelobe level as a function of the original m-sequence code length and the asymptotically complementary set size.

9.19 **Cross-complementary sets derived from the Ipatov code**
   The Ipatov code of length 24 and its reference code are given by (see Chapter 6)

   $\{1 \ \ 1 \ \ 1 \ -1 \ -1 \ -1 \ \ 1 \ -1 \ -1 \ -1 \ -1 \ -1 \ \ 1 \ -1 \ -1$
   $\ \ 1 \ -1 \ \ 1 \ \ 1 \ -1 \ -1 \ -1 \ \ 1 \ -1\}$

and

$$\{5 \quad 11 \quad 11 \quad -7 \quad -7 \quad -7 \quad 5 \quad -7 \quad -7 \quad -7 \quad -7 \quad -7 \quad 5 \quad -7$$
$$-7 \quad 11 \quad -7 \quad 11 \quad 5 \quad -7 \quad -7 \quad -7 \quad 11 \quad -7\}$$

(a) Verify that the periodic cross-correlation of the two codes is ideal.
(b) Calculate the aperiodic cross-correlation of the 24 pairs formed by cyclically shifting both codes and show that it is cross-complementary.

9.20 Almost-complementary quadriphase sets
Luke (2001) introduced a family of almost-perfect quadriphase sequences of length $N = p^J + 1 = 2 \pmod 4$ (p odd prime, $J = 1, 2, \ldots$) that have a periodic correlation function which vanishes for all delays except 0 (mod $N/2$). An example of such a sequence for $N = 18$ is

$$\{-1 \quad 1 \quad j \quad 1 \quad 1 \quad 1 \quad j \quad 1 \quad -1 \quad -1 \quad 1 \quad j \quad 1 \quad -j \quad 1 \quad j \quad 1 \quad -1\}$$

(a) Verify that the periodic autocorrelation of the sequence is almost perfect. What is the value of the peak sidelobe?
(b) Use the 18 cyclic shifts of the almost-perfect sequence to write an almost complementary set. What is the value of the peak sidelobe of the almost-complementary set pulse train?

9.21 Periodic complementary binary sequences
In CW radar applications the matched filter response is described by the periodic autocorrelation function rather than aperiodic autocorrelation.
(a) Check that the Golay binary sets of length 2, 10, and 26 are also periodic complementary.
(b) Prove that periodic-complementary sequences include aperiodic complementary sequences as a special case.
(c) Give an example of a periodic complementary set that is not aperiodic complementary (try to avoid the trivial case).

9.22 Delay and add receiver
In many radar applications the target may become visible for a duration that is of the order of the integration time (e.g., a cannon shell crossing the radar antenna mainlobe). In these cases it is essential for the radar to be in sync with the target appearance in order to get maximal signal power.
(a) Assume that the waveform transmitted is a train of three phase-coded complementary pulses described by the complementary set (A) 0 π π, (B) 0 $2\pi/3$ $7\pi/3$, (C) 0 $\pi/3$ $5\pi/3$. Verify that this is a complementary set.
(b) Assume that the three pulses are transmitted continuously and only one period is used in the receiver as a reference. Plot the periodic

correlation function of the pulse train assuming that the reference pulse train is ABC.

- (c) Repeat the calculation of the periodic correlation function assuming that the reference train is BCA and CAB. Compare the PACFs of the three receivers.
- (d) Assume that a three-channel receiver is used, all three correlations are performed simultaneously, and the outputs are properly delayed and added such that the peak response of all channels is aligned. Draw a schematic diagram of the receiver proposed and write a mathematical expression for its output.
- (e) Show that the delay and add receiver output is the sum of the three PACFs calculated in parts (b) and (c). What can be said on the recurrent lobe level?

9.23 Zero cross-correlation complementary code (examples)
- (a) Calculate the three-channel delay and add the receiver response when the phase code used is given by (A) 0 0 0, (B) 0 $4\pi/3$ $2\pi/3$, (C) 0 $2\pi/3$ $4\pi/3$. Compare the response to the one obtained in Problem 9.22(e).
- (b) The family of phase codes having the property demonstrated in part (a) are termed *zero cross-correlation* (ZCC) *complementary*. Show that the Frank code is also ZCC complementary.
- (c) A P4 code of length N^2 can be arranged into an $N \times N$ matrix when the code is read by concatenating the matrix rows. Show that the matrix rows form a ZCC complementary code.

9.24 General form of zero cross-correlation complementary code (after Gerlach and Kretschmer, 1990)
Consider an $N \times N$ code matrix $\mathbf{C}$ where an element of $\mathbf{C}$ is defined by $\lambda^m d_{i+1} W^{Mmi}$ $m, i = 0, 1, \ldots, N-1$, where $W = \exp(j2\pi/N)$, $\lambda \in \{1, W, W^2, \ldots, W^{N-1}\}$, $d_1, d_2 \cdots d_{N-1}$ are arbitrary complex numbers, and M is an integer relatively prime to N.
- (a) Prove that the rows of the matrix $\mathbf{C}$ form a ZCC complementary set.
- (b) Show that the Frank code is a special case of a ZCC complementary code. What are the values of $\lambda, d_1, d_2 \cdots d_{N-1}$ in this case?
- (c) Show that the P4 code is also a special case of a ZCC complementary code.

REFERENCES

Budisin, S. Z., New complementary pairs of sequences, *Electronics Letters*, vol. 26, no. 13, June 1990, pp. 881–883.

Costas, J. P., A study of a class of detection waveforms having nearly ideal range–doppler ambiguity properties, *Proceedings of the IEEE*, vol. 72, no. 8, August 1984, pp. 996–1009.

Frank, R. L., Polyphase complementary codes, *IEEE Transactions on Information Theory*, vol. IT-26, no. 6, October 1980, pp. 641–647.

Gerlach, K., and F. F. Kretschmer, Jr., *General Forms and Properties of Zero Cross Correlation Radar Waveforms*, NRL Report 9120, January 30, 1990.

Golay, M. J. E., Multislit spectroscopy, *Journal of the Optical Society of America*, vol. 39, June 1949, pp. 437–444.

Golay, M. J. E., Complementary series, *IRE Transactions on Information Theory*, vol. IT-7, April 1961, pp. 82–87.

Kajiwara, A., Stepped-FM pulse radar for vehicular collision avoidance, *Electronics and Communications in Japan*, pt. 1, vol. 82, no. 6, June 1999, pp. 1–7.

Kretschmer, F. F., Jr., and K. Gerlach, Low sidelobe radar waveforms derived from orthogonal matrices, *IEEE Transactions on Aerospace and Electronic Systems*, vol. 27, no. 1, January 1991, pp. 92–102.

Levanon, N., Stepped-frequency pulse-train radar signal, *IEE Proceedings on Radar, Sonar and Navigation*, vol. 149, no. 6, December 2002, pp. 297–309.

Levanon, N., and E. Mozeson, Nullifying ACF grating lobes in stepped-frequency train of LFM pulses, *IEEE Transactions on Aerospace and Electronic Systems*, vol. 39, no. 2, April 2003, pp. 694–703.

Luke, H. D., Almost perfect quadriphase sequences, *IEEE Transactions on Information Theory*, vol. 47, no. 6, September 2001, pp. 2607–2608.

Mozeson, E., and N. Levanon, Removing autocorrelation sidelobes by overlaying orthogonal coding on any train of identical pulses, *IEEE Transactions on Aerospace and Electronic Systems*, vol. 39, no. 2, April 2003, pp. 583–603.

Popovic, B. M., Complementary sets based on sequences with ideal periodic autocorrelation, *Electronics Letters*, vol. 26, no. 18, 1990, pp. 1428–1430.

Popovic, B. M., Generalized chirp-like polyphase sequences with optimum correlation properties, *IEEE Transactions on Information Theory*, vol. 38, no. 4, July 1992, pp. 1406–1409.

Rabideau, D. J., Nonlinear synthetic wideband waveforms, *Proceedings of the IEEE Radar Conference*, Los Angeles, May 2002, pp. 212–219.

Rohling, H., and E. Lissel, 77 GHz sensor for car applications, *Proceedings of the IEEE 1995 International Radar Conference*, Alexandria, VA, May 1995, pp. 373–379.

Rohling, H., and R. Mende, Method for operating a radar system, U.S. patent 6,147,638, November 14, 2000.

Shapiro, H. S., Extremal problems for polynomials and power series, M.Sc. thesis, MIT, 1951.

Sivaswamy, R., Multiphase complementary codes, *IEEE Transactions on Information Theory*, vol. IT-24, no. 5, September 1978a, pp. 546–552.

Sivaswamy, R., Digital and analog sub-complementary sequences for pulse compression, *IEEE Transactions on Aerospace and Electronic Systems*, vol. AES-14, no. 2, March 1978b, pp. 343–350.

Tseng, C. C., and Liu, C. L., Complementary sets of sequences, *IEEE Transactions on Information Theory*, vol. IT-18, no. 5, September 1972, pp. 644–652.

Vannicola, V. C., T. B. Hale, M. C. Wicks, and P. Antonik, Ambiguity function analysis for the chirp diverse waveform (CDW), *Proceedings of the IEEE 2000 International Radar Conference*, Alexandria, VA, May 8–12, 2000, pp. 666–671.

Wehner, D. R., *High Resolution Radar*, Artech House, Norwood, MA, 1987.

Zulch, P., M. Wicks, B. Moran, S. Surova, and J. Byrens, A new complementary waveform technique for radar signals, *Proceedings of the IEEE 2002 International Radar Conference*, Long Beach, CA, April 22–25, 2002, pp. 35–40.

10

CONTINUOUS-WAVE SIGNALS

Continuous-wave (CW) radar is one of the earliest forms of radar, and it is also in widespread use today. The proximity fuze, developed at the Applied Physics Laboratory of Johns Hopkins University before and during World War II, is a prominent example of a CW radar. The CW concept still dominates the proximity fuze industry. The CW approach is also found in short-range radar applications such as radar altimeters and atmospheric probing radars. The CW signal is also useful in velocity-measuring radars such as airborne Doppler navigation radars, artillery muzzle velocity, and police radars. Relative to a pulsed signal, a continuous-wave signal is more difficult to intercept by an unprepared receiver, where by *unprepared* we imply that the receiver does not contain a filter (or correlator) matched to the signal expected. For this reason, CW waveforms are sometimes referred to as *low probability of intercept* (LPI) *waveforms*.

As its name implies, CW radar emits a continuous signal. It must therefore receive the returned signal while it is transmitting. An inherent problem is how to distinguish or separate between the signal transmitted and the signal received. In the simple proximity fuze, the return signal is always Doppler shifted (due to the projectile's high relative velocity). This Doppler shift is the mean for separating between the transmitted and reflected signals. Other CW radars utilize modulation to implement separation. Usually, it is also necessary to employ two well-isolated antennas, one for transmitting and one for receiving, to help further separate the signal transmitted from the signal reflected.

Modulation of the continuous-wave signal is also the key for range resolution. Modulation provides marking on the time axis, necessary for measuring delay.

Radar Signals, By Nadav Levanon and Eli Mozeson
ISBN 0-471-47378-2 Copyright © 2004 John Wiley & Sons, Inc.

As we have learned in early chapters, the range resolution is related inversely to the spectral width. It is also well known that the range error is inversely proportional to the root-mean-square bandwidth. An unmodulated CW signal has zero bandwidth and therefore an infinitely wide range window or unbounded range error.

10.1 REVISITING THE PERIODIC AMBIGUITY FUNCTION

Doppler resolution is inversely related to the signal duration that is processed coherently by the receiver. Clearly, that duration cannot be infinitely long. To simplify the analysis we assume that if the carrier signal is modulated by a periodic waveform with period T_r, the receiver is matched to an integer number of periods, that is, to a signal of duration NT_r (see Fig. 10.1). The tool for analyzing the response of such a receiver is the periodic ambiguity function (PAF), introduced in Section 3.6. The PAF serves CW signals in a role similar to that served by the regular ambiguity function for finite energy signals. A brief revisit of the PAF follows.

The reference signal $r(t)$ is constructed from N coherent periods and can be described as

$$r(t) = \text{Re}[u_N(t) \exp(j2\pi f_c t)] \qquad (10.1)$$

Its complex envelope is

$$u_N(t) = \frac{1}{\sqrt{N}} \sum_{n=1}^{N} u_n[t - (n-1)T_r] \qquad (10.2)$$

Henceforth, the term *signal* will refer to the complex envelope of the signal. We first assume that the N periods are identical: namely, $u_n(t) = u_1(t)$. Adding a weight window on receive will violate this assumption. The complex envelope of the received signal (at zero relative delay and zero Doppler) is not limited to N periods and will be described as an infinite periodic signal whose envelope is

$$u(t) = \sum_{n=-\infty}^{\infty} u_1[t - (n-1)T_r] \qquad (10.3)$$

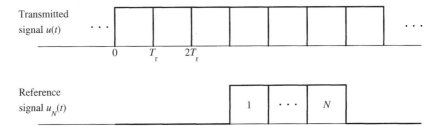

FIGURE 10.1 Processing a periodic CW signal.

Doppler-induced phase shifts of the received signal will be handled by the periodic ambiguity function. Other than those phase shifts, equation (10.3) assumes that the target return exhibits constant phase during the dwell. We now return to Fig. 10.1, where the matched receiver acts to correlate a received periodic signal $u(t)$ of many ($>N$) identical periods with a reference periodic signal $u_N(t)$ of exactly N identical (and same) periods. The normalized response of this processor in the presence of Doppler shift is given by the periodic ambiguity function (a slightly different version from the one in Chapter 3):

$$|\chi_{NT}(\tau, \nu)| = \left| \frac{1}{NT_r} \int_0^{NT_r} u(t - \tau) u^*(t) \exp(j2\pi \nu t) \, dt \right| \quad (10.4)$$

Of special interest is the single-period reference signal ($N = 1$), which yields

$$|\chi_T(\tau, \nu)| = \left| \frac{1}{T_r} \int_0^{T_r} u(t - \tau) u^*(t) \exp(j2\pi \nu t) \, dt \right| \quad (10.5)$$

It can be shown (Freedman and Levanon, 1994; Getz and Levanon, 1995) that a very simple and important relationship exists between equations (10.4) and (10.5):

$$|\chi_{NT}(\tau, \nu)| = |\chi_T(\tau, \nu)| \left| \frac{\sin N \pi \nu T_r}{N \sin \pi \nu T_r} \right| \quad (10.6)$$

Equation (10.6) suggests that it is sufficient to calculate the single-period PAF (10.5) and then multiply it by the function $|(\sin N \pi \nu T_r)/(N \sin \pi \nu T_r)|$ to get the N-period PAF. The multiplying function is a function of the Doppler shift ν but not of the delay τ. Its version for $N = 8$ was plotted in Fig. 3.5.

One of the outcomes of (10.6) is that the cuts at Doppler shifts that are multiples of $1/T_r$ are independent of N: namely,

$$\left| \chi_{NT}\left(\tau, \frac{k}{T_r}\right) \right| = \left| \chi_T\left(\tau, \frac{k}{T_r}\right) \right|, \quad k = 0, \pm 1, \pm 2, \ldots \quad (10.7)$$

Of those cuts, the zero-Doppler cut of the PAF $|\chi_{NT}(\tau, 0)| = |\chi_T(\tau, 0)|$ is the magnitude of the periodic autocorrelation function PACF of the signal. The zero-delay cut of the PAF is obtained by setting $\tau = 0$ in (10.4). It yields

$$|\chi_{NT}(0, \nu)| = \left| \frac{1}{NT_r} \int_0^{NT_r} |u(t)|^2 \exp(j2\pi \nu t) \, dt \right| \quad (10.8)$$

When the signal is only phase- and/or frequency-modulated, $|u(t)| = 1$. Using that fact in (10.5), (10.6), and (10.8) yields a universal cut, independent of other details of the signal:

$$|\chi_{NT}(0, \nu)| = \left| \frac{\sin N \pi \nu T_r}{N \pi \nu T_r} \right|, \quad |u(t)| = 1 \quad (10.9)$$

Because of the periodicity of the signal, that cut reappears every T_r:

$$|\chi_{NT}(nT_r, \nu)| = |\chi_{NT}(0, \nu)|, \qquad n = \pm 1, \pm 2, \ldots \qquad (10.10)$$

More symmetry and periodicity properties of the PAF were discussed by Freedman and Levanon (1994). Of special interest are CW radar signals that exhibit an ideal (perfect) periodic autocorrelation function (PACF). Many of those signals were discussed in Chapter 6. Since the zero-Doppler cut of the PAF is the magnitude of the PACF, all those signals exhibit the same ideal zero-Doppler cut of the PAF. Two examples are shown in the next section.

10.2 PAF OF IDEAL PHASE-CODED SIGNALS

One of the advantages of CW periodic signals is that they can make use of the fact that there are signals with an ideal (or perfect) periodic autocorrelation function (zero sidelobes). There is no such thing as a sidelobe-free aperiodic autocorrelation function. Appendix 10A describes a simple test for determining if a phase-coded sequence yields ideal PACF. Many ideal signals were discussed in detail in Chapter 6, among them the Frank code, Zadoff–Chu codes, P4 code, and Golomb's biphase code. In this section we demonstrate the PAF of the last two signals because they exhibit opposite behaviors. The significance of the number of periods N, processed coherently, will also be demonstrated.

We begin with a P4 signal of length M. One period of the modulation signal is divided into M bits, each of duration $t_b = T_r/M$. The complex envelope of one period is

$$u_1(t) = \sum_{m=1}^{M} u_{1m}[t - (m-1)t_b], \qquad 0 \leq t \leq T_r \qquad (10.11)$$

The bits are phase modulated by the phase sequence $\{\phi_m\}$ of length M:

$$u_{1m}(t) = \begin{cases} \exp(j\phi_m), & 0 \leq t \leq t_b \\ 0, & \text{elsewhere,} \end{cases} \qquad m = 1, 2, \ldots, M \qquad (10.12)$$

The phase sequence of P4 (one of several different versions) is given by

$$\phi_m = \frac{\pi}{M}(m-1)^2 - \pi(m-1), \qquad m = 1, 2, \ldots, M \qquad (10.13)$$

The periodic ambiguity functions for $M = 16, N = 1$ and for $M = 16, N = 8$ are given in Figs. 10.2 and 10.3, respectively. The zero-Doppler cut (in dB) of either PAF is given in Fig. 10.4. Both PAFs, and obviously the PACF in Fig. 10.4, demonstrate the sidelobe-free zero-Doppler cut. The main difference between Figs. 10.2 and 10.3 is the improved Doppler resolution when $N = 8$. The first Doppler null moved from $\nu = 1/T_r$ in Fig. 10.2 to $\nu = 1/(8T_r)$ in Fig. 10.3.

298 CONTINUOUS-WAVE SIGNALS

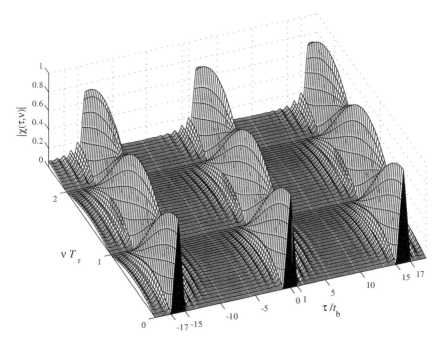

FIGURE 10.2 PAF of P4 phase-coded CW signal with $M = 16$ bits and $N = 1$ period.

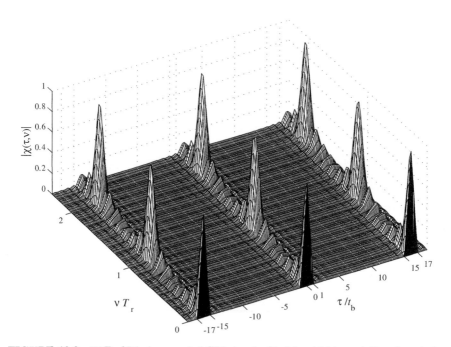

FIGURE 10.3 PAF of P4 phase-coded CW signal with $M = 16$ bits and $N = 8$ periods.

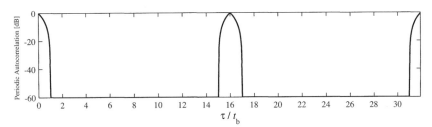

FIGURE 10.4 PACF of P4 phase-coded CW signal with $M = 16$ bits.

What is unique to P4 phase coding, and can be seen more clearly in Fig. 10.3, is that there is very little difference between the PAF's zero-Doppler cut and the PAF's cut at $|\chi_{NT}(\tau, 1/T_r)|$. Both are sidelobe-free. The main difference is that the $|\chi_{NT}(\tau, 1/T_r)|$ cut is shifted to the right by t_b. The shifts continue at higher Doppler multiples of $1/T_r$. The shift of exactly t_b is due to a unique property of P4 and P3. When a phase staircase is added, which results in an accumulation of 2π over one sequence, the new code is a one-bit cyclic shift of the original code (plus a constant phase). Using (10.11), note for P4 that

$$\phi_m + \frac{2\pi}{M}(m-1) = \phi_{m+1} + \pi\left(1 - \frac{1}{M}\right) \qquad (10.14)$$

A Doppler shift of $1/T_r$ adds a linear-phase ramp that accumulates 2π over one sequence (whose duration is T_r). A linear phase ramp is not exactly a staircase, but for a Doppler shift of $\pm 1/T_r$ or $\pm 2/T_r$, the difference is negligible, and the effect is a shift in the location of the peak of t_b, with negligible change in its triangular shape. At much higher multiples of $1/T_r$, the shifts are accompanied by some shape changes, as noted in (Getz and Levanon, 1995), and are as shown in Fig. 10.7. We will come back to this shift property when we consider methods to create delay-shifted "matched" filters to P4 and to other signals with a similar property. Except for those shifts, the PAF in Fig. 10.3 resembles the "bed of nails" PAF of a coherent train of identical, unmodulated pulses.

Another family of CW signals with ideal PACF is that of Golomb's biphase signals (Levanon and Freedman, 1992), described in Section 6.4. We consider a Golomb signal of length 15 whose phase sequence is

$$\{\phi_m\} = \{0 \quad 0 \quad 0 \quad 0 \quad \alpha \quad \alpha \quad \alpha \quad 0 \quad \alpha \quad \alpha \quad 0 \quad 0 \quad \alpha \quad 0 \quad \alpha\},$$
$$\alpha = \cos^{-1}(-\tfrac{7}{8}) \qquad (10.15)$$

The PAFs of Golomb 15 signal for $N = 1$ and $N = 8$ processed periods are shown in Figs. 10.5 and 10.6. Contrary to the ridges seen in the PAF ($N = 1$) of P4, in the PAF of the Golomb biphasic signal the sidelobe volume is spread more uniformly. When N is increased, what remains in the PAF is the volume around $\nu = k/T_r$ ($k = 0, \pm 1, \pm 2, \ldots$). In the P4 case, that creates peaks (Fig. 10.3), whereas in the Golomb case, it creates ridges that are parallel to the delay axis (Fig. 10.6).

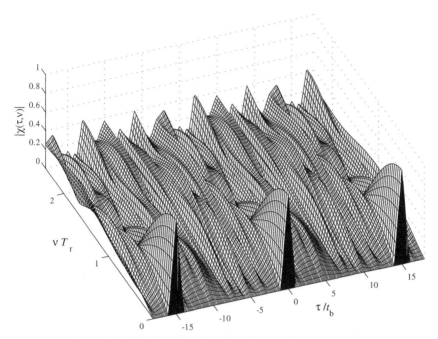

FIGURE 10.5 PAF of Golomb biphasic CW signal with $M = 15$ bits and $N = 1$ periods.

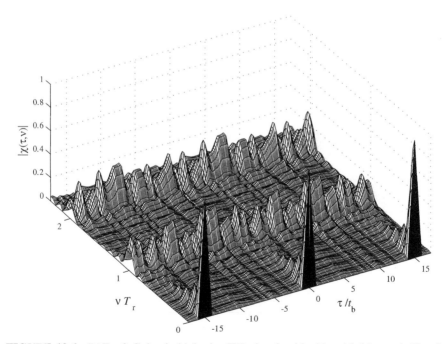

FIGURE 10.6 PAF of Golomb biphasic CW signal with $M = 15$ bits and $N = 8$ periods.

Increasing N, the number of periods that are processed coherently, squelches the sidelobes between the PAF ridges at $\nu = k/T_r$ ($k = 0, \pm 1, \pm 2, \ldots$). However, it is clearly evident from Figs. 10.3 and 10.6 that some Doppler sidelobes remain. As in other signals, Doppler sidelobes can be reduced by adding an amplitude weight window along the time axis. In the next section we analyze Doppler sidelobe reduction using weight windows.

10.3 DOPPLER SIDELOBE REDUCTION USING WEIGHT WINDOWS

In a finite-length signal the weight window could have been split between the transmitter and receiver. In a continuous periodic signal only the finite reference signal can be modified by a weight function $w(t)$. Automatically, that converts the receiver from a matched receiver to a mismatched receiver (with a corresponding degradation in SNR and some decrease in Doppler resolution). The analysis of the effect of a weight window on the delay–Doppler response was introduced by Getz and Levanon (1995) and is summarized below. The reference signal $u^*(t)$ in (10.4) is divided into a product of three functions: $r(t)$ [periodic with the same period as $u(t)$], $w(t)$ (an aperiodic weight function), and $p(t)$ (a rectangular window of duration NT),

$$p(t) = \begin{cases} 1, & 0 \leq t < NT \\ 0, & \text{elsewhere} \end{cases} \tag{10.16}$$

The delay–Doppler response of the mismatched receiver becomes

$$|\psi(\tau, \nu)| = \left| \int_{-\infty}^{+\infty} u(t - \tau) r(t) p(t) w(t) \exp(j 2\pi \nu t) \, dt \right| \tag{10.17}$$

Since (10.17) is the Fourier transform of two products (except for the missing negative sign in the exponential), it can be described by the convolution (denoted $\otimes$) of two Fourier transforms,

$$|\psi(\tau, \nu)| = \left| \int_{-\infty}^{+\infty} u(t - \tau) r(t) \exp(j 2\pi \nu t) \, dt \right|$$
$$\otimes \left| \int_{-\infty}^{+\infty} p(t) w(t) \exp(j 2\pi \nu t) \, dt \right| \tag{10.18}$$

With the first transform, since both $u(t)$ and $r(t)$ are infinitely long and periodic with period T_r, the Fourier transform of their product (for any τ) can be shown (Getz and Levanon, 1995) to be a series of delta functions at $\nu = n/T_r, n = 0, \pm 1, \pm 2, \ldots$:

$$\int_{-\infty}^{+\infty} u(t - \tau) r(t) \exp(j 2\pi \nu t) \, dt = \sum_{n=-\infty}^{+\infty} \delta\left(\nu - \frac{n}{T_r}\right) g_n(\tau) \tag{10.19}$$

where
$$g_n(\tau) = \frac{1}{T_r} \int_0^{T_r} u(t-\tau) r(t) \exp(j2\pi nt/T_r)\, dt \qquad (10.20)$$

The second integral in (10.18) is the Fourier transform of the product of the rectangular window and the weight function:

$$W(\nu) = \int_{-\infty}^{+\infty} p(t) w(t) \exp(j2\pi \nu t)\, dt = \int_0^{NT_r} w(t) \exp(j2\pi \nu t)\, dt \qquad (10.21)$$

Finally, the delay Doppler response of the weighted correlation receiver is obtained from the convolution between (10.19) and (10.21), yielding

$$|\psi(\tau,\nu)| = \sum_{n=-\infty}^{+\infty} g_n(\tau) W\left(\nu - \frac{n}{T}\right) \qquad (10.22)$$

The significance of this equation is that at any given coordinate (τ, ν) the delay–Doppler receiver response is determined by contributions from $g_n(\tau)$ and by the weight function. The set of functions $g_n(\tau)$, determined by (10.20), depend on the transmitted signal modulation that is used in $u(t)$ and in the reference signal $r(t)$ [if $r(t)$ is different from $u^*(t)$, as is the case in Ipatov's signal, described in Section 6.5]. Getz and Levanon (1995) developed a general expression of $g_n(\tau)$ for phase-coded signals when one period of the signal is described by

$$u_{1m}(t) = \begin{cases} \exp(j\phi_m), & 0 \le t \le t_b \\ 0, & \text{elsewhere}, \end{cases} \qquad m = 1, 2, \ldots, M \qquad (10.23)$$

and one period of the reference signal is described by

$$r_{1m}(t) = \begin{cases} b_m \exp(-j\phi_m), & 0 \le t \le t_b \\ 0, & \text{elsewhere}, \end{cases} \qquad m = 1, 2, \ldots, M \qquad (10.24)$$

The delay is described by

$$\tau = \tau_0 + p t_b \qquad (10.25)$$

where $0 \le \tau \le t_b$ and p is an integer. The resulting expression is

$$g_n(\tau_0 + p t_b) = \frac{1}{M} \sum_{m=1}^{M} b_m \exp\left[j 2\pi (m-1)\frac{n}{M}\right]$$
$$\cdot \begin{cases} \dfrac{\tau_0}{t_b} \dfrac{\sin(\pi n\, \tau_0/T_r)}{\pi n\, \tau_0/T_r} \end{cases}$$
$$\times \exp\left[j\left(\phi_{M+m-p-1} - \phi_m + \frac{\pi n\, \tau_0}{T_r}\right)\right] \qquad (10.26)$$

$$+ \left(1 - \frac{\tau_0}{t_b}\right) \frac{\sin[\pi n(t_b - \tau_0)/T_r]}{\pi n(t_b - \tau_0)/T_r}$$
$$\times \exp\left[j\left(\phi_{M+m-p} - \phi_m + \frac{\pi n(t_b + \tau_0)}{T_r}\right)\right]\bigg\}$$

Note that for a negative delay, $g_n(\tau)$ can be described by its complement to T_r, since

$$g_n(-\tau) = g_n(T_r - \tau) \tag{10.27}$$

Figure 10.7 displays $|g_n(\tau)|$ of a P4 signal for $n = -1, 0, 1, \ldots, 16$. Note that for all n the width remains $2t_b$; however, the shape changes as n increases. For all signals with ideal (sidelobe-free) $|g_0(\tau)|$ it was shown (Ipatov et al., 1984) that

$$\sum_{m=0}^{M-1} |g_n(mt_b)|^2 = |g_0(0)|^2 \left(\frac{\sin(n\pi/M)}{n\pi/M}\right)^2 \tag{10.28}$$

Regarding the transform of the weight window, three important amplitude-weighting windows can be described by selecting the parameter c in the expression

$$p(t)w(t) = \frac{1}{NT_r}\left(1 - \frac{1-c}{c}\cos\frac{2\pi t}{NT_r}\right), \quad 0 \leq t \leq NT_r, \quad \text{zero elsewhere} \tag{10.29}$$

For uniform, Hann, and Hamming weight windows, c is selected as $c = 1, 0.5,$ and 0.53836, respectively. Using (10.21) to transform $p(t)w(t)$ yields

$$W(\nu) = \frac{\sin \pi \nu NT_r}{\pi \nu NT_r}\left(1 + \frac{(1-c)(\nu NT_r)^2}{c[1-(\nu NT_r)^2]}\right)\exp(j\pi\nu NT_r) \tag{10.30}$$

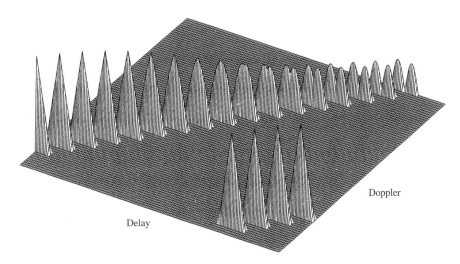

FIGURE 10.7 $|g_n(\tau)|$ of a P4 signal for $n = -1, 0, 1, \ldots, 16$.

The exponent in (10.30) results from the fact that the weight function is not centered at $t = 0$. Note that when $(\nu N T_r)^2 = 1$, equation (10.30) yields $W(\nu) = -(1-c)/2c$. Another observation regarding (10.30) is that when N is an integer,

$$W\left(\frac{n}{T_r}\right) = 0, \qquad n = \pm 1, \pm 2, \ldots \tag{10.31}$$

Equation (10.31) implies, for example, that on the delay axis, namely for $\nu = 0$, only $g_0(\tau)$ contributes to (10.22). In general, for $\nu = k/T_r$ (k an integer), only the respective $g_k(\tau)$ need to be considered in (10.22). For other values of ν (not multiples of $1/T_r$), the neighboring $g_k(\tau)$ should be considered. How many neighbors make a significant contribution depends on the rate of decay of $W(\nu)$.

Although amplitude weighting reduces the Doppler sidelobe levels, the mainlobe in Doppler is also spread. The effect will be demonstrated with a uniform weight window for the two examples (P4 and Golomb) that were plotted in Figs. 10.3 and 10.6. The corresponding delay–Doppler responses using a Hamming window are presented in Figs. 10.8 and 10.9. The amplitude and phase evolution of the reference P4 signal with Hamming weight are plotted in Fig. 10.10.

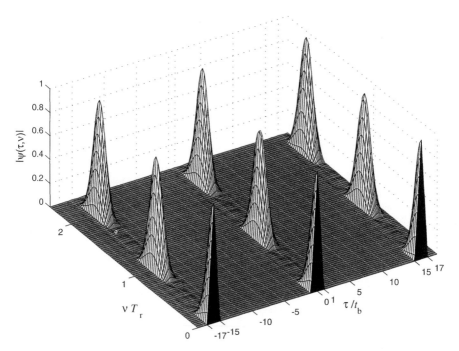

FIGURE 10.8 Delay–Doppler response of P4 phase-coded CW signal with $M = 16$ bits and $N = 8$ periods, Hamming window.

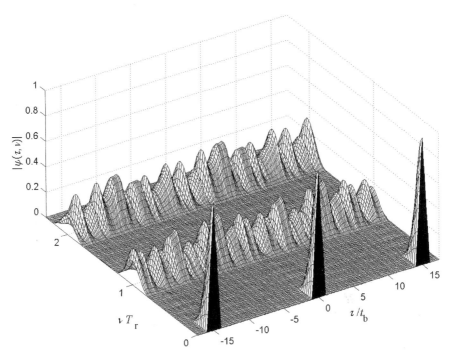

FIGURE 10.9 Delay–Doppler response of Golomb biphase CW signal with $M = 15$ bits and $N = 8$ periods, Hamming window.

Figure 10.8 strongly resembles the delay–Doppler response of a coherent train of unmodulated pulses with interpulse weighting. Yet it is produced by a CW signal (Levanon, 1993). The main difference is in the delay shift of the peaks by kt_b at the $v = k/T_r$ Doppler cut. That shift is the key for a method to produce a delay–Doppler response that is shifted by multiples of t_b, discussed in the next section.

10.4 CREATING A SHIFTED RESPONSE IN DOPPLER AND DELAY

In some applications of CW radar, it is desired to preset a response window at a given delay and Doppler rather than providing continuous range and Doppler measurements. In the case of a P4 signal, adding a stepwise phase ramp to the reference signal can control both the Doppler and delay coordinates of the response peak. Consider a phase step (per bit) of $\Delta\phi = 2\pi d/M$. As long as $|d| \leq \frac{1}{2}$, the response peak will shift in Doppler within $|vT_r| \leq \frac{1}{2}$. For integer d the response will shift in delay by dt_b. When $d = k + \alpha$, $k = 0, \pm 1, \pm 2, \ldots$, $|\alpha| \leq \frac{1}{2}$, the response peak will shift in delay by kt_b and in Doppler by α/T_r. Figure 10.11 demonstrates the delay–Doppler response when $d = 1$: namely, $\Delta\phi = 2\pi/M$. A delay shift of 1 bit is clearly evident.

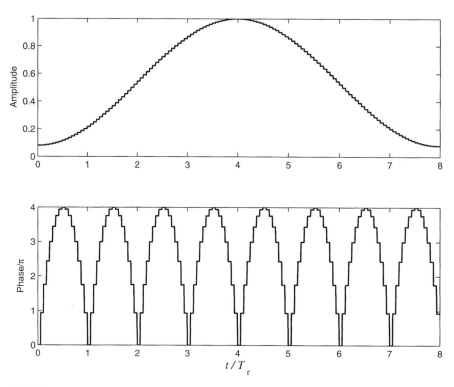

FIGURE 10.10 Amplitude (top) and phase (bottom) of the P4 reference signal with $M = 16$ bits and $N = 8$ periods, Hamming window.

Figure 10.12 demonstrates the delay–Doppler response when $d = 1/4$: namely, $\Delta\phi = \pi/(2M)$. A Doppler shift of $1/(4T_r)$ is clearly evident. Figures 10.11 and 10.12 were obtained with a P4 signal, Hamming weight window, and $N = 8$ periods. In the Golomb signal case, the peak response can be moved in Doppler by choosing $|d| \leq \frac{1}{2}$ (see Fig. 10.13). However, a delay shift cannot be implemented by choosing $d = \pm 1$. This choice will destroy the range resolution by bringing the ridge of $v = \pm 1/T_r$ (see Fig. 10.9) to the zero-Doppler cut of the delay–Doppler response.

10.5 FREQUENCY-MODULATED CW SIGNALS

Early CW radars used frequency modulation (FM) to define range windows. A survey of FM CW signals and processors and important reprints appear in (Barton, 1978; Skolnik, 1980). The processors in those radars were not always matched filters. They usually involved mixing the signal received with a variant of the signal transmitted. Our analysis is based on the periodic ambiguity function (PAF). In many analog FM waveforms it is possible to obtain closed-form expressions of

FREQUENCY-MODULATED CW SIGNALS

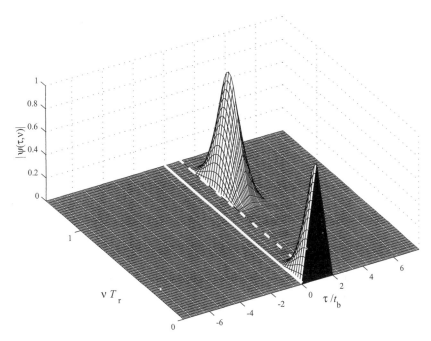

FIGURE 10.11 Delay–Doppler response of P4 phase-coded CW signal with $M = 16$ bits and $N = 8$ periods, Hamming window, added phase step per bit $\Delta\phi = 2\pi/M$.

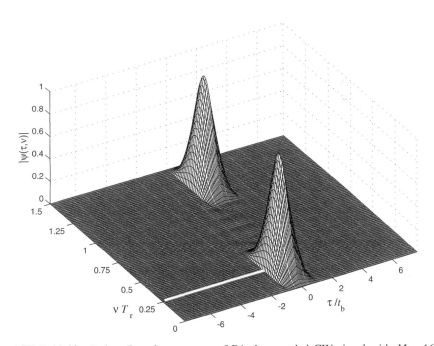

FIGURE 10.12 Delay–Doppler response of P4 phase-coded CW signal with $M = 16$ bits and $N = 8$ periods, Hamming window, added phase step per bit $\Delta\phi = \pi/(2M)$.

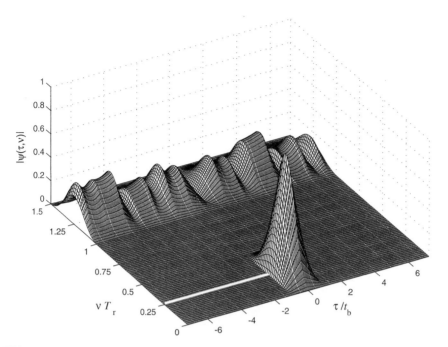

FIGURE 10.13 Delay–Doppler response of Golomb phase-coded CW signal with $M = 15$ bits and $N = 8$ periods, Hamming window, added phase step per bit $\Delta\phi = \pi/(2M)$.

the cuts of the PAF at multiples of the inverse of the modulation period:

$$\left| \chi_{NT} \left(\tau, \frac{n}{T_r} \right) \right| = \left| \chi_T \left(\tau, \frac{n}{T_r} \right) \right| = |g_n(\tau)|, \qquad n = 0, \pm 1, \pm 2, \ldots \ldots \quad (10.32)$$

The last equality holds if the reference signal is identical to the signal transmitted.

To simplify development of closed-form expressions, we start from yet another version of the PAF,

$$|\chi_T(\tau, \nu)| = \left| \frac{1}{T_r} \int_0^{T_r} u\left(t + \frac{\tau}{2}\right) u^*\left(t - \frac{\tau}{2}\right) \exp(j2\pi\nu t)\, dt \right| \quad (10.33)$$

Its cuts at $\nu = n/T_r = nf_m$, without the magnitude symbol are

$$\chi_T\left(\tau, \frac{n}{T_r}\right) = \chi_T(\tau, nf_m) = g_n(\tau)$$

$$= \frac{1}{T_r} \int_0^{T_r} u\left(t + \frac{\tau}{2}\right) u^*\left(t - \frac{\tau}{2}\right) \exp(j2\pi nf_m t)\, dt \quad (10.34)$$

10.5.1 Sawtooth Modulation

The first modulation waveform to be discussed is sawtooth modulation. The complex envelope is given by

$$u_1(t) = \begin{cases} \exp\left[j\pi\dfrac{\Delta f}{T}\left(t - \dfrac{T_r}{2}\right)^2\right], & 0 \leq t < T_r \\ 0, & \text{elsewhere} \end{cases} \qquad (10.35)$$

$$u(t) = \sum_{n=-\infty}^{\infty} u_1(t - nT_r) \qquad (10.36)$$

Using (10.35) and (10.36) in (10.34) yields

$$\begin{aligned} g_n(\tau) = &\frac{1}{T_r}\int_0^\tau \exp\left[-j\pi\frac{\Delta f}{T_r}\left(t - \frac{T_r}{2}\right)^2\right] \\ &\times \exp\left[j\pi\frac{\Delta f}{T_r}\left(t + \frac{T_r}{2} - \tau\right)^2\right]\exp\left(j\pi\frac{n}{T_r}t\right)dt \\ &+ \frac{1}{T_r}\int_\tau^{T_r}\exp\left[-j\pi\frac{\Delta f}{T_r}\left(t - \frac{T_r}{2}\right)^2\right] \\ &\times \exp\left[j\pi\frac{\Delta f}{T_r}\left(t - \frac{T_r}{2} - \tau\right)^2\right]\exp\left(j\pi\frac{n}{T_r}t\right)dt \end{aligned} \qquad (10.37)$$

Solving the integrals, we get

$$g_n(\tau) = \exp\left(j\pi\frac{n}{T_r}\tau\right)\left[(-1)^n\left(1 - \frac{\tau}{T_r}\right)\frac{\sin(\pi n - \alpha)}{\pi n - \alpha} + \frac{\tau}{T_r}\frac{\sin\alpha}{\alpha}\right] \qquad (10.38)$$

where $0 \leq \tau < T_r$ and

$$\alpha = \frac{\pi\tau}{T_r}\left[\Delta f T_r\left(1 - \frac{\tau}{T_r}\right) + n\right] \qquad (10.39)$$

For τ outside the defined range, the periodicity can be applied:

$$g_n(\tau) = g_n(\tau + kT_r), \qquad k = \pm 1, \pm 2, \ldots \qquad (10.40)$$

A plot of $|g_n(\tau)|$, $n = 0, 1, 2$, appears in Fig. 10.14 for the case $\Delta f T_r = 20$. Recall that $\chi_{NT}(\tau, n/T_r) = g_n(\tau)$ for any N. Note the delay shift as n increases, resembling the P4 response. The delay–Doppler response (when using an overall amplitude-weighted reference) was calculated numerically for Hamming weight

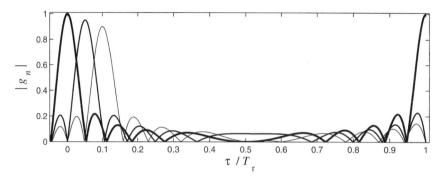

FIGURE 10.14 $|g_n(\tau)|$, $n = 0$ (thick), 1, 2 (thin) for sawtooth FM with $\Delta f T_r = 20$.

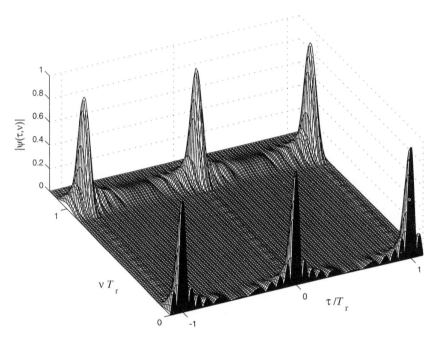

FIGURE 10.15 Delay–Doppler response of sawtooth FM with $N = 16$, $\Delta f T_r = 20$, Hamming weight.

and $N = 16$ periods and is presented in Fig. 10.15. We see that the overall weight window reduced the Doppler sidelobes but had no effect on the delay sidelobes.

As in an LFM pulse, it is possible to use the linear relationship between the time along a single period and the instantaneous frequency to reshape the spectrum through intraperiod amplitude weighting. Figure 10.16 shows a reference signal

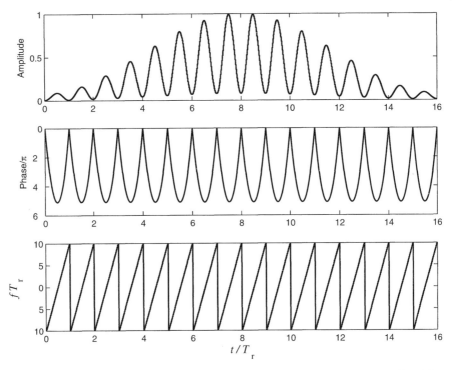

FIGURE 10.16 Sawtooth FM, $N = 16$, $\Delta f T_r = 20$, Hamming weight (overall and intraperiod).

used in such a processor. A block diagram of a possible implementation was described in (Levanon and Getz 1994). The top subplot in Fig. 10.16 describes 16 Hamming windows, each with a length of one modulation period, multiplied by a single Hamming window with a length of 16 periods. Such a weight window results in a significant SNR loss compared to a uniform window. However, it reduces both Doppler and delay sidelobes, as shown in Fig. 10.17. On the other hand, comparing Figs. 10.8 and 10.17 demonstrates that despite the weighting, the remaining delay sidelobes in sawtooth FM CW are still high relative to the sidelobe-free case of a P4 phase-coded CW signal.

10.5.2 Sinusoidal Modulation

The next modulation waveform to be discussed is sinusoidal modulation. The complex envelope is

$$u(t) = \exp\left[j\frac{\Delta f}{2f_m}\cos 2\pi f_m t\right], \qquad f_m = \frac{1}{T_r} \qquad (10.41)$$

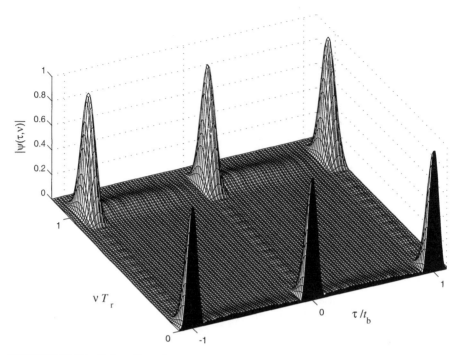

FIGURE 10.17 Delay–Doppler response of sawtooth FM with $N = 16$, $\Delta f T_r = 20$, Hamming weight (overall and intraperiod).

where Δf is the peak-to-peak frequency deviation and f_m is the modulation frequency. Using (10.41) in (10.34) yields

$$\chi_T(\tau, nf_m) = \frac{1}{T_r} \int_0^{T_r} \exp\left\{ j \frac{\Delta f}{2f_m} \cos\left[2\pi f_m \left(t + \frac{\tau}{2} \right) \right] \right\} \\ \cdot \exp\left\{ -j \frac{\Delta f}{2f_m} \cos\left[2\pi f_m \left(t - \frac{\tau}{2} \right) \right] \right\} \exp(j 2\pi n f_m t)\, dt \quad (10.42)$$

which reduces to

$$\chi_T(\tau, nf_m) = \frac{1}{T_r} \int_0^{T_r} \exp\left(-j \frac{\Delta f}{f_m} \sin \pi f_m \tau \sin 2\pi f_m t \right) \exp(j 2\pi n f_m t)\, dt \quad (10.43)$$

Defining

$$\alpha = \frac{\Delta f}{f_m} \sin \pi f_m \tau \quad (10.44)$$

and changing the integration variable to

$$x = 2\pi f_m t - \pi \quad (10.45)$$

yields

$$\chi_T(\tau, nf_m) = \frac{(-1)^n}{2\pi} \int_{-\pi}^{\pi} \exp\{j[\alpha \sin(x) + nx]\} \, dx = (-1)^n J_{-n}(\alpha) = J_n(\alpha) \tag{10.46}$$

or

$$\chi_T(\tau, nf_m) = g_n(\tau) = J_n(\alpha) \tag{10.47}$$

where $J_n(\alpha)$ is the Bessel function of order n of the argument α. Note that as the delay spans one period $0 \leq \tau \leq T_r$, α changes from 0 to $\Delta f/f_m$ (when $\tau = T_r/2$) and back to 0 (when $\tau = T_r$).

The delay–Doppler response of a sinusoidal FM, with $\Delta f/f_m = 10$, is displayed in Fig. 10.18. The receiver is matched to eight periods but is amplitude weighted using a Hamming window. The Doppler span extends beyond $\nu = 2f_m$; hence three ridges are shown, corresponding to $n = 0, 1, 2$. Note that $|J_0(\alpha)|$ exhibits a first null when $\alpha = 2.405$. Thus, the zero-Doppler cut, which follows $|J_0(\alpha)|$, should exhibit a first null at a normalized delay of

$$\tau f_m = \frac{1}{\pi} \sin^{-1}\left(2.405 \frac{f_m}{\Delta f}\right) = \frac{1}{\pi} \sin^{-1}\left(2.405 \frac{1}{10}\right) = 0.0773 \approx 0.08 \tag{10.48}$$

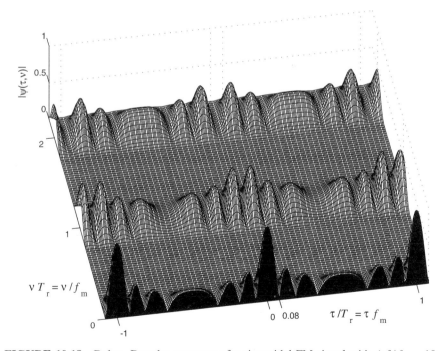

FIGURE 10.18 Delay–Doppler response of a sinusoidal FM signal with $\Delta f/f_m = 10$, $N = 8$ periods, Hamming window.

The delay–Doppler response ridges at $v/f_m = 1$ and 2, also seen in Fig. 10.18, behave according to $|J_1(\alpha)|$ and $|J_2(\alpha)|$, respectively. Namely, the higher the order of the Bessel function, the farther the first peak is from zero delay.

The higher-order Bessel response can be applied to the zero-Doppler cut by adding a phase ramp to the reference signal. For example, a phase ramp that completes 6π radians within one cycle of the modulation frequency, namely within $T_r = 1/f_m$, will bring the $J_3(\alpha)$ response to the zero-Doppler cut. This is demonstrated in Fig. 10.19. Note that while the peak value of $J_0(\alpha)$ is 1, the peak value of $J_3(\alpha)$ is only 0.43. Hence, a significant loss is entailed.

The $|J_n(\alpha)|$ delay response of a sinusoidal FM CW signal for any n suffers from relatively high sidelobes, hence is not very useful for range measurements. On the other hand, its Doppler resolution and sidelobes are determined mostly by the number of periods N processed coherently and by the amplitude weight function. Thus, the Doppler resolution can be made as good as with any other signal.

Sinusoidal FM can therefore be used for velocity measurements as long as the expected range (delay) falls within the broad delay lobe of the $|J_n(\alpha)|$ response. For this reason and for its relative simplicity, sinusoidal FM found use in early airborne Doppler navigation radars. In such a radar there is basically only one major reflecting target—the ground—and it is the Doppler shift (velocity) that is

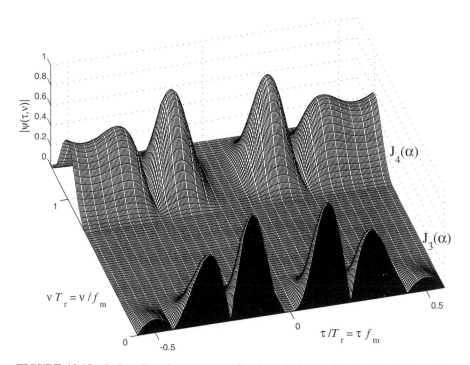

FIGURE 10.19 Delay–Doppler response of a sinusoidal FM signal with $\Delta f/f_m = 10$. A phase ramp of 6π per cycle was added to the reference signal. $N = 8$ periods, Hamming window.

sought, not the delay (range). The main competition to the distant ground return are reflections from the aircraft frame. Because those are reflected from very short distances their intensity can be very large. For this reason a response that attenuates returns at near zero delay, such as the $J_3(\alpha)$ response, is preferred over a response that peaks at zero-delay, such as the $J_0(\alpha)$ response. Recall that the $J_n(\alpha)$ response is obtained by using a reference complex envelope shifted from zero frequency to nf_m. Suggested use of higher harmonics of the modulating frequency can be found in an early U.S. patent (Varian et al., 1948).

10.5.3 Triangular Modulation

The instantaneous frequency of a triangular FM (Fig. 10.20) can be described by

$$f(t + kT_r) = f(t) = \begin{cases} \dfrac{2\Delta f}{T_r}\left(t - \dfrac{T_r}{4}\right), & 0 \leq t \leq \dfrac{T_r}{2} \\ \dfrac{-2\Delta f}{T_r}\left(t - \dfrac{3T_r}{4}\right), & \dfrac{T_r}{2} < t \leq T_r \end{cases} \quad (10.49)$$

Using (10.49), the PAF can be calculated numerically. Figure 10.21 displays approximately two periods of the PAF for $N = 1$, $\Delta f T_r = 20$, and without weighting. A zoom around the origin is displayed in Fig. 10.22. Note the bifurcation of the mainlobe, which begins at Doppler values close to $\nu \approx 1/T_r$. The bifurcation results from the two opposite slopes of the frequency modulation, creating two PAF ridges in opposite slopes. The height of the two ridges is about half the value of the PAF at the origin.

In Fig. 10.22 note also that the first null of the zero-Doppler cut is at $\tau = 1/\Delta f$, which agrees with the pulse compression obtained in conventional linear-FM pulse signals. When the number N of periods processed coherently increases, the delay–Doppler response will be squelched between the values of $\nu = n/T_r$. Figure 10.23 displays the delay–Doppler response of a triangular FM CW signal, with $\Delta f T_r = 20$, $N = 8$ periods processed coherently, and with Hamming weighting on receive. The amplitude, phase, and frequency of the reference signal are plotted in Fig. 10.24.

As far as creating a filter matched to a nonzero Doppler shift, triangular FM is no different from any other modulation waveform. What is required is to add a phase ramp to the reference signal. To create a filter matched to a Doppler shift of

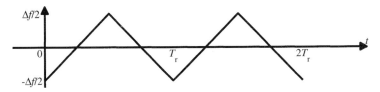

FIGURE 10.20 Triangular FM.

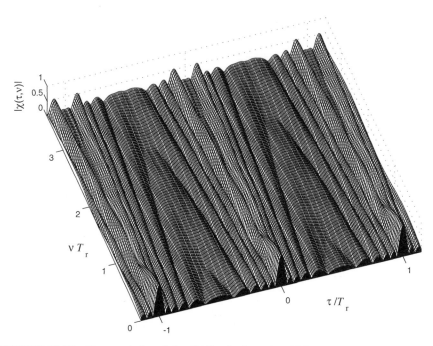

FIGURE 10.21 Two periods of the PAF of triangular FM, $N = 1$, $\Delta f T_r = 20$, no weighting.

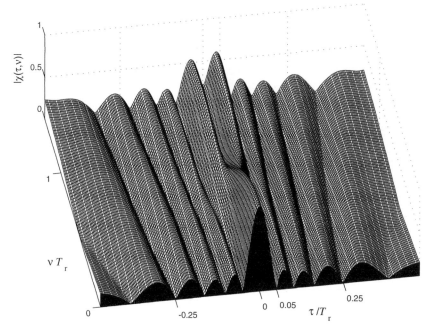

FIGURE 10.22 Zoom on the PAF of triangular FM, $N = 1$, $\Delta f T_r = 20$, no weighting.

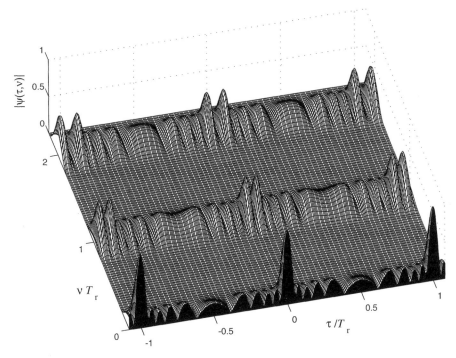

FIGURE 10.23 Delay–Doppler response of triangular FM, $N = 8$, $\Delta f T_r = 20$, Hamming weighting.

$v = d/T_r$ the ramp should complete a phase shift of $2\pi d$ every T_r. Figure 10.25 displays the amplitude and phase of triangular FM with $\Delta f T_r = 20$, $N = 16$, Hamming weight, and a phase ramp factor of $d = \frac{1}{8}$. The resulted delay–Doppler response is plotted in Fig. 10.26. Because a weight function practically doubles the Doppler lobe width, the lowest Doppler shift to which the filter could be matched while still exhibiting a null at zero Doppler is $v_{\min} = 2/NT_r$, where N is the number of modulation periods processed coherently. This is the case in Fig. 10.26. If the zeroth-order filter is undesirable because it peaks at $\tau = 0$, a higher-order "matched" filter can be created [e.g., the second-order filter, matched to $v = 1/(8T_r)$, can be created by adding a phase ramp with a slope of $\Delta\phi/T_r = -4\pi + 2\pi/8$]. The resulting delay–Doppler response is shown in Fig. 10.27.

The significance of creating a delayed response can be questioned because a delayed response can be implemented by delaying the appearance of the reference signal in the correlator relative to the transmitted signal. Indeed, this is the case in matched filter processing, and there the match reference is likely to be used, yielding a zero-order response. However, there are simple CW radars where the reference signal is a diverted fraction of the signal transmitted (without delay). An example is given in the next section. In such a receiver the reference signal

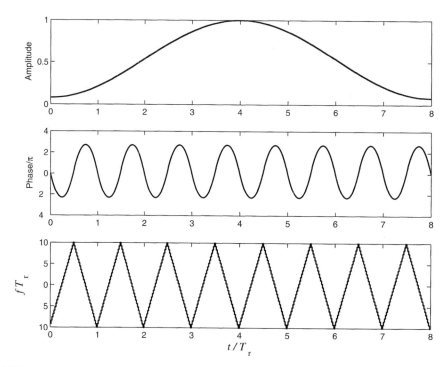

FIGURE 10.24 Reference triangular FM signal with $\Delta f T_r = 20$, $N = 8$, Hamming weight.

cannot be delayed, and a delayed response can be created only through the use of a higher-order response.

10.6 MIXER IMPLEMENTATION OF AN FM CW RADAR RECEIVER

The results in preceding sections apply to a matched filter implementation of the receiver. Many FM CW radars implement a simpler receiver, whose basic function is to mix (multiply) the signal received (delayed) with an attenuated version of the signal transmitted (Fig. 10.28). This kind of processing is an early version of stretch, discussed in earlier chapters. The mixer output is passed first through a low-pass filter to remove signal components centered about twice the carrier frequency. It is then passed through a narrow bandpass filter centered at the nth multiple of the inverse of the modulation period $n/T_r = nf_m$, where $n = 0, 1, 2, \ldots$. Additional processing then follows. In this section we show that the results regarding the cuts of the PAF apply to the output of such mixing plus harmonic filtering.

Let the transmitted signal at time $t + \tau/2$ be

$$s\left(t + \frac{\tau}{2}\right) = \text{Re}\left\{u\left(t + \frac{\tau}{2}\right) \exp\left[j\omega_c\left(t + \frac{\tau}{2}\right)\right]\right\} \qquad (10.50)$$

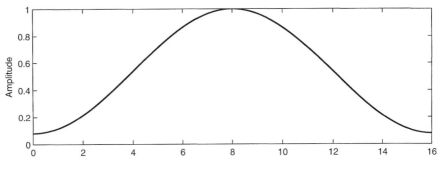
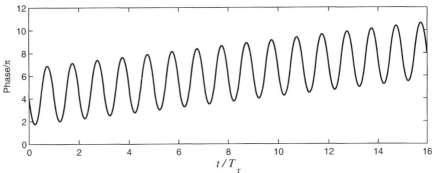

FIGURE 10.25 Reference triangular FM signal with $\Delta f T_r = 20$, $N = 16$, Hamming weight. A phase ramp of $\Delta\phi/T_r = 2\pi/8$ was added.

Assuming a stationary point target and ignoring attenuations, the signal received is identical to the signal transmitted, but delayed by τ. The signal $x(t)$ in Fig. 10.28, which follows the wide LPF that removes the $2\omega_c$ components, can be described as

$$x(t) = \frac{1}{2}\text{Re}\left\{u\left(t+\frac{\tau}{2}\right)u^*\left(t-\frac{\tau}{2}\right)\exp(j\omega_c\tau)\right\} \quad (10.51)$$

Note that $j\omega_c\tau$ is a fixed phase term. Regarding the other component of the complex envelope of $x(t)$, note that since $u(t)$ is periodic with period T_r, the product $u(t+\tau/2)u^*(t-\tau/2)$ is also periodic. If $x(t)$ is available for a very long time, its complex envelope can be described by a Fourier sum,

$$u\left(t+\frac{\tau}{2}\right)u^*\left(t-\frac{\tau}{2}\right) = \sum_{n=-\infty}^{\infty} F_n \exp(jn\omega_m t), \quad \omega_m = \frac{2\pi}{T_r} \quad (10.52)$$

and

$$F_n = \frac{1}{T_r}\int_0^{T_r} u\left(t+\frac{\tau}{2}\right)u^*\left(t-\frac{\tau}{2}\right)\exp(-jn\omega_m t)\,dt \quad (10.53)$$

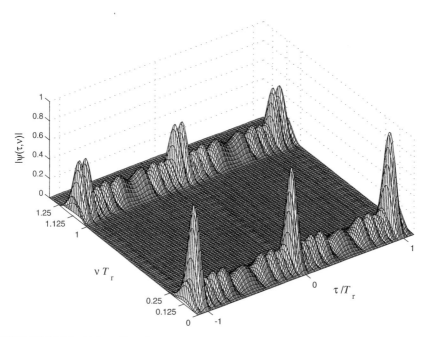

FIGURE 10.26 Delay–Doppler response of triangular FM signal with $\Delta f T_r = 20$, $N = 16$, Hamming weight. A phase ramp of $\Delta\phi/T_r = 2\pi/8$ was added to create a filter matched to $\nu = 1/(8T_r)$.

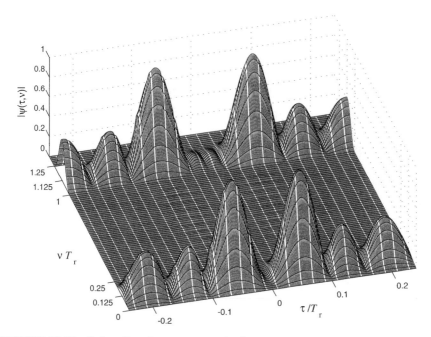

FIGURE 10.27 Delay–Doppler response of triangular FM signal with $\Delta f T_r = 20$, $N = 16$, Hamming weight, and a phase ramp of $\Delta\phi/T_r = -4\pi + 2\pi/8$.

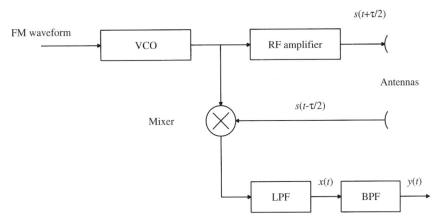

FIGURE 10.28 Generic block diagram of a simple FM CW radar.

Comparing (10.53) with (10.34), we note that

$$F_n = \chi_T(\tau, -nf_m) = \chi_T\left(\tau, -\frac{n}{T_r}\right) \qquad (10.54)$$

The equivalent of saying that $x(t)$ is available for a very long time is saying that the number of modulation periods processed coherently is very large: namely, $N \to \infty$. If the narrow bandpass filter (BPF) is, indeed, very narrow, it will extract only two terms out of the Fourier sum in (10.52), the nth and $-n$th terms. The signal at the output of the narrow BPF is thus

$$y(t) = \frac{1}{2}\mathrm{Re}\left\{\left[\chi_T\left(\tau, \frac{n}{T_r}\right)\exp(-jn\omega_m t) \right.\right.$$
$$\left.\left. +\chi_T\left(\tau, \frac{-n}{T_r}\right)\exp(jn\omega_m t)\right]\exp(j\omega_c \tau)\right\} \qquad (10.55)$$

We have thus demonstrated the close ties between the nth cut of the periodic ambiguity function and the real signal at the output of a narrow BPF centered around nf_m. Note that several different bandpass filters can be implemented using a bank of filters or DFT. The output of each filter will peak at a different delay. As mentioned earlier with regard to triangular FM, the output of the nth filter will peak when

$$\alpha = \pi\Delta f \tau = \frac{n\pi}{2} \qquad (10.56)$$

namely, when

$$\tau = \frac{n}{2\Delta f} \qquad (10.57)$$

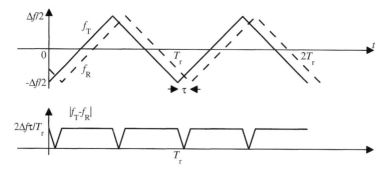

FIGURE 10.29 Frequency difference in triangular FM.

Some intuition regarding the result in (10.57) can be obtained with the help of Fig. 10.29. Its top part presents the instantaneous frequencies: f_T of the transmitted signal and f_R of the signal received (delayed). Its lower part displays the magnitude of the frequency difference (the *beat frequency*), which is the frequency of the signal $x(t)$ in Fig. 10.28.

Note that the instantaneous beat frequency is periodic, with period T_r. (The magnitude looks periodic with period $T_r/2$.) If it was not periodic, frequency analysis using DFT could have yielded any value according to the expression

$$|f_T - f_R| = f_B = 2\Delta f \frac{\tau}{T_r} = 2\Delta f \tau f_m \qquad (10.58)$$

Because of the periodicity and the very long duration ($N \to \infty$), in reality the spectrum analysis of the beat signal can yield only frequency values that are multiples of $f_m = 1/T_r$. The strongest multiple n will be the one closest to

$$\text{round}\left\{\frac{f_B}{f_m}\right\} = \text{round}\{2\Delta f \tau\} = n \qquad (10.59)$$

which agrees with (10.57).

When the signal reflected is Doppler shifted by f_D, the instantaneous frequencies are as displayed in Fig. 10.30. The spectrum (Fig. 10.31) of the beat signal shows that the peaks at nf_m are split into pairs of peaks at $nf_m \pm f_D$. The BPF centered on a particular nf_m needs to be wide enough to pass those Doppler components. Estimating the Doppler shift (and in particular its sign) requires a more complex receiver than the one shown in Fig. 10.28.

In addition to the FM waveforms above, there are other basic modulation waveforms, such as the square wave. More complex waveforms can be constructed using superposition of pairs of basic waveforms (Couch, 1973). Adjusting the amplitude ratios of the summed waveforms can optimize the delay sidelobe reduction.

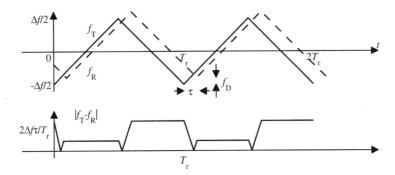

FIGURE 10.30 Frequency difference in triangular FM with Doppler.

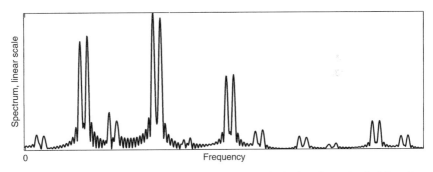

FIGURE 10.31 Mixer output spectrum, triangular FM, signal received delayed and Doppler shifted.

APPENDIX 10A: TEST FOR IDEAL PACF

A simple test can determine if a phase-coded signal with a phase sequence $\{\phi_m\}$, $m = 0, 1, 2, \ldots, M - 1$ exhibits an ideal periodic autocorrelation function (PACF). The test was described by Scholtz and Welch (1978). If the sequence is ideal, it must obey the Fourier transform equation

$$\frac{1}{M}\begin{bmatrix} 1 & 1 & 1 & \cdots & 1 \\ 1 & w & w^2 & \cdots & w^{M-1} \\ 1 & w^2 & w^4 & \cdots & w^{2(M-1)} \\ \vdots & & & \cdots & \vdots \\ 1 & w^{M-1} & w^{2(M-1)} & \cdots & w^{(M-1)^2} \end{bmatrix} \begin{bmatrix} \exp(j\phi_0) \\ \exp(j\phi_1) \\ \exp(j\phi_2) \\ \vdots \\ \exp(j\phi_{M-1}) \end{bmatrix} = \begin{bmatrix} \exp(j\alpha_0) \\ \exp(j\alpha_1) \\ \exp(j\alpha_2) \\ \vdots \\ \exp(j\alpha_{M-1}) \end{bmatrix} \quad (10A.1)$$

where

$$w = \exp\left(-j\frac{2\pi}{M}\right) \quad (10A.2)$$

and there are no restrictions on the resulting $\{\alpha_m\}$, $m = 0, 1, 2, \ldots, M - 1$.

The rationale that led to (10A.1) is as follows. The vector on the right-hand side of (10A.1) is the Fourier transform (spectrum) of the code sequence. A vector whose elements are the magnitude square of the respective spectrum elements, is a vector representing the power spectrum. An inverse Fourier transform of the power spectrum will yield the autocorrelation. If we want a perfect autocorrelation vector $[1 \ 0 \ 0 \ \cdots \ 0]^T$, the power spectrum has to be a unit vector $[1 \ 1 \ \cdots \ 1]^T$. That implies a unit magnitude of each element of the spectrum vector [on the right-hand side of (10A.1)].

PROBLEMS

10.1 PAF theory
Prove equations (10.9), (10.14), and (10.31).

10.2 PAF of Frank code
Plot the PAF of a Frank-coded signal of length 16:
(a) When the number of periods processed is $N = 1$.
(b) When $N = 8$.

10.3 PAF and PACF of periodic LFM
Plot the PACF and the PAF of a periodic LFM signal in which $BT_r = 20$ (B is the total frequency deviation and T_r is the repetition period), and the number of periods processed coherently in the receiver is $N = 8$.

10.4 PAF and PACF of staircase LFM
Consider a periodic LFM signal in which each period T_r is constructed from M staircases each of duration $T_s = T_r/M$, and the frequency spacing between stairs is $\Delta f = B/(M - 1)$. The number of periods processed coherently in the receiver is N (see Fig. P10.4).
(a) Plot the PAF and PACF for the case of $BT_r = 20$, $N = 8$, when $M = 16$.
(b) Repeat part (a) for $M = 8$.

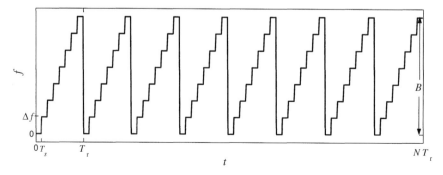

FIGURE P10.4 Frequency evolution of the reference signal.

(c) Repeat part (a) for $M = 6$.
(d) Repeat part (a) for $M = 3$.
(e) Comment and explain the differences. (*Hint*: Find $T_s \Delta f$ for each case.)

10.5 Interperiod weight window in periodic CW
Explain the problem(s) (that are unique to an interperiod weight window) caused by implementing part of the interperiod weight window in the transmitter.

10.6 Mismatch SNR loss of amplitude weighting on receive
Calculate the SNR loss caused by implementing inter- and intraperiod (full) Hamming weight (only in the receiver) to the signal discussed in Problem 10.5. Use the parameters $N = 8$ periods, $M = 16$ steps per period.

10.7 PAF Doppler cuts of triangular FM
Develop the $\chi(\tau, nf_m) = g_n(\tau)$ expressions of a triangular FM CW signal.

10.8 PAF Doppler cuts of square-wave FM
Develop the $\chi_T(\tau, nf_m) = g_n(\tau)$ expressions of a square-wave FM CW signal.

10.9 PAF of square-wave FM
Plot the PAF of a square-wave FM CW signal for the case $\Delta f T_r = 20$, $N = 8$. Extend the Doppler axis of the PAF as far as $\nu = 2.2/T_r$. Zoom the delay axis over $|\tau| \leq T_r/2$.

10.10 Delay–Doppler response of sawtooth FM with weighting on receive
Plot the periodic delay–Doppler response of 16 periods processed coherently of a sawtooth-FM periodic signal with $\Delta f T_r = 20$:

(a) No weighting.
(b) On receipt, interperiod Hamming weighting.
(c) On receipt, intraperiod Hamming weighting.
(d) On receipt, inter- and intraperiod Hamming weighting.

MATLAB m-files cw_sgnl.m and cross_ambfn2.m in the Appendix at the end of the book could be used. Extend the delay scale over two periods and the Doppler scale as far as $1.5/T_r$.

10.11 Delay–Doppler response of periodic P4 with weighting on receive
Repeat Problem 10.10(a) and (b) using a 30-element P4 signal.

10.12 Delay–Doppler response of periodic Golomb biphase with weighting on receipt
Repeat Problem 10.10(a) and (b) using a 15-element Golomb biphase signal.

10.13 Delay–Doppler response of periodic Frank code with weighting on receive

Repeat Problem 10.10 (a) and (b) using a 16-element Frank code.

10.14 Delay–Doppler response of Doppler-shifted receiver

Repeat Problem 10.13 (b) when the receiver is matched to a Doppler shift of $1/(8T_r)$. (*Hint*: In cw_sgn1.m use Dphi*M/pi = 0.25.)

10.15 Range response of sinusoidal-FM Doppler radar

A CW Doppler navigation radar has to operate over a slant range of 1500 to 4500 m. The signal is sinusoidal FM with a deviation of Δf. The radar receiver utilizes the third harmonic of the modulating frequency f_m. Find the combination of Δf and f_m that will yield the best (high and flat) response over the range specified. Plot the response using linear scales. (Ignore the range-dependent path attenuation.)

REFERENCES

Barton, D. K., *CW and Doppler Radar*, Artech House, Bedford, MA, 1978.

Couch, L. W., Effects of modulation nonlinearity on the range response of FM radars, *IEEE Transactions on Aerospace and Electronic Systems*, vol. AES-9, no. 4, July 1973, pp. 598–606.

Freedman, A., and N. Levanon, Properties of the periodic ambiguity function, *IEEE Transactions on Aerospace and Electronic Systems*, vol. AES-30, no. 3, July 1994, pp. 938–941.

Getz, B., and N. Levanon, Weight effects on the periodic ambiguity function, *IEEE Transactions on Aerospace and Electronic Systems*, vol. AES-31, no. 1, January 1995, pp. 182–193.

Ipatov, V. P., et al., Boundaries of the sidelobes of a periodic discrete signal in a broad Doppler band, *Radio Engineering and Electronic Physics*, vol. 29, February 1984, pp. 25–32.

Levanon, N., CW alternatives to the coherent pulse train: signals and processors, *IEEE Transactions on Aerospace and Electronic Systems*, vol. AES-29, no. 1, January 1993, pp. 250–254.

Levanon, N., and A. Freedman, Periodic ambiguity function of CW signals with perfect periodic autocorrelation, *IEEE Transactions on Aerospace and Electronic Systems*, vol. AES-28, no. 2, April 1992, pp. 387–395.

Levanon, N., and B. Getz, Comparison between linear FM and phase-coded CW radars, *IEE Proceedings: Radar, Sonar and Navigation*, vol. 141, no. 4, August 1994, pp. 230–240.

Scholtz, R. A., and L. R. Welch, Group characters: sequences with good correlation properties, *IEEE Transactions on Information Theory*, vol. IT-24, no. 5, September 1978, pp. 537–545.

Skolnik, M. I., *Introduction to Radar Systems*, 2nd ed., McGraw-Hill, New York, 1980.

Varian, R. H., W. W. Hansen, and J. R. Woodyard, Object detecting and locating system, U.S. patent 2,435,615, February 10, 1948.

11

MULTICARRIER PHASE-CODED SIGNALS

In this chapter we introduce a family of multicarrier frequency-, and phase-modulated pulse compression waveforms with thumbtack ambiguity functions. The basic building blocks described in previous chapters (single pulse, pulse train, CW, phase/frequency coding or modulation, etc.) are used and generalized to the multicarrier case. Signal parameters are optimized such that the resulting ambiguity function, complex envelope, and spectrum have favorable properties.

The multicarrier signal was introduced by the authors (Levanon 2000a,b, 2001; Levanon and Mozeson, 2002; Mozeson and Levanon, 2003) and named *multicarrier phase coded* (MCPC). Every multicarrier phase-coded pulse consists of N sequences (phase codes) transmitted *simultaneously* on N carriers. Each sequence contains M phase-modulated bits (chips). The frequency difference between two adjacent carriers is set equal to the inverse of the bit duration t_b yielding orthogonal frequency-division multiplexing (OFDM; see Box 11A).

The general expression for the MCPC pulse complex envelope $g(t)$ is

$$g(t) = \sum_{n=1}^{N} \sum_{m=1}^{M} w_n a_{n,m} s[t - (m-1)t_b] \exp\left[j2\pi \left(n - \frac{N+1}{2}\right)\frac{t}{t_b}\right] \quad (11.1)$$

where w_n is the complex weight associated with the nth carrier, $a_{n,m}$ is the mth element of the sequence modulating carrier n ($|a_{n,m}| = 1$), and $s(t) \equiv 1$ for $0 \leq t < t_b$ and zero elsewhere. Note that $g(t)$ is zero for $t < 0$ or $t > Mt_b$ ($T = Mt_b$ is the pulse length). The structure of the MCPC signal is illustrated schematically in Fig. 11.1.

Radar Signals, By Nadav Levanon and Eli Mozeson
ISBN 0-471-47378-2 Copyright © 2004 John Wiley & Sons, Inc.

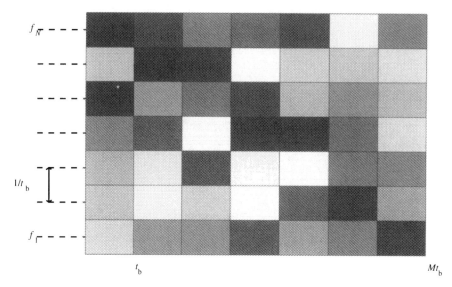

FIGURE 11.1 Structure of a multicarrier phase-coded pulse.

Figure 11.2 shows the complex envelope (magnitude and phase) of a randomized MCPC pulse with $N = 35$ carriers and $M = 25$ chips in each carrier. The normalized autocorrelation function (ACF) and normalized power spectrum density (PSD) are shown in Fig. 11.3. Figure 11.4 shows the ambiguity function (AF). As can easily be observed from Fig. 11.3, the signal's frequency spectrum occupies a width B given by $N/t_b = MN/T$. The signal's time–bandwidth product (TB) can thus be expressed as the product of the number of carriers (N) and the number of chips modulating each carrier (M). Recall that this is consistent with the single-carrier phase-coded signal, where the frequency spectrum bandwidth was shown to equal $1/t_b = M/T$ and the time–bandwidth product was M.

As can easily be noted from the AF plot in Fig. 11.4 and the AF zero-Doppler cut (ACF) in Fig. 11.3, as the time–bandwidth product ($TB = MN$) is higher and when there are no regularities in the randomized pulse structure, the ambiguity function has a shape closer to a thumbtack shape. The delay resolution of the thumbtack ambiguity function with the time–bandwidth product of MN is $T/(MN) = t_b/N$ and the Doppler resolution is $1/T$ (the resolution cell size is thus $1/MN$). The average sidelobe power level relative to the central peak is $1/(MN)$.

A major drawback of the multicarrier signal is its varying amplitude (see Fig. 11.2). If the signal generator contains a power amplifier, it becomes desirable to reduce the peak-to-mean envelope power ratio (PMEPR) as much as possible to avoid operating the power amplifier at a large input backoff (used mainly to avoid out-of-band interference). The power loss involved in not using the power amplifier to its full output power all the time is $-10\ \log_{10}(\text{PMEPR})$. An additional loss is in using a linear power amplifier instead of a saturated amplifier, which is usually more power efficient.

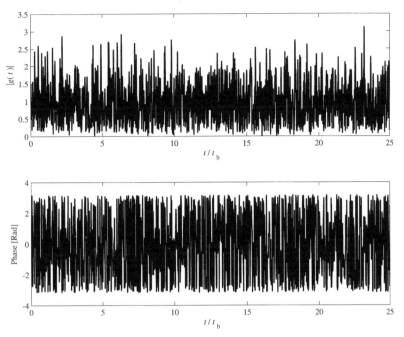

FIGURE 11.2 Complex envelope (magnitude and phase) of a 35-carrier 25-bit MCPC pulse.

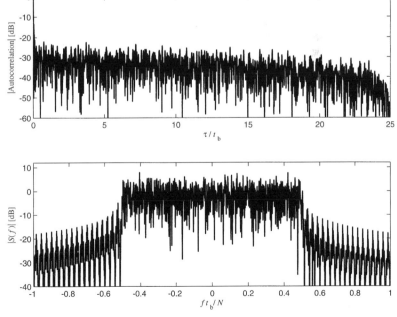

FIGURE 11.3 Autocorrelation (top) and spectrum (bottom) of a 35-carrier 25-bit MCPC pulse.

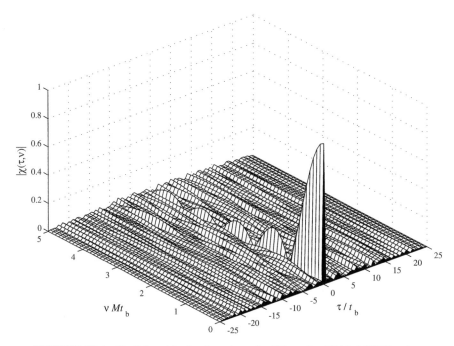

FIGURE 11.4 Partial ambiguity function of a 35-carrier 25-bit MCPC pulse.

The orthogonality of the multicarrier phase-coded signal implies that over a bit duration one carrier has no effect on the others (see Box 11A). Hence, if each carrier is of unit power, the mean power of the N carriers must be N. The instantaneous peak power during a bit can be at most N^2 (obtained when all carriers are with an instantaneous identical phase). When the carriers are amplitude modulated, the mean power of the N carriers is $\sum |w_n|^2$, while the instantaneous peak power during a bit can be at most $(\sum |w_n|)^2$. We conclude, therefore, that in general (randomized multicarrier phase-coded signal), the PMEPR is at most $(\sum |w_n|)^2 / \sum |w_n|^2$, which can be shown to be in general at least N, and equal to N if and only if all carriers are of equal energy (i.e., no carrier amplitude modulation).

Thus, in the process of designing a multicarrier phase-coded pulse the sequences used for the carrier modulation should be picked with extra care such that the signal results in tolerable PMEPR as well as good ambiguity properties.

BOX 11A: Orthogonal Frequency-Division Multiplexing

OFDM is a method of transmitting data simultaneously over multiple equally spaced carrier frequencies, using Fourier transform processing for modulation and demodulation. The method has been proposed for many types of radio systems, such as wireless local area networks and digital audio and video broadcasting. By dividing the bandwidth into many small orthogonal frequencies (efficiently achievable using the fast Fourier transform), the data can be transmitted across multiple narrowband channels having overlapping frequency spectra.

Orthogonality is a mathematical concept derived from the vector representation of time-dependent waveforms. Any two vectors are orthogonal if the cosine of the angle between them is zero (i.e., they are perpendicular to each other). Waveforms are tested for orthogonality in a similar manner. If the waveforms to be compared are laid out on the time axis and the average of the integral of the products of pairs of values for all instances of time extending over their common period is taken, and this average is found to be zero, the waveforms are said to be orthogonal. Thus, orthogonality becomes a broader concept than perpendicularity.

For narrowband waveforms it can be shown that the procedure described above is equivalent to using the complex envelope of the two waveforms, with the exception that the conjugate of one of the signals should be used. In general, two time-dependent narrowband waveforms represented by their complex envelopes $u_n(t)$ and $u_m(t)$ are orthogonal if

$$\frac{1}{t_b} \int_0^{t_b} u_n(t) u_m^*(t) \, dt = 0 \quad (11A.1)$$

for $m \neq n$ over the interval t_b.

In OFDM, the various time-dependent waveforms are selected to lie on carriers separated by the inverse of the signal duration t_b and are given by

$$u_n(t) = \begin{cases} \exp\left(j2\pi n \frac{t}{t_b}\right), & 0 \leq t < t_b \\ 0, & \text{otherwise} \end{cases} \quad (11A.2)$$

Note that for each subcarrier a rectangular pulse shaping is applied. Orthogonality of these waveforms is shown simply by using (11A.2) in (11A.1), yielding

$$\frac{1}{t_b} \int_0^{t_b} u_n(t) u_m^*(t) \, dt = \frac{1}{t_b} \int_0^{t_b} \exp\left[j2\pi(n-m)\frac{t}{t_b}\right] dt = \begin{cases} 1, & m = n \\ 0, & m \neq n \end{cases} \quad (11A.3)$$

since the integrand is 1 for $m = n$ and for $m \neq n$ the integral is over full periods of cosine and sine functions. Each subcarrier can be modulated independently with the complex modulation symbol s_n, where the subscript n refers to the number of the subcarrier. Thus within the signal duration the following waveform is formed:

$$g(t) = \sum_{n=1}^{N} s_n u_n(t) \tag{11A.4}$$

Due to the rectangular pulse shaping of the signal, the spectra of the subcarriers are $(\sin x)/x$ functions with a first null at the inverse of the signal duration t_b. In a practical application the OFDM signal $g(t)$ is generated in a first step as a discrete-time signal in the digital signal processing part of the transmitter. As the bandwidth of the OFDM system is N/t_b, the signal must be sampled with the sampling time t_b/N. The samples of the signal are written as g_k, $k = 1, 2, \ldots, N$, and can be calculated as

$$g_k = \frac{1}{\sqrt{N}} \sum_{n=1}^{N} s_n \exp\left(j2\pi k \frac{n}{N}\right) \tag{11A.5}$$

The equation describes the inverse discrete Fourier transform, which is typically realized by an inverse FFT (IFFT).

The spectra of the subcarriers overlap, but the subcarrier signals are mutually orthogonal and the modulation symbols s_n can be recovered by a correlation with $[u_n(t)]^*$ [the conjugate of $u_n(t)$]. Alternatively, the correlation at the receiver can be realized as the DFT or an FFT, respectively:

$$\hat{s}_n = \frac{1}{\sqrt{N}} \sum_{k=1}^{N} r_k \exp\left(-j2\pi k \frac{n}{N}\right) \tag{11A.6}$$

where r_k is the kth sample of the received signal $r(t)$.

Equation (11A.6) implies perfect synchronization at the receiver, which is an important topic in OFDM transmission systems, since time and frequency synchronization errors disturb the orthogonality of the subcarriers. For the application described here (radar), the sensitivity to exact synchronization is beneficial since the receiver knows what were the "data" transmitted and measures the delay (time) and Doppler (frequency) shifts between the signal transmitted and the echo received.

11.1 MULTICARRIER PHASE-CODED SIGNALS WITH LOW PMEPR

The problem of minimizing the peak/mean envelope power ratio (PMEPR) of a multicarrier (or multitone) waveform has been addressed mostly in the context of instrumentation and measurement or in the context of OFDM communication systems. In the field of instrumentation and measurement, it is common to test a system by injecting a predetermined multicarrier signal into the system's input and from looking at the output, learn about the system transfer function. It is often required that the PMEPR of the input signal be as low as possible (e.g., to avoid saturations or other nonlinearity of the system being tested).

To be able to use multicarrier signals in real radar systems it is often necessary to find not a few but families of multicarrier signals that have favorable PMEPR and attractable range–Doppler behavior. The design principle is based on restricting the search to families of multicarrier waveforms that are robust in having low PMEPR while changing the waveform-free parameters such that the other signal properties meet the design requirements.

Two multicarrier families are described. The first design is based on using an identical sequence (IS) to modulate all carriers. The PMEPR is lowered by proper amplitude and phase modulating of the various carriers (the same carrier phase and amplitude is used for all bits). The second design is based on using consecutive ordered cyclic shifts (COCS) of a chirplike ideal sequence to modulate all carriers (see Chapter 6 for details on the variants and properties of the chirplike ideal sequences).

11.1.1 PMEPR of an IS MCPC Signal

When using identical sequences (IS) to modulate all carriers in a multicarrier phase-coded pulse or CW signal, the real envelope (i.e., the magnitude of the complex envelope) of the multicarrier signal is identical for all bits, independent of the sequence used to modulate all carriers and is a function only of the carrier amplitude and phase weighting used. The problem of minimizing the peak-to-mean envelope power ratio (PMEPR) becomes a problem of minimizing the PMEPR of a multicarrier bit.

With all carriers modulated by the same sequence, we can write the complex envelope of the complex mth multicarrier bit as

$$g_m(t) = a_m \sum_{n=1}^{N} w_n \exp\left[j2\pi \left(n - \frac{N+1}{2}\right)\frac{t}{t_b}\right] \quad (m-1)t_b \leq t < mt_b \quad (11.2)$$

where w_n is the complex weight associated with the nth carrier and a_m is the mth element of the sequence modulating all carriers ($|a_m| = 1$). Note that $g_m(t)$ is zero for $t < (m-1)t_b$ or $t \geq mt_b$ (t_b is the bit length). The real-valued multicarrier signal is derived from (11.2) according to

$$\tilde{g}_m(t) = \text{Re}\{g_m(t)\exp(j2\pi f_c t)\} \quad (m-1)t_b \leq t < mt_b \quad (11.3)$$

where f_c is the central carrier frequency.

The *crest-factor* (CF) of the complex multicarrier signal $g_m(t)$ is defined by

$$\text{CF} = \frac{\max_{(m-1)t_b \leq t < mt_b}\{|g_m(t)|\}}{\sqrt{(1/t_b)\int_{(m-1)t_b}^{mt_b}|g_m(t)|^2\,dt}} \quad (11.4)$$

Since $|a_m| = 1$, the CF is the same for all bits (not a function of m). The *peak-to-mean envelope power ratio* (PMEPR) is defined as the square of the crest factor. Note that what is usually of interest is the peak-to-average power ratio (PAPR), defined as the ratio of the peak power to the average power of the real-valued multicarrier signal, and can be lower than the PMEPR since the actual peak of the real signal can be lower than the peak of the complex envelope. However, for low-bandwidth signals (i.e., $f_c \gg N/t_b$) the PMEPR gives a good approximation for finding the true PAPR.

In the example shown in Fig. 11.5, the peak envelope power is 1.16, while the peak signal power is 1.12. The mean envelope power is 0.62, yielding a PMEPR value of 1.87 (1.16/0.62). The mean signal power is approximately the mean envelope power yielding a PAPR value of 1.84, which is slightly lower than the PMEPR. Several phasing schemes for designing a carrier initial phase such that the PMEPR of a multicarrier signal is low were given in closed form by Newman (1965), Schroeder (1970a), Boyd (1986), Friese (1997), and Narahashi and Nojima (1997). The advantage of closed-form solutions is that the construction rule allows for easily generating multicarrier signals with a varying number of carriers. Their common disadvantage is that the resulting PMEPR is low, but not really close to the optimum value achievable. The effect of using different carriers phasing methods on the multicarrier bit amplitude and PMEPR is demonstrated in Box 11B.

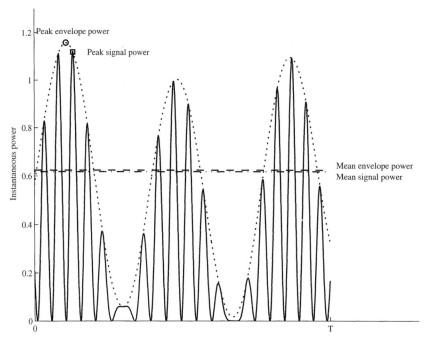

FIGURE 11.5 Peak envelope power and peak power of a varying-amplitude narrowband waveform.

BOX 11B: Closed-Form Multicarrier Bit Phasing with Low PMEPR

Assume a single multicarrier bit based on N carriers and duration t_b. The phasing scheme usually refereed to as *Newman phases* is given by

$$\theta_n = \frac{(n-1)^2}{N}\pi, \qquad n = 1, 2, \ldots, N \qquad (11\text{B}.1)$$

where the amplitudes of all carriers are assumed identical. Numerical investigation of the crest factor of the Newman phases shows that it is always very small, about 4.6 dB for moderate N, and decreasing slightly for larger N.

Schroeder (1970a,b) introduced a carrier phasing approach based on the simple intuitive concept concerning the asymptotic relationship between the power spectra of frequency-modulated signals and their instantaneous frequency described in Chapter 5. Using Schroeder's method, the carrier phase values as a function of the carrier amplitude weights are given by

$$\theta_n = -2\pi \sum_{l=1}^{n-1}(n-l)|w_l|^2, \qquad n = 1, 2, \ldots, N \qquad (11\text{B}.2)$$

where $|w_n|$ are the carrier amplitudes and are normalized such that $\sum |w_n|^2 = 1$. When all carrier amplitude weights are identical (no amplitude weighting—flat spectrum), the weight phases are given by

$$\theta_n = \pi\frac{n^2}{N}, \qquad n = 1, 2, \ldots, N \qquad (11\text{B}.3)$$

Note that the Schroeder phases (11B.3) and the Newman phases (11B.1) give rise to the same PMEPR. The two methods differ only in a linear-phase term that has no effect on the PMEPR.

A third method, proposed by Narahashi and Nojima (1997), also applies only to the flat spectrum case but gives better results than the Newman phases. The Narahashi–Nojima phases are given by

$$\theta_n = \frac{(n-1)(n-2)}{N-1}\pi, \qquad n = 1, 2, \ldots, N \qquad (11\text{B}.4)$$

The PMEPRs of multicarrier bits having equal carrier amplitudes and phase modulated using the three methods described above are summarized in Table 11.1. Note that the PMEPR is lower or equal to 2 (= 3.01 dB) for the Narahashi–Nojima phasing. For tone numbers less than or equal to 29, the Narahashi–Nojima phasing method gives a lower PMEPR value than do Newman and Schroeder phasings, while for higher values of N, Newman and

TABLE 11.1 PMEPR Comparison with No Carrier Amplitude Weighting

N	Newman and Schroeder	Narahashi–Nojima	N	Newman and Schroeder	Narahashi–Nojima
2	2.0000	2.0000	22	1.8687	1.8542
3	2.3333	1.6667	23	1.7895	1.9029
4	2.0000	1.8171	24	1.8619	1.8544
5	1.7336	1.9667	25	1.8324	1.8914
6	2.0152	1.8101	26	1.8419	1.8839
7	1.9277	1.8618	27	1.8532	1.8713
8	1.7936	1.9455	28	1.8149	1.8945
9	1.9541	1.7916	29	1.8565	1.8466
10	1.8478	1.9034	30	1.8089	1.8927
11	1.8546	1.9165	31	1.8486	1.8640
12	1.9172	1.8289	32	1.8347	1.8828
13	1.7636	1.9223	64	1.8200	1.8703
14	1.8878	1.8667	128	1.8190	1.8544
15	1.8665	1.8858	256	1.8193	1.8418
16	1.8278	1.9111	512	1.8150	1.8364
17	1.8869	1.8365	1,024	1.8163	1.8297
18	1.7937	1.9098	16,384	1.8156	1.8194
19	1.8681	1.8685	32,768	1.8156	1.8182
20	1.8518	1.8868	65,536	1.8156	1.8175

Schroeder phasings are better. Finally, note that as the tone number increases, PMEPR variation decreases and its value converges to 1.8156 (2.59 dB).

In the case of varying carrier amplitudes, Schroeder phasing gives better results than either the Newman or the Narahashi–Nojima phasing methods. In that case, the PMEPR is a function of both N and carrier amplitude.

The Schroeder carrier phasing approach is the only one of those presented in Box 11B that takes carrier amplitude weighting into consideration. Slightly extending the definition of the original Schroeder phasing, the carrier phase values as a function of the carrier amplitude weights are given in their most general form by

$$w_n = |w_n| \exp\left[-j2\pi \sum_{l=1}^{n-1}(n-l)|w_l|^2 - j2\pi\lambda\frac{n}{N} - j\varphi_1\right], \qquad n = 1, 2, \ldots, N$$

(11.5)

where w_n are normalized such that $\sum |w_n|^2 = 1$, φ_1 is an arbitrary phase with no effect on PMEPR, and $0 \leq \lambda < N$. Note that the additional linear term in n, with λ acting as the slope parameter, constitutes a pure cyclic delay to the multicarrier bit complex envelope and does not change the PMEPR. However, changing λ does affect the correlation function of the multicarrier signal and can be used as a design parameter.

When all carrier amplitude weights are identical (no amplitude weighting—flat spectrum), the weights are given by

$$w_n = \frac{1}{\sqrt{N}} \exp\left(-j\pi \frac{n^2}{N} - j2\pi\lambda \frac{n}{N} - j\varphi_1\right), \qquad n = 1, 2, \ldots, N \qquad (11.6)$$

Figure 11.6 presents an example of the multicarrier bit complex envelope using various phasing techniques on a 15-carrier bit, with carriers amplitude-modulated following a square root of the Hamming window rule. The figure also shows the case of identical carrier phases, which yields the worst PMEPR. Note the significant improvement introduced by adding carrier phasing. The PMEPR without carrier phasing is 13, and the PMEPR using the various carrier phasing methods is 2.15 (Newman), 1.91 (Narahashi–Nojima), and 1.55 (Schroeder). Adding a linear phase term in the form of $n\pi$ (where n is the carrier index) to the multicarrier bit with no carrier phasing could be used to shift the image cyclically in

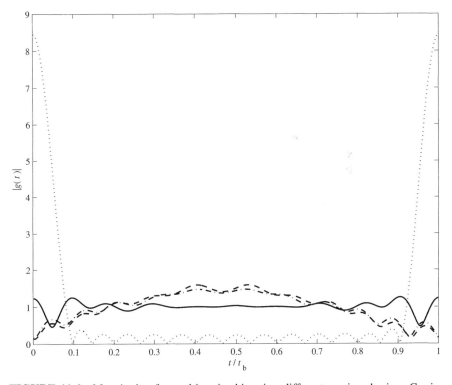

FIGURE 11.6 Magnitude of a multicarrier bit using different carrier phasing. Carrier amplitudes are modulated using a square root of Hamming weight. Dotted, identical carrier initial phase; dashed, Newman phases; dash–dotted, Narahashi–Nojima phases; solid, Schroeder phases.

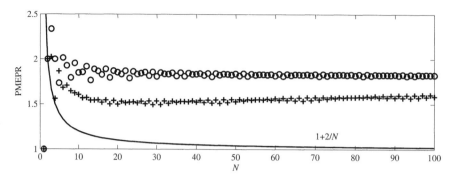

FIGURE 11.7 PMEPR of a multicarrier bit. Carrier initial phase calculated using Schroeder's method. Circles, identical carrier amplitudes; plus signs, carriers' amplitude-modulated using a square root of Hamming weight.

Fig. 11.6 such that the bit exhibits low values, close to the starting and ending points ($t = 0$ and $t = t_b$), and the peak is obtained at the center of the bit.

The PMEPR of a multicarrier bit with no carrier amplitude weight (circles) and with a square root of Hamming carrier amplitude weight (plus signs) as a function of N (number of carriers) using Schroeder's method is shown in Fig. 11.7. The curve in Fig. 11.7 is a lower bound on the PMEPR when no carrier amplitude weight is used (which is $1 + 2/N$). Note that the closed-form solution yields a relatively low PMEPR but it is far from the theoretical lower bound.

To lower the PMEPR further, some iterative algorithms can be used. An example of such algorithms is the time-frequency switching and clipping approach described by Van-Der Ouderaa et al. (1998). The results obtained with this algorithm were reported to be excellent compared with other iterative algorithms. The freeware MATLAB procedure `msinclip.m` (a part of the frequency-domain identification toolbox) can be used for implementing the algorithm. The iterative clipping algorithm implemented in MATLAB is designed to lower the crestfactor of a baseband signal. To lower the PMEPR of the complex envelope of the signal, it is necessary to add a carrier term artificially.

An example of the envelope (absolute value and phase) of a single MCPC bit with $N = 25$ carriers using a square root of Hamming carrier amplitude weighting and Schroeder (solid) or iterative clipping (200 iterations, dashed) carrier phasing is given in Fig. 11.8. The PMEPR is 1.51 for Schroeder carrier phasing and 1.22 using iterative clipping. The single-bit phase exhibits a close-to-quadratic shape; note that in Fig. 11.8 only the deviation from the quadratic term of the phase was plotted.

We could now proceed and describe a multicarrier pulse, CW signal, and pulse train with low PMEPR based on using the carrier phasing approach described here. Instead, a different method for designing low PMEPR multicarrier signals is presented next. Examples of how to use both methods for designing MCPC signal are addressed later.

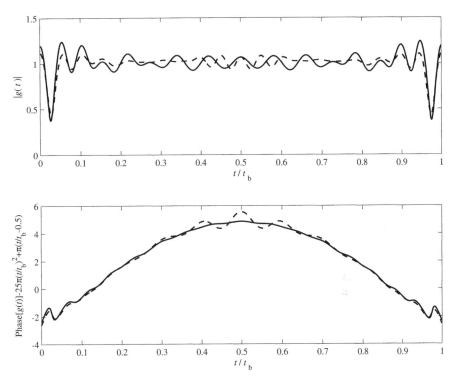

FIGURE 11.8 Magnitude (top) and phase (bottom) of a single multicarrier bit employing 25 carriers. Carrier amplitude is modulated using a square root of Hamming weight. Solid, Schroeder phases; dashed, iterative clipping (200 iterations).

11.1.2 PMEPR of an MCPC Signal Based on COCS of a CLS

In Section 11.1.1 we demonstrated how to lower the PMEPR by using carrier phases with quadratic dependence on n. Another design, yielding close to quadratic carrier phase dependence, is now introduced. The design is based on modulating all N carriers with consecutive ordered cyclic shifts (COCS) of an ideal chirplike sequence (CLS) of length M. The MCPC design will be named MCPC based on a COCS of a CLS or just a COCS of a CLS MCPC.

When using COCS of any ideal sequence (not necessarily a chirplike ideal sequence), $N = M$, and no carrier weighting, the N sequences modulating all N carriers form a complementary set. It can be shown that for this constellation the autocorrelation function is zero for integer multiples of the bit duration t_b. In general, N can be different from M, and for any M there are several (nonequivalent) ideal sequences with chirplike phase behavior. For any N and M the PMEPR of a MCPC pulse based on COCS of a chirplike ideal sequence depends on the variant of the ideal sequence used, the direction of the consecutive ordered cyclic shifts (upward or downward), and the initial phase (or cyclic shift) of the ideal sequence used.

The PMEPR as a function of N and M using a P4 sequence (basic variant of a CLS) is shown in Fig. 11.9. The minimal PMEPR using the optimal chirplike ideal sequences (different variant) can take even lower values. Note that the area where $N \cong M$ has, on average, lower PMEPR values than for other selections of N and M (i.e., $N \gg M$ or $M \gg N$). Even so, low PMEPR values (and sometimes even lower than for $N = M$) are possible for $N \neq M$ (e.g., for low values of N).

For $N = M$ the chirplike ideal sequence that yields the lowest PMEPR is P4. The PMEPR for $N = M$ approached 1.825 when $N = M$ is large and is lower than 2.04 for $N = M > 4$. Note also that the local minimum around $N = M$ is quite flat (the second derivative is low), so we can easily state that for $N/M \cong 1$ (N close to M) and for large compression ratios (large N and M), the PMEPR using the consecutive order cyclic shifts of P4 will give a PMEPR close to 1.825.

The results displayed in Fig. 11.9 were limited to the situation where no carrier weighting is used. Using carrier amplitude weighting changes the PMEPR and in some situations will also change the optimal chirplike ideal sequence for a specific selection of N and M. Figure 11.10 shows the minimal PMEPR (optimal chirplike ideal sequence) as a function of N and M, where the carriers are amplitude modulated using a square root of Hamming weight.

Comparing Fig. 11.10 (square root of Hamming carrier amplitude weight) with Fig. 11.9 (no carrier amplitude weight) reveals that using carrier amplitude weighting changes the area where PMEPR is globally minimal (i.e., instead of $N = M$ with no carrier weighting, we get a low area for $N \cong 1.5M$ using a square root of Hamming carrier amplitude weight). In hindsight this is no surprise, since amplitude-modulating the carriers implies a lower effective number of carriers.

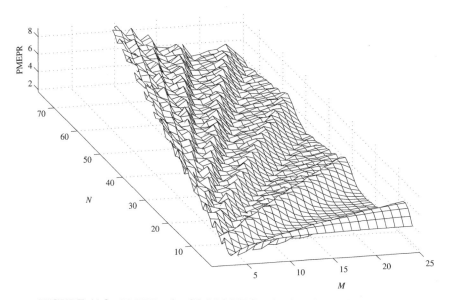

FIGURE 11.9 PMEPR of a COCS MCPC pulse based on a P4 sequence.

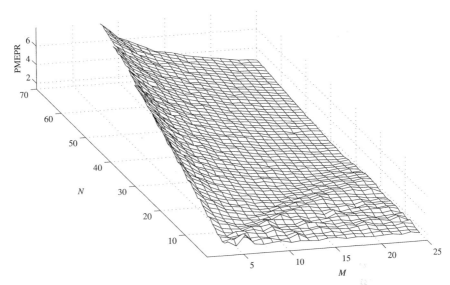

FIGURE 11.10 Minimal PMEPR of a COCS MCPC pulse based on the optimal CLS. Carrier amplitudes are modulated using a square root of Hamming weight.

Amplitude-modulating the carriers has another important effect. The PMEPR values obtained with carrier amplitude weighting for the optimal selection of N are lower (all around 1.5) than the PMEPR values of the optimal selection of N ($N = M$) with no carrier amplitude weighting (PMEPR around 1.8).

We are now ready to design a low PMEPR MCPC pulse, CW signal, and pulse train using the identical sequence approach described in Section 11.1.1 and the consecutive ordered cyclic shifts of a chirplike ideal sequence approach described above.

11.2 SINGLE MCPC PULSE

It can be shown (see Problem 11.3 for details) that the autocorrelation function (ACF) magnitude of a single MCPC pulse can be written explicitly as

$$|R(it_b + \eta)| = \left| \sum_{l=1}^{N} \exp\left[j2\pi l \frac{\eta}{t_b} \right] \right.$$

$$\left. \times \sum_{n=1}^{N} w_n w_l^* \left[I_1 \sum_{m=1}^{M} a_{n,m} a_{l,m-i-1}^* + I_2 \sum_{m=1}^{M} a_{n,m} a_{l,m-i}^* \right] \right| \quad (11.7)$$

where $\tau = it_b + \eta$, i is an integer, $0 \leq \eta < t_b$, and where I_1 and I_2 are given by

$$I_1 = \eta \operatorname{sinc}(\beta) \exp(j\beta) \quad \text{where} \quad \beta = \pi(n-l)\frac{\eta}{t_b} \quad \text{and} \quad I_2 = t_b \, \delta(n-l) - I_1$$
$$(11.8)$$

and where we define $a_{n,m}$ as zero for "illegal" values of m (i.e., $m > M$ or $m < 1$). Note that the sum over m is the aperiodic cross-correlation of the sequences modulating carrier n and carrier l at delay i and $i+1$.

11.2.1 Identical Sequence

Using (11.7), a MCPC pulse based on a single sequence a_m modulating all carriers can be shown to have an autocorrelation function (ACF) magnitude given by

$$|R(it_b + \eta)| = \left| \sum_{l=1}^{N} \exp\left[j2\pi l \frac{\eta}{t_b} \right] \right.$$
$$\left. \times \sum_{n=1}^{N} w_n w_l^* \left[I_1 \sum_{m=1}^{M} a_m a_{m-i-1}^* + I_2 \sum_{m=1}^{M} a_m a_{m-i}^* \right] \right| \quad (11.9)$$

where $\tau = it_b + \eta$, i is an integer, $0 \leq \eta < t_b$, and where I_1 and I_2 were given in (11.8) and a_m is defined to be zero for illegal values of m (i.e., $m > M$ or $m < 1$).

Note that the ACF is a function of the complex carrier weight and the sequence used for modulating all carriers. Note also that any carrier phasing used is not unique since adding a linear carrier phase term given by $\exp(-j2\pi\lambda n/N)$ results in no change to the PMEPR but can change the ACF sidelobe pattern. The free parameters of design are $N-1$ carrier weights (although complex, the phase and amplitude are related through the phasing method used to lower the PMEPR), λ (the linear carrier phase term slope), and $(M-1)$ sequence phase values.

The ACF magnitude at integer multiples of the bit length $\tau = it_b$ is given by substituting $\eta = 0$ in (11.9), yielding

$$|R(it_b)| = \left| t_b \sum_{l=1}^{N} \sum_{n=1}^{N} w_n w_l^* \delta(n-l) \sum_{m=1}^{M} a_m a_{m-i}^* \right|$$
$$= t_b \sum_{n=1}^{N} |w_n|^2 \left| \sum_{m=1}^{M} a_m a_{m-i}^* \right| = t_b \left| \sum_{m=1}^{M} a_m a_{m-i}^* \right| \quad (11.10)$$

where we assumed that the signal is normalized such that average power is 1 ($\sum |w_n|^2 = 1$). Note that the sum over m is the aperiodic autocorrelation function of the sequence a_m and that the result is independent of the complex carrier weight.

Recall that for a single carrier signal the ACF between integer multiples of the bit duration t_b is obtained by connecting the values at integer multiples of t_b using straight lines (see Chapter 6). In the multicarrier case we still have a weighted sum of the values at the peaks [see equation (11.9)], but the connection is no longer linear. Minimum peak sidelobe is obtained by using minimum aperiodic correlation peak sidelobe codes for a given length M (e.g., the Barker code

SINGLE MCPC PULSE

introduced in Chapter 6). Note that the ACF peaks minimal level is a function of M (and not the compression ratio MN).

An example showing the autocorrelation function of a MCPC pulse based on a Barker sequence of length 13 with $N = 7$ carriers or $N = 15$ carriers is shown in Fig. 11.11. The carriers were not amplitude-modulated and the carrier phase was calculated using Schroeder's method with no additional linear phase ($\lambda = 0$). The figure also shows the ACF of the single-carrier signal ($N = 1$). For the three signals the PMEPR value was 1 (the single-carrier case), 1.928 (for $N = 7$), and 1.87 (for $N = 15$). Note that the ACF at integer multiples of t_b is independent of the number of carriers used (N). Increasing N yields narrower peaks and lower sidelobes between the peaks.

The power spectral density (PSD) of the 15-carrier 13-bit MCPC pulse based on modulating all carriers with a 13-element Barker sequence is shown in Fig. 11.12. Note the relatively flat central area of the frequency spectrum, which is a result

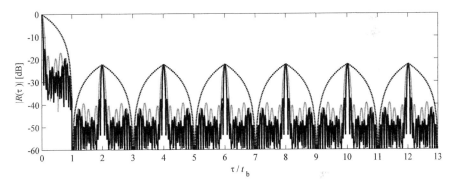

FIGURE 11.11 ACF of a 13-bit MCPC pulse based on modulating all carriers with a Barker sequence. All carriers have equal amplitude. The initial carrier phase is set using Schroeder phases. Dashed line, $N = 1$; dotted line, $N = 7$; solid line, $N = 15$.

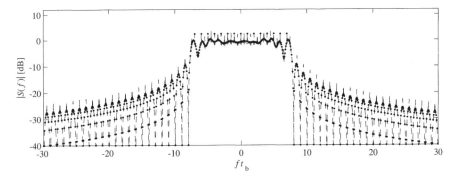

FIGURE 11.12 Frequency spectrum of a 15-carrier 13-bit MCPC pulse based on modulating all carriers with a Barker sequence. The spectrum was calculated every $1/T$ (emphasized) and every $1/16T$. All carriers have equal amplitudes. Schroeder carrier phasing.

of not amplitude-modulating the carriers. The spectrum shown in Fig. 11.12 was calculated by taking the DFT of the signal with zero padding the signal to 16 times its original length (dashed) and without zero padding (darker dots). Thus, the power spectrum is given for points spaced $1/(16T) = 1/(16Mt_b)$ apart. Note that changing M or the sequence type has little effect on the spectrum shape. Increasing N widens the spectrum mainlobe (normalized to the bit length).

The effect of carrier amplitude weighting on the ACF is demonstrated in Fig. 11.13. The figure shows the ACF (top) and ACF mainlobe area (bottom) of the MCPC pulse having $N = 26$ carriers, all modulated by the same 13-element Barker sequence. The dotted line shows the case where all carriers have equal amplitude, while the solid line shows the result of amplitude modulating the carriers. A square root of Hamming carrier amplitude weight was used. For both cases the Schroeder phasing method was used for minimizing the PMEPR. The resulting PMEPR is 1.84 when equal carrier amplitudes are used and 1.52 when amplitude-modulating the carriers. In both cases no additional linear carrier phase term was added ($\lambda = 0$).

Note how weighting increases the mainlobe and secondary lobes widths while lowering the sidelobe level close to the mainlobe and especially close to the center

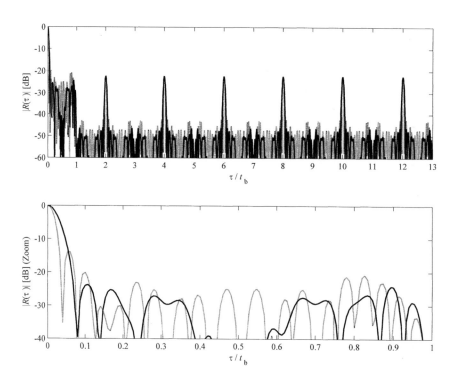

FIGURE 11.13 ACF (top) and ACF mainlobe zoom (bottom) for a 26-carrier 13-bit MCPC pulse based on modulating all carriers using a Barker sequence. Schroeder carrier phasing ($\lambda = 0$). Dashed, equal carrier amplitudes; solid, carriers amplitude modulated using a square root of Hamming weight.

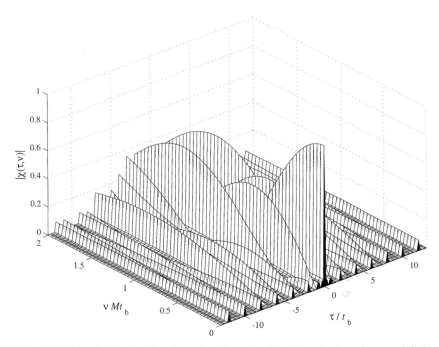

FIGURE 11.14 Ambiguity function of a 26-carrier MCPC pulse based on modulating all carriers with a 13-bit Barker sequence. Carriers were amplitude modulated using a square root of Hamming weight. Schroeder phases used for a carrier's initial phase.

of the bit. Using iterative clipping instead of Schroeder's method to calculate the carrier phase gives lower PMEPR and slightly different ACF values. Many different numerical methods can be used for designing optimal carrier amplitude weights such that the sidelobe level is lowered and the mainlobe width is limited. All those methods cannot, however, lower the sidelobe level at integer multiples of the bit duration. The partial ambiguity function of the MCPC pulse based on modulating all 26 carriers with the same 13-element Barker sequence is shown in Fig. 11.14. Carriers are amplitude-weighted using a square root of Hamming amplitude weighting, and the carrier phase is calculated using Schroeder's method to lower PMEPR.

Note that the first recurrent lobe in Doppler is expected at $\nu = 1/t_b$ ($\nu M t_b = M = 13$) and is not shown in Fig. 11.14. The level of the peak is not high due to the carrier weighting used (amplitude and phase). Note also that most of the ambiguity volume is concentrated in a strip around zero delay ($-t_b \leq \tau \leq t_b$) and around delays equal to integer multiples of the bit duration (including the odd multiples of the bit length for which the ACF is zeroed).

11.2.2 MCPC Pulse Based on COCS of a CLS

Figure 11.15 shows the ACF (top) and magnitude (bottom) of a MCPC pulse based on COCS of a CLS with 15 elements (bits). Two possible MCPC designs

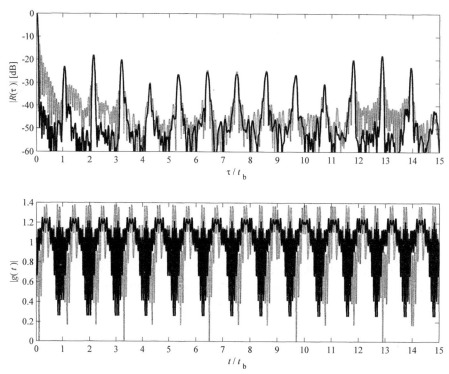

FIGURE 11.15 ACF (top) and magnitude (bottom) of a 15-bit COCS of P4 MCPC pulse. Dashed, $N = 14$, equal carrier amplitudes; solid: $N = 21$, carrier amplitude-modulated using a square root of Hamming weight.

are shown: (1) a MCPC pulse with identical carrier amplitudes (dashed), and (2) a MCPC pulse with carriers amplitude modulated using a square root of Hamming weight (solid). In either case the number of carriers (N) is selected such that the PMEPR is minimized ($N = 14$ in the identical carrier amplitude case and $N = 21$ with carriers amplitude modulated using a square root of Hamming weight). In both cases the effective bandwidth is very close and the ACF mainlobe width is almost identical. However, as can be seen from the figure, increasing N and amplitude-modulating the carriers yields lower ACF sidelobes and lower PMEPR.

For both cases the P4 version of the CLS was used. In both cases, this selection yields minimum PMEPR and minimum PSLL (for some different CLS, the PMEPR can be the same but the PSLL different). The PMEPR was 1.78 for the first case (identical carrier amplitudes) and 1.48 with amplitude modulating of the carriers. Both MCPC pulses were consecutive downward ordered (i.e., $a_{N,1} = a_{N-1,2} = a_{N-2,3} = \cdots$). For some cases a different CLS variant can be selected (suboptimal from PMEPR point of view) to yield lower PSLL with only little increase in the PMEPR. Note also that using a suboptimal value of N can result in a similar decrease to PSLL with only little increase in the PMEPR.

The autocorrelation shape of the signal with amplitude-modulated carriers exhibits a shape similar to a shape without carrier amplitude weighting but with a reduction in the sidelobe level between the peaks and an increase in the secondary peaks width (the secondary peaks level remains the same). Using a square root of Hamming carrier amplitude weight mainly lowers the sidelobes within the first bit.

Note that in case $N = M$ and all carriers are of identical amplitude, the N sequences modulating all N carriers form a complementary set and the ACF is zero at integer multiples of the bit duration t_b. When $N \neq M$ or when the carriers are amplitude-modulated, the (amplitude-weighted) sequences no longer form a complementary set and the ACF is not necessarily zero at integer multiples of the bit duration t_b.

Increasing M (and N) while keeping the general signal structure the same (i.e., $N \cong M$ for the equal-carrier-amplitude case or $N \cong 1.5M$ for a square root of Hamming carrier amplitude weighting) changes only the details of sidelobe pattern (not the overall picture) and has a limited effect on the sidelobe peak value. The main influence is a reduction in the integrated sidelobe level and in the mainlobe and sidelobes width, which is lowered as the sequence length (M) and the number of carriers N increases.

Figure 11.16 shows the PSLL (excluding near sidelobes in the first bit) and PMEPR when $N = M$ and using P4 sequence (minimum PMEPR) as a function of M. All carriers were assumed to have equal amplitudes. Note how the PMEPR approaches 1.825 and the PSLL is lowered when M is increased. The PSLL is approximately given by $-10 \log (4M)$. Examining the behavior of the PSLL and PMEPR with carrier amplitude weighting reveals that the PMEPR is lower with carrier amplitude weighting (higher N) and that the PSLL behavior is the same as with identical carrier amplitudes. Note that from observing the PSLL for $N = M$ (identical carrier amplitudes) for different variants of the CLS, it can be realized that P4 gives not only the minimal PMEPR but also the minimal PSLL.

Figure 11.17 shows the frequency spectra of the MCPC pulses based on COCSs of a chirplike ideal sequence. The top part of Fig. 11.17 shows the case of

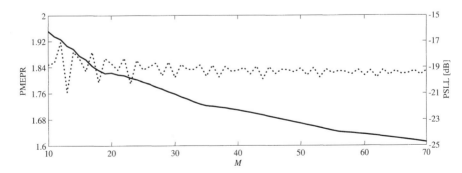

FIGURE 11.16 PSLL (excluding the first bit) and PMEPR (dotted) of a MCPC pulse based on COCS of P4. $M = N$. Identical carrier amplitudes.

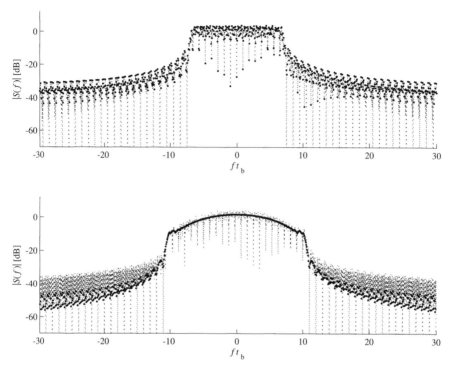

FIGURE 11.17 Frequency spectra of a MCPC pulse based on COCS of a 15-bit CLS calculated every $1/T$ (emphasized) and every $1/16T$. Top, $N = 14$, equal carrier amplitudes; bottom, $N = 21$, carrier amplitude modulated using a square root of Hamming weight.

14 equal-amplitude carriers. The lower part of Fig. 11.17 shows the case of the 21 amplitude-modulated carriers (a square root of a Hamming window was used). Both signals have their carriers phase-modulated using the COCS of a P4 sequence of length 15. The spectrum shown in Fig. 11.17 was calculated by taking the DFT of a signal, with zero padding the signal to 16 times its original length (dashed) and without zero padding (darker dots). Thus the power spectrum is given for points spaced $1/(16T) = 1/(16Mt_b)$ apart. Although not shown here, the spectrum samples at integer multiples of $1/T$ are independent of the ideal sequence used (this will be clarified when discussing the CW MCPC signal).

Notice the relatively flat central part of the spectrum (shaped to a square root of Hamming weight in the lower part of the figure) and the relatively low sidelobes. The spectrum sidelobes are much lower than those of a 15×13 MCPC pulse based on modulating all carriers with the same Barker sequence (see Fig. 11.12). Note that using a different CLS variant (instead of P4) yields a frequency spectrum picture that is not as clean as the one shown in Fig. 11.17. Note the payment in using a square root of Hamming weight to lower the ACF sidelobes, while maintaining the same ACF mainlobe width, is in the increase in PSD mainlobe width (increased number of carriers), although the spectrum sidelobe level remains practically the same. The COCS signal exhibits a power spectrum drop at every $\sim 1/t_b$ in the central part of the power spectrum (flat area) and nulls every $1/t_b$ outside the area of the mainlobe. The nulls outside the mainlobe are shifted $1/2t_b$ relative to the nulls in the mainlobe area.

Finally, note that a square root of Hamming weight or any other classic windowing method is not optimal in the classical sense when used to modulate the amplitudes of the MCPC pulse carriers. In the search for an optimal weight for the MCPC pulse, one has to use an exhaustive search or simply try some classical weights (optimize the free parameters, such as β for a Kaiser window or the sidelobe pedestal for Chebyshev) and select the window that most closely suits a given application.

The structure and properties of the ambiguity function of a multicarrier pulse based on COCS of a chirplike ideal sequence are demonstrated using Fig. 11.18. The figure shows the ambiguity function of a COCS of a CLS MCPC pulse for $0 \leq \nu \leq N/T$ and $|\tau| < Mt_b$. The MCPC pulse used for the plot is based on modulating 15 identical amplitude carriers with downward consecutive ordered cyclic shifts of a 15-element P4 sequence.

The diagonal ridge passing through the origin is a result of using the P4 chirplike ideal sequence. Using a different variant of the chirplike sequence yields

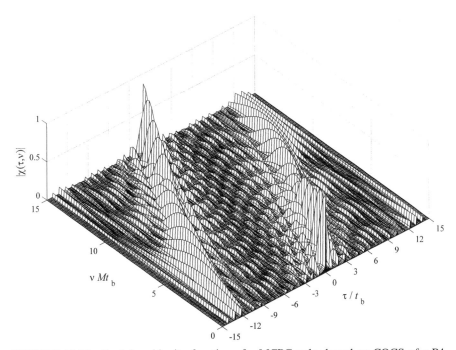

FIGURE 11.18 Partial ambiguity function of a MCPC pulse based on COCS of a P4. $N = M = 15$. Equal carrier amplitudes.

a different diagonal ridge angle. As a result of using COCS, the recurrent lobe in Doppler located at $\nu = 15/T = 1/t_b$ is shifted in delay by t_b relative to the peak in zero Doppler. Since a downward COCS structure was used for the plot, the Doppler recurrent lobe at $\nu = 1/t_b$ is shifted to the left in Fig. 11.18 (negative delay), while using an upward COCS would result in a shift in the opposite direction (positive delay).

11.3 CW (PERIODIC) MULTICARRIER SIGNAL

In the CW case the MCPC signal defined in (11.1) is repeated continuously. The actual complex envelope of the signal is the periodic continuation of the basic period. Recall that in CW radar applications, the matched-filter delay response is determined by the periodic autocorrelation function (PACF). It is possible to show that when N is odd or M is even and the sequence modulating all carriers is any ideal sequence, the magnitude of the PACF is given by

$$|R(\tau)| = \begin{cases} Mt_b \left| \int_0^{1-\tau/t_b} W(x) W^* \left(x + \frac{\tau}{t_b} \right) dx \right|, & 0 \leq |\tau| < t_b \pmod{Mt_b} \\ 0, & t_b \leq |\tau| < (M-1)t_b \pmod{Mt_b} \end{cases} \quad (11.11)$$

where $W(x)$ is the inverse Fourier series of w_n [defined in (11.1)] and is periodic with period 1. Notice that the PACF is independent of the type of sequence used as long as the sequence exhibits ideal periodic autocorrelation. For brevity this design will be named IIS (identical ideal sequence) and we will refer to a multicarrier phase-coded CW signal when all carriers are phase modulated by the same ideal sequence as an IIS MCPC CW signal.

As for the IIS MCPC pulse, a linear phase term (λ acting as the slope parameter) can be added to the carrier phase without an effect on the PMEPR. The free parameter λ is then used for minimizing the PSLL once the PMEPR was set (by changing the carrier amplitudes and phase). The PACF for the optimal $\lambda = 4.525$ (minimal PSLL) and identical carrier amplitudes (flat spectrum) is shown in Fig. 11.19 for $N = 25$. Schroeder carrier phasing was used to lower the PMEPR to 1.83. Setting $\lambda = 4.525$ gives a PSLL of -16.35 dB and a mainlobe width (measured at the point on the mainlobe slopes where the mainlobe equals the PSLL) is $0.88t_b/N$. The IIS MCPC CW was assumed to have $M = 16$ bits when one complete period of the PACF was drawn.

Amplitude-modulating the carriers widens the mainlobe but can help lower the sidelobe level for large N. Numerical methods of designing optimal PSLL, PMEPR, and mainlobe width can be used. One weight function that yields acceptable results with relatively low computational effort for various values

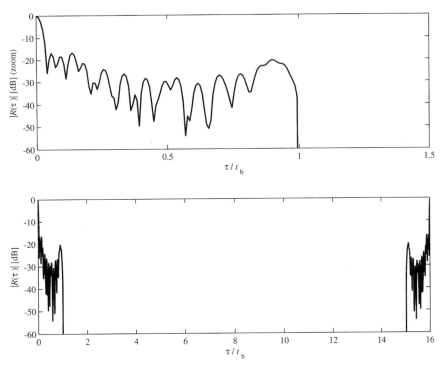

FIGURE 11.19 PACF first bit (top) and full PACF (bottom) of a 25-carrier 16-bit IIS MCPC CW signal. Equal carrier amplitudes. Schroeder carrier phasing with minimal PSLL ($\lambda = 4.525$).

of N and a prescribed mainlobe width is the Taylor weight function raised to a power α, given by

$$|w_n| = \left| \left\{ 1 + \sum_{k=0}^{K-1} c_{k+1} \cos\left[\frac{2\pi}{N}\left(n - \frac{N+1}{2}\right) k \right] \right\}^{\alpha} \right|, \qquad n = 1, 2, \ldots, N \tag{11.12}$$

where c_k ($1 \leq k \leq K$) are the optimized weight coefficients, α is the weight power (also optimized), and K is the number of the phase weights. Note that the optimization is on $K + 2$ free parameters instead of N parameters. Increasing the number of carriers (N) for a fixed mainlobe width yields lower PSLL.

In general, optimizing the carrier amplitudes where the carrier phase is calculated using any phasing method will not guarantee that there does not exist a different carrier weight with lower PSLL and possibly even lower PMEPR (with the same mainlobe width). To find lower PMEPR or PSLL, various iterative optimization methods can be used. One example is to use a two-step iterative scheme, where the first step is to use a numerical gradient method for finding carrier weights with lower PSLL (by changing carrier complex weights), and the second step is to lower PMEPR using a single clipping step (or Schroeder phasing). The optimization steps are repeated until a stop criterion is met. Note that in each step a local minimum to a different minimization problem with different initial values is found.

An example of the optimization process is shown in Fig. 11.20 for $N = 25$ and a designed mainlobe width of $2t_b/N$. The optimization scenario was restricted to symmetric weights (i.e., $w_1 = w_{25}$, $w_2 = w_{24}$, etc.). The initial weight used for the plot yields a PSLL of -25.5 dB and a PMEPR of 1.5 (shown by the square in Fig. 11.20). An optimization step where PSLL is minimized (moving to the left in Fig. 11.20) is shown by dotted lines. The optimization step where the carrier phasing is calculated (a single clipping step and optimizing λ) is marked by solid lines (moving down on Fig. 11.20). Note that it is possible to choose to stop after any iteration. Note also that the PSLL or PMEPR optimization methods are the same for all iterations; alternatively, one could try to adapt the type of optimization to the resulting PSLL and PMEPR.

Figure 11.21 shows iteration steps in which the PSLL and PMEPR were such that no other iteration step yields both lower PMEPR and lower PSLL. Note how all lines intersect in the vicinity of a PSLL of -43 dB and a PMEPR of 1.4.

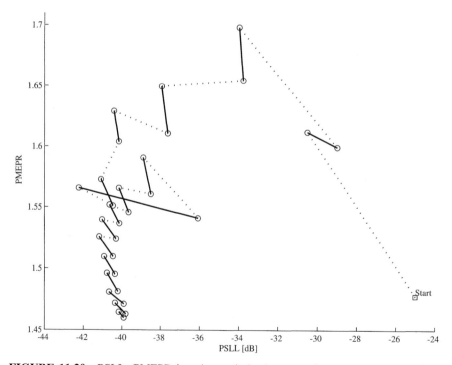

FIGURE 11.20 PSLL–PMEPR iterative optimization steps for a 25-carrier IIS MCPC CW signal. PSLL is minimized using gradient optimization of the carrier complex weights (dotted). PMEPR is minimized using a single clipping step (solid). Designed mainlobe width is $2t_b/N$.

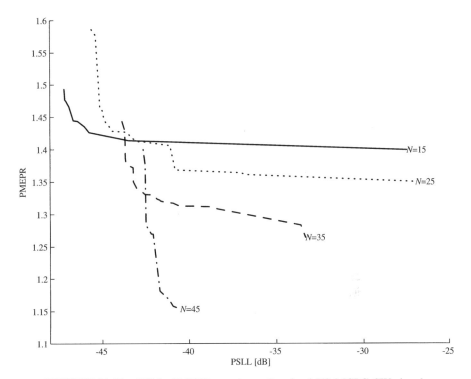

FIGURE 11.21 PSLL–PMEPR contours of optimal IIS MCPC CW signal.

For higher values of N, lower PMEPR can be obtained, but with relatively high PSLL (higher than -43 dB). Obtaining lower PSLL is possible when N is lower but with a price of increasing the PMEPR (above 1.4).

An example of a PSLL- and PMEPR-optimized CW MCPC is shown in Fig. 11.22. The solid lines in Fig. 11.22 show the PACF of a 25-carrier IIS MCPC CW signal with carrier amplitude and phase modulated using the weight given in Table 11.2. The weight selected yields a PMEPR of 1.46 and a PSLL of -42.85 dB. Note that all bits have the same magnitude and that both the PACF and magnitude are independent of the identical sequence used for the construction. The value of M used for the plot was 16. The value of M selected affects only the size of the zero-ACF sidelobe gap between the PACF first and last bits. Figure 11.23 shows the absolute value and phase of the carrier weight used for Fig. 11.22. The amplitudes and phases in Table 11.2 should be read from left to right, top to bottom. Only the first to the center elements are listed; the remaining elements are symmetrical.

The PACF and magnitude of a MCPC CW based on consecutive ordered cyclic shifts (COCS) of a P4 sequence are also shown in Fig. 11.22 (dotted lines). Note that the PACF is not a function of the ideal sequence type, but the signal magnitude is. Using the specific variant of CLS (P4) selected for the plot yields the lowest PSLL and PMEPR. The sequence length (M) is 16 and the number of

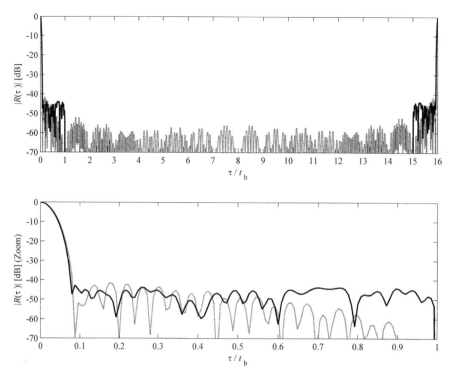

FIGURE 11.22 PACF of a 25-carrier 16-bit MCPC CW signal. ACF mainlobe width set to $2t_b/N$. Solid, IIS MCPC, optimized carriers amplitude and phase; dotted, COCS of a CLS MCPC, carriers amplitude modulated using a square root of Hamming weight.

carriers (N) is 25. The carriers were amplitude-modulated using the square root of a Hamming window, and no additional carrier phase was introduced (on top of the phase introduced by the cyclic shifts of P4). The peak-to-mean envelope power ratio is 1.66, and the PACF PSLL is -41.6 dB.

The frequency spectrum of the periodic MCPC pulse is obtained by sampling the frequency spectrum of the pulsed version at multiples of $1/Mt_b$. Figure 11.24 shows the frequency spectra of a conventional single-carrier phase-coded signal and the two multicarrier variants (IIS MCPC CW and COCS of a CLS MCPC CW). The three signals were designed to give the same effective bandwidth (same PACF mainlobe width) and to have the same energy. The COCS of a CLS MCPC CW (solid) and the IIS MCPC CW (dotted) are both based on modulating 25 carriers with 16 bits each. Both multicarrier signals use a 16-element P4 sequence for the modulation. The single-carrier CW signal, phase-coded using a 200-element P4 sequence, has a $(\sin x)/x$ frequency spectrum, shown in Fig. 11.24 by the dashed line.

Note that for the multicarrier signals, $MN = 400$; thus, the 25 carriers were weighted to double the PACF mainlobe width such that it matches the single-carrier CW signal PACF (having a compression ratio of 200). The COCS of

TABLE 11.2 PSLL–PMEPR Carrier Weights Found Using a Numerical Search

N	PSLL	PMEPR	Carrier Amplitudes				Carrier Phase (deg)		
15	45.7711	1.4262	0.0979	0.1395	0.1966	0.2466	225.3291	276.0215	324.4146
			0.2895	0.3231	0.3459	0.3569	15.5203	76.7491	159.3598
							287.1053		101.0686
	43.4958	1.4137	0.0979	0.1395	0.1966	0.2466	227.5197	276.6462	324.3254
			0.2895	0.3231	0.3459	0.3569	14.9890	76.5365	159.8468
							286.7583		101.2564
25	43.8500	1.4601	0.0746	0.0893	0.1164	0.1367	37.8384	5.3206	336.9453
			0.1621	0.1840	0.2057	0.2233	311.1725	277.5807	243.2525
			0.2401	0.2521	0.2655	0.2718	202.0755	149.9241	76.6448
			0.2695				331.2418	212.9400	89.5602
							288.8825		
	40.8108	1.3684	0.0605	0.0810	0.1101	0.1308	353.0705	320.5062	289.8441
			0.1585	0.1850	0.2091	0.2251	264.2552	231.6907	196.4908
			0.2405	0.2511	0.2704	0.2755	154.6376	103.1894	30.7458
			0.2716				283.9350	166.3384	42.6926
							241.1956		
35	40.5535	1.3125	0.0551	0.0674	0.0815	0.1001	12.9432	45.8554	60.2763
			0.1148	0.1269	0.1365	0.1543	85.2793	109.0830	137.8351
			0.1660	0.1767	0.1921	0.1952	166.0243	211.1545	298.2143
			0.2073	0.2241	0.2221	0.2215	62.6985	141.8614	195.9987
			0.2302			0.2265	246.2286	336.2650	159.0019
							287.9341	46.6190	215.9444
	43.8604	1.4431	0.0751	0.0774	0.0854	0.1024	266.2930	289.9654	314.4874
			0.1150	0.1273	0.1436	0.1569	333.7195	355.8814	17.3640
			0.1680	0.1757	0.1900	0.1953	42.0062	68.3669	100.1482
			0.2063	0.2151	0.2180	0.2220	140.1083	194.3263	272.8975
			0.2267			0.2261	8.1099	100.6538	196.5404
							334.3121	133.8321	291.6906
45	41.1715	1.1677	0.0355	0.0412	0.0508	0.0677	131.6231	149.9123	166.5481
			0.0830	0.0961	0.1037	0.1119	186.4523	205.2363	224.8374
			0.1250	0.1324	0.1406	0.1543	244.6320	261.6153	282.6144
			0.1611	0.1659	0.1717	0.1811	306.0950	332.6862	5.7138
			0.1918	0.1935	0.1950	0.2062	53.6612	118.8589	203.8495
			0.1992	0.2051		0.2075	281.8542	353.8931	78.6399
							204.6199	348.4874	120.8483
							273.1983		76.1462
	42.4925	1.2835	0.0471	0.0468	0.0582	0.0728	208.4063	240.3182	257.5081
			0.0853	0.0981	0.1021	0.1131	274.1558	292.7084	306.3941
			0.1263	0.1360	0.1413	0.1498	326.1055	345.5451	8.4544
			0.1603	0.1653	0.1757	0.1788	29.5518	57.2396	91.7492
			0.1874	0.1932	0.1949	0.2014	139.8323	203.1658	284.7230
			0.2028	0.2012		0.2071	4.4137	77.6961	163.6365
							287.0401	66.7995	200.5068
							358.3404		162.6618

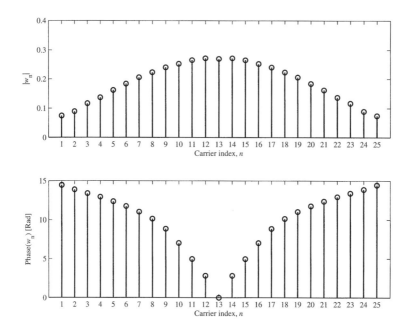

FIGURE 11.23 Optimized carrier amplitudes (top) and phase (bottom) of a 25-carrier IIS MCPC CW having $2t_b/N$ ACF mainlobe width.

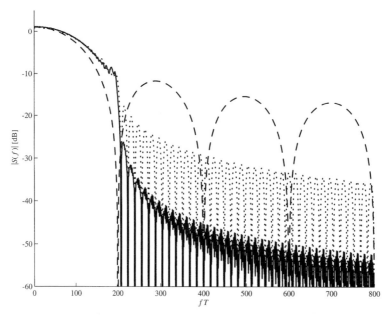

FIGURE 11.24 Frequency spectra of equal energy and ACF mainlobe width signals. Solid: 25-carrier 16-bit MCPC CW signal based on COCS of an ideal sequence; Dashed, single-carrier signal phase coded using 200-element ideal sequence; dotted, 25-carrier 16-bit IIS MCPC CW signal.

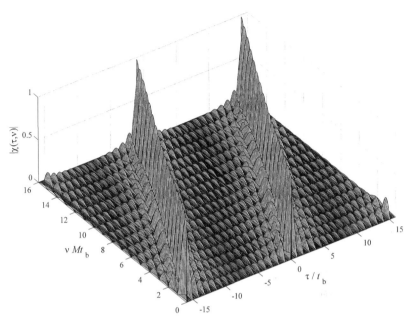

FIGURE 11.25 PAF of a 25-carrier COCS of a 16-bit P4 MCPC CW signal. Carriers amplitude-modulated using a square root of Hamming weight.

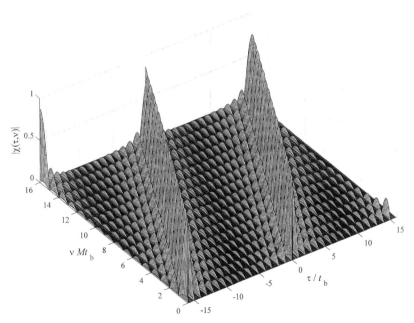

FIGURE 11.26 PAF of a 25-carrier IIS MCPC CW signal (16-bit P4 used). Optimized carriers amplitude and phase. ACF mainlobe width set to $2t_b/N$.

a CLS MCPC CW amplitude weighting rule was a square root of a Hamming window. For the IIS MCPC CW, the optimal carrier amplitude and phase given in Table 11.2 were used. Note that using suboptimal weights (e.g., a square root of Hamming amplitudes and Schroeder phases) yields higher-frequency spectrum sidelobes. Finally, note that the resulting frequency spectra are not a function of the ideal sequence used for the construction (since the PACF is also not a function of the specific ideal sequence).

The periodic ambiguity function of the 25-carrier 16-bit MCPC CW signal based on COCS of a P4 sequence and the 25-carrier 16-bit IIS MCPC CW signal based on a P4 sequence are shown in Figs. 11.25 and 11.26. The figures show two periods of the PAF ($|\tau| < Mt_b$) and are limited in Doppler up to the first recurrent lobe at $\nu = 1/t_b$. Note that using P4 as the ideal sequence yields a diagonal PAF. Note also that for the COCS of P4 MCPC CW signal the recurrent lobe in Doppler is shifted in delay. The shift direction in Fig. 11.25 is one bit to the left since a downward COCS was used (using an upward COCS yields a shift to the right).

11.4 TRAIN OF DIVERSE MULTICARRIER PULSES

The third type of multicarrier signal presented here is a train of diverse MCPC pulses separated at a constant interval T_r. Three types of MCPC pulse train will be analyzed. The three types differ in the method in which the carriers are phase and amplitude modulated. The first and second methods are a direct continuation of the MCPC pulse using identical and consecutive ordered cyclic shifts of a basic sequence. The third method is applicable only to a train of MCPC pulses and is based on using mutually orthogonal complementary sets to phase-modulate the carriers.

11.4.1 ICS MCPC Diverse Pulse Train

The first modulation method is based on using an identical modulation phase for all carriers. The complex envelope of a train of P diverse MCPC pulses where all carriers are modulated by the same phase is given by

$$g_P(t) = \sum_{p=1}^{P} \sum_{n=1}^{N} \sum_{m=1}^{M} w_n a_{m,p} s[t - (m-1)t_b - (p-1)T_r]$$

$$\times \exp\left[j2\pi \left(n - \frac{N+1}{2}\right) \frac{t - (p-1)T_r}{t_b}\right] \qquad (11.13)$$

where $a_{m,p}$ is the mth element of a sequence modulating all carriers in pulse number p ($|a_{m,p}| \equiv 1$ for all m and p), w_n is the complex weight associated with the nth carrier, T_r is the pulse repetition interval, t_b is the bit length (pulse length is Mt_b), and $s(t) \equiv 1$ for $0 \le t < t_b$ and zero elsewhere.

The design is based on using identical complementary sets to modulate all carriers and is thus termed an identical complementary set MCPC pulse train or, for brevity, an ICS MCPC pulse train. The pth sequence of the set modulates all the carriers of the pth pulse. Recall that for the CW MCPC we require that $M(N+1)$ be even. We now require that $T_r/t_b(N+1)$ is an even integer (that T_r/t_b is an even integer or T_r/t_b is odd and N is odd). It can be shown that the normalized correlation function of the MCPC pulse train for $0 \leq \tau \leq T_r - Mt_b$ (the mainlobe area) when $T_r > 2Mt_b$ (the duty cycle is below $\frac{1}{2}$) and the set of P sequences (with M elements each) form a complementary set is given by

$$R_P(\tau) = \begin{cases} \left| \sum_{l=1}^{N} \sum_{n=1}^{N} w_n w_l^* [t_b \delta(n-l) - \tau] \right. \\ \left. \times \operatorname{sinc}\left[\pi(n-l)\frac{\tau}{t_b} \right] \exp\left[j\pi(n+l)\frac{\tau}{t_b} \right] \right|, & 0 \leq \tau < t_b \\ 0, & t_b \leq \tau < T_r - Mt_b \end{cases} \quad (11.14)$$

where $\sum |w_n|^2 = 1$. Note that the normalized correlation function in the area of the mainlobe is not a function of M or P but only of N and the complex carrier weights.

As in the CW case, many phasing methods can be used to lower the PMEPR of the diverse MCPC pulse train (e.g., Schroeder phasing). An example showing the periodic autocorrelation function (PACF) of a diverse MCPC pulse train based on modulating 16 carriers with identical complementary sets is shown in Fig. 11.27. The carriers were phase and amplitude modulated to yield a PACF mainlobe width of $2t_b/N$ and low PMEPR (a square root of Hamming amplitude weight and Schroeder phases were used). The complementary set used to modulate all carriers is based on all cyclic shifts of a chirplike ideal sequence of length 8. The sequences were permutated, between the pulses, to avoid high recurrent lobes in range. The permutation used for the plot was {5 6 2 7 4 8 1 3}, and the duty cycle used for the plot was $\frac{1}{3}$.

Note the relatively high correlation sidelobes in the first bit (-20 dB). The PMEPR of the ICS MCPC pulse train is 1.48 (here the peak/mean envelope power ratio is calculated for all the pulses in the train and is identical to the PMEPR of a single bit in any of the pulses). Increasing N while increasing the normalized mainlobe width (the absolute mainlobe width remains the same) lowers the sidelobe level. For example, increasing N to 25 and the designed mainlobe width to $3t_b/N$ results in a PSLL of -30 dB. However, increasing the number of carriers while keeping the carrier spacing constant will result in a higher effective bandwidth. Keeping the bandwidth constant implies increasing t_b (i.e., lowering M). Iterative and numerical methods similar to the ones demonstrated for the CW case can be used for finding optimal carrier weights yielding lower PMEPR and lower PSLL.

The PAF of the eight-pulse 16-carrier 8-bit ICS MCPC diverse pulse train based on all cyclic shifts of an 8-bit P4 sequence is shown in Fig. 11.28. The

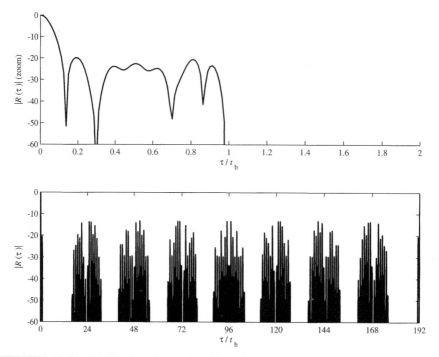

FIGURE 11.27 PACF of an eight-pulse 16-carrier 8-bit ICS MCPC diverse pulse train based on all cyclic shifts of an 8-bit P4 sequence ({ 5 6 2 7 4 8 1 3} pulse permutations). 33% duty cycle. Carrier amplitude modulated using a square root of Hamming weight. Schroeder carrier phasing.

figure extends in Doppler up to the first recurrent lobe ($1/T_r$) and is limited in delay to the first recurrent lobe (the other six are not shown). Note that the sidelobe level off the zero-Doppler axis is very similar to the one obtained for zero Doppler (PACF). This is due primarily to the pulse permutation used. Using consecutive ordered pulses could yield high sidelobe levels off zero Doppler.

11.4.2 COCS of a CLS MCPC Diverse Pulse Train

The second multicarrier diverse pulse train design is based on using consecutive ordered cyclic shifts of a chirplike ideal sequence to modulate all carriers in all pulses. The pulse diversity is obtained by using a different cyclic carrier order in each pulse such that for any carrier the sequences modulating all pulses are different. The PMEPR is low, due to the inherent COCS structure of the pulse (no additional carrier phasing is used).

The PACF for consecutive ordered cyclic shifts (COCS) of an ideal sequence MCPC pulse train is demonstrated in Fig. 11.29. The number of pulses in the train (P) used for the plot was eight, The sequence length (M) is 8 bits and the number of carriers (N) is 16. A square root of Hamming weight was used

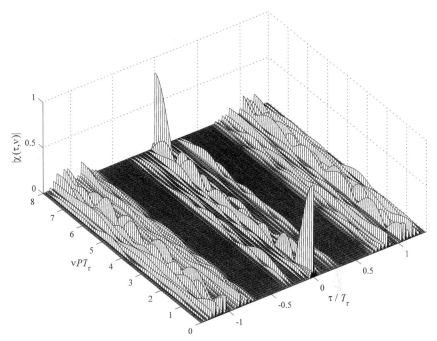

FIGURE 11.28 Partial PAF of an eight-pulse 16-carrier 8-bit ICS MCPC diverse pulse train based on all cyclic shifts of an 8-bit P4 sequence ({5 6 2 7 4 8 1 3} pulse permutations); 33% duty cycle. Carrier amplitude modulated using a square root of Hamming weight. Schroeder carrier phasing.

for carrier amplitude weighting, and no additional carrier phasing was used. The complementary set, pulse permutation, and duty cycle were selected identical to those used in Fig. 11.27 for plotting the PACF of the ICS MCPC diverse pulse train. Note that in Fig. 11.29 the property of zero PACF for $t_b \leq |\tau| \leq Mt_b$ is lost, but the PACF sidelobes within the PACF first bit are much lower than in Fig. 11.27. The PMEPR of the COCS MCPC pulse train is 1.9 instead of 1.48 (for an ICS MCPC diverse pulse train). The partial PAF of the COCS MCPC pulse train is shown in Fig. 11.30.

Note that for the COCS MCPC diverse pulse train the number of carriers N cannot be increased without limit while keeping the PMEPR low. Specific values of N (such as $N = M$ with identical carrier amplitudes or $N = 2M$ with a square root of the Hamming carrier amplitude modulation rule) give considerably lower PMEPR than do other selections of N.

11.4.3 MOCS MCPC Pulse Train

The third type of diverse MCPC pulse train design is based on using (different) *mutually orthogonal complementary sets* (MOCS) to modulate the different carriers. Two complementary sets are mutually orthogonal when the sum of the

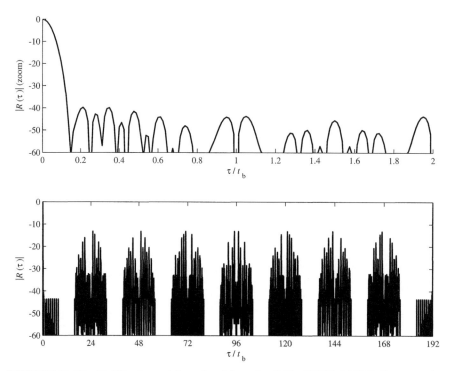

FIGURE 11.29 PACF of an eight-pulse 16-carrier 8-bit COCS MCPC diverse pulse train based on an 8-bit P4 sequence. Carriers amplitude-modulated using a square root of Hamming weight. {5 6 2 7 4 8 1 3} pulse permutations; 33% duty cycle.

cross-correlation between the sequences in one complementary set with the corresponding sequences in the second set is zero for all shifts (Tseng and Liu, 1972). The initial phase is set, using numerical methods, such that the PMEPR is low. Unlike the ICS design, where calculating the initial carrier phase is independent of the complementary set used, the carrier initial phase calculation for the MOCS design is a function of the type of MOCSs used for the construction. Using mutually orthogonal complementary sets yields zero PACF sidelobes for $t_b \leq |\tau| \leq Mt_b$.

An example of a MCPC pulse train design using mutually orthogonal complementary sets is shown in Fig. 11.31. The MOCS MCPC pulse train used for the plot consists of $N = 16$ carriers. In each pulse the same block of eight binary mutually orthogonal complementary sets are used to phase modulate the lower and upper eight carriers. Note that without additional carrier phasing (identical for all bits in all pulses) the PMEPR is maximized since bit 7 in all pulses exhibits maximal amplitude.

The PACF of the 16-carrier MOCS MCPC diverse pulse train is shown in Fig. 11.32. The carrier amplitudes were weighted using a square root of a Hamming window and the initial carrier phase was optimized numerically to minimize

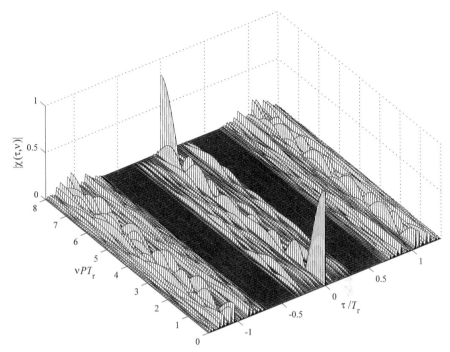

FIGURE 11.30 Partial PAF of an eight-pulse 16-carrier 8-bit COCS MCPC diverse pulse train based on an 8-bit P4 sequence. Carriers amplitude-modulated using a square root of Hamming weight. {5 6 2 7 4 8 1 3} pulse permutations; 33% duty cycle.

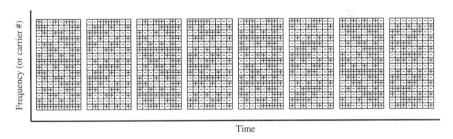

FIGURE 11.31 Structure of a 16-carrier eight-pulse MOCS MCPC pulse train based on two blocks of eight-set eight-sequence 8-bit binary MOCS (the initial phase of each carrier, used to reduce PMEPR, is not shown).

PMEPR (yielding a PMEPR of 1.99). Note the low sidelobes within the first bit and the lower and narrower recurrent lobes (compare to Figs. 11.27 and 11.29). The mainlobe width is the same as for the ICS MCPC pulse train (Fig. 11.27) and the COCS MCPC pulse train (Fig. 11.29).

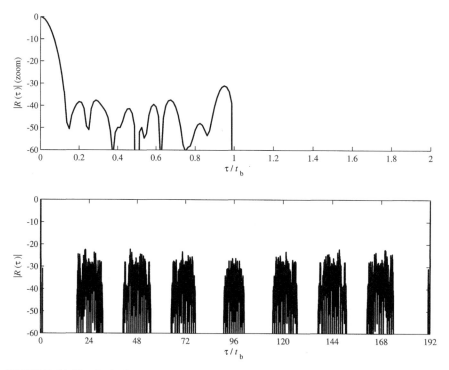

FIGURE 11.32 Periodic correlation function of an eight-pulse 16-carrier 8-bit MOCS MCPC pulse train based on two blocks of binary eight-set eight-sequence 8-bit MOCS. Carriers amplitude-modulated using a square root of Hamming weight. Carrier phasing is {0° −30° −161° −19° −107° 151° 36° −169° 30° 140° −119° −144° 36° 86° −51° −73°}; 33% duty cycle.

11.4.4 Frequency Spectra of MCPC Diverse Pulse Trains

The frequency spectra of the MOCS MCPC pulse train, COCS MCPC pulse train, ICS MCPC pulse train, and a single-carrier phase-coded diverse pulse train are compared in Fig. 11.33. The ideal sequence used for construction of the COCS MCPC signal was a chirplike ideal sequence with $M = 8$ bits. The complementary set used for the ICS MCPC pulse train is based on all cyclic shifts of the eight-element chirplike ideal sequence, while the complementary set used for the single-carrier signal is based on eight cyclic shifts of a 64-element chirplike ideal sequence (shifted at integer multiples of eight elements). All signals are designed to exhibit approximately the same PACF mainlobe width ($T/64$). The single-carrier complementary set has zero PACF sidelobes starting at $\tau = T/64$, the ICS and MOCS MCPC diverse pulse trains PACF sidelobes extend as far as $\tau = T/8$, and the COCS MCPC diverse pulse train PACF sidelobes extend as far as $\tau = T$. The pulses are permutated to avoid high recurrent lobes in range. The permutation used for the plot was {5 6 2 7 4 8 1 3} for all pulse trains. The duty cycle was $\frac{1}{3}$. Note that the MOCS MCPC pulse train and the COCS MCPC diverse pulse train exhibit the lowest-frequency spectrum sidelobes. Note also that all signals except the MOCS MCPC pulse train exhibit spikes in their frequency spectrum mainlobe area due to the ordered nature of the designs.

11.5 SUMMARY

Design principles of a multicarrier phase-coded pulse, continuous-wave, and diverse pulse train with low peak-to-mean envelope power ratio were shown. The two basic building blocks yielding low PMEPR are the identical sequence design and the consecutive ordered cyclic shifts of a chirplike sequence design. The main two advantages of the identical sequence design over the consecutive ordered cyclic shifts design are (1) lower PMEPR and (2) autocorrelation sidelobes extending as far as t_b instead of Mt_b. The two main drawbacks of the identical sequence design are (1) higher correlation sidelobes in the first bit and (2) a less bandwidth efficient spectrum. Both the consecutive ordered cyclic

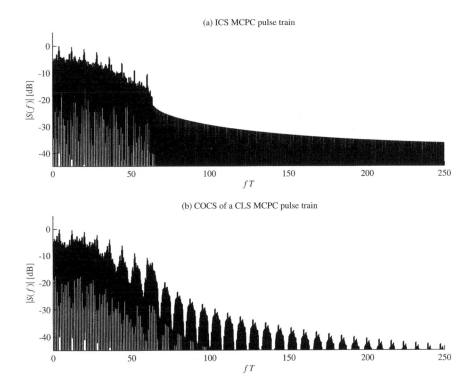

FIGURE 11.33 Frequency spectra of equal PACF mainlobe width and power signals: (a) 16-carrier eight-pulse 8-bit ICS MCPC diverse pulse train; (b) 16-carrier eight-pulse 8-bit COCS of a CLS MCPC diverse pulse train.

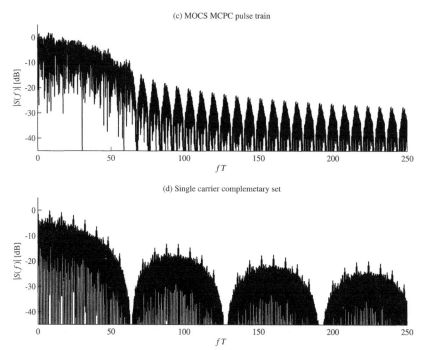

FIGURE 11.33 Frequency spectra of equal PACF mainlobe width and power signals: (c) 16-carrier eight-pulse 8-bit MOCS MCPC pulse train; (d) single-carrier eight-pulse complementary 64-bit chirplike sequences.

shifts design and the identical sequence design are considerably more bandwidth efficient than the single-carrier complementary phase-coded signal having the same autocorrelation mainlobe width.

For a diverse MCPC pulse train, a third design approach based on using mutually orthogonal complementary sets (MOCS) also exists. It was shown that the ACF of a MOCS MCPC pulse train benefits both from the large zero sidelobe gap of the ICS MCPC pulse train and the low first bit sidelobes of the COCS MCPC pulse train. The frequency spectrum of the MOCS MCPC pulse train has low sidelobes and is free of some of the peaks that can be found in the frequency spectrum of the other MCPC pulse trains. The PMEPR of the MOCS MCPC pulse train is higher than the one obtained for the other optional designs of a MCPC diverse pulse train but still is not too high.

The single-carrier signal has two clear advantages over the multicarrier signals: (1) fixed envelope and (2) zero correlation sidelobes. The main drawback of the single-carrier phase-coded signal is the $(\sin x)/x$ frequency spectrum, implying higher out-of-band interference and losses. The multicarrier signals can be compared to a triathlon athletic event (running, swimming, and bicycling). Although the MCPC signal does not have the lowest PMEPR, the lowest ACF sidelobes, or the highest-efficiency frequency spectrum, it is relatively difficult to find signals that outperform it when all three aspects are considered. Another advantage of multicarrier signals is the added dimension for introducing diversity between signals. This is useful in reducing mutual interference between radars operating in close proximity.

PROBLEMS

11.1 Waveform orthogonality

(a) Show that the $u(t)$ and $v(t)$ defined below are orthogonal.

$$u(t) = \begin{cases} 1, & t \in [0, t_b) \\ 0, & t \notin [0, t_b) \end{cases} \qquad v(t) = \begin{cases} \dfrac{1}{2}, & t \in \left[0, \dfrac{t_b}{2}\right) \\ -\dfrac{1}{2}, & t \in \left[\dfrac{t_b}{2}, t_b\right) \\ 0, & t \notin [0, t_b) \end{cases}$$

(b) Assume that $u(t)$ is used to amplitude modulate a signal with carrier frequency f_u and that $v(t)$ modulates f_v. What are the conditions on f_u and f_v such that the two modulated signals are orthogonal?

(c) How will your answer change if f_u and f_v are much higher than $1/t_b$ (narrowband signals)?

11.2 PACF and ACF of MCPC diverse pulse trains

We define an MCPC diverse pulse train as

$$g_P(t) = \sum_{p=1}^{P} \sum_{n=1}^{N} \sum_{m=1}^{M} w_n a_{n,m,p} s[t - (m-1)t_b - (p-1)T_r]$$

$$\times \exp\left[j2\pi\left(n - \frac{N+1}{2}\right)\frac{t - (p-1)T_r}{t_b}\right]$$

where all the symbols follow the conventions used in the chapter body.

(a) Show that when $T_r/t_b(N+1)$ is an even integer (T_r/t_b is an odd integer and N is odd or that T_r/t_b is even), the *aperiodic autocorrelation function* of the MCPC diverse pulse train is given by

$$\left| \sum_{l=1}^{N} \exp\left[j2\pi l \frac{\eta}{t_b}\right] \sum_{n=1}^{N} w_n w_l^* \left\{ I_1 \sum_{p=r+1}^{P} \sum_{m=1}^{M} a_{n,m,p} a_{l,m-i_r-1,p-r}^* \right.\right.$$

$$\left.\left. + I_2 \sum_{p=r+1}^{P} \sum_{m=1}^{M} a_{n,m,p} a_{l,m-i_r,p-r}^* \right\} \right|$$

where we assume $a_{n,m,p}$ as zero for unallowed values of m, n, or p (e.g., $m > M$ or $m < 1$), r is the integer part of T_r in τ, i_r is the integer part of t_b in $\tau - rT_r$ and η is the remainder of τ after division with t_b ($\tau = rT_r + i_r t_b + \eta$), and where

$$I_1 = \eta \operatorname{sinc} \beta \exp(j\beta) \quad \text{where} \quad \beta = \pi(n-l)\frac{\eta}{t_b}$$

$$\text{and} \quad I_2 = t_b\, \delta(n-l) - I_1$$

(b) Show that under the same assumptions as in part (a), the *periodic* correlation function of the MCPC diverse pulse train is given by

$$\left| \sum_{l=1}^{N} \exp\left(j2\pi l \frac{\eta}{t_b}\right) \sum_{n=1}^{N} w_n w_l^* \left\{ I_1 \sum_{p=1}^{P} \sum_{m=1}^{M} \right.\right.$$

$$\left.\left. \times\, a_{n,m,p} a_{l,m-i_r-1,[p-r]_P}^* + I_2 \sum_{p=1}^{P} \sum_{m=1}^{M} a_{n,m,p} a_{l,m-i_r,[p-r]_P}^* \right\} \right|$$

11.3 MCPC diverse pulse train PACF and ACF mainlobe area

(a) Use the results of Problem 11.2 to explicitly write the periodic and aperiodic autocorrelation function of the MCPC pulse train in the area around zero delay.

(b) Show that the expressions for the PACF and the aperiodic ACF are identical. Explain why.

11.4 MOCS and ICS MCPC pulse trains PACF and ACF mainlobe area

(a) Use the results of Problem 11.3 to show that the mainlobe area correlation function of a MCPC pulse train based on modulating all carriers with an identical complementary set (ICS MCPC pulse train) is given by

$$\left| PM \sum_{l=1}^{N} \sum_{n=1}^{N} w_n w_l^* [t_b \delta(n-l) - \tau] \operatorname{sinc}\left[\pi(n-l)\frac{\tau}{t_b}\right] \right.$$

$$\left. \times \exp\left[j\pi(n+l)\frac{\tau}{t_b}\right] \right|$$

during the first bit and zero otherwise.

(b) Use the results of Problem 11.3 again to show that the mainlobe area correlation function of a MOCS MCPC pulse train is given by

$$\left| PM(t_b - \tau) \sum_{n=1}^{N} |w_n|^2 \exp\left[j2\pi n \frac{\tau}{t_b}\right] \right|$$

during the first bit and zero otherwise. Note that the normalized ACF shape in the area of the mainlobe is not a function of M or P but only of N and the complex carrier amplitudes.

(c) Note the similarity between the results of parts (a) and (b), although the designs are different. The shape (absolute value) of the correlation function of the MOCS MCPC pulse train for $0 \leq \tau < t_b$ is composed of two multiplied components: (1) a negative ramp with a maximum at $\tau = 0$ and zero for $\tau = t_b$, and (2) a shape, symmetric around $\tau = t_b/2$, given by the inverse Fourier transform of $|w_n|^2$. The correlation function of the ICS MCPC pulse train can also be decomposed into two factors. Write these two factors and explain why the PSLL of the ICS MCPC cannot be lowered dramatically below 1/M.

11.5 COCS of a CLS MCPC diverse pulse train PACF and ACF mainlobe area

(a) Use the results of Problem 11.3 to show that correlation function of a COCS MCPC pulse train in the mainlobe area is given by

$$\left| P \sum_{l=1}^{N} \exp\left(j 2\pi l \frac{\eta}{t_b}\right) \sum_{n=1}^{N} w_n w_l^* \right.$$

$$\left. \times \left[(M - i_0 - 1) I_1 \delta(n - l \pm i_0 \pm 1) + (M - i_0) I_2 \delta(n - l \pm i_0) \right] \right|$$

where the $\pm$ sign is used to distinguish between consecutive upward and consecutive downward cyclic shifts (minus sign for upward ordered shifts), $\tau = i_0 t_b + \eta$, i_0 is an integer, $0 \leq \eta < t_b$, and where I_1 and I_2 are given by

$$I_1 = \eta \text{ sinc } \beta \exp j\beta \quad \text{where} \quad \beta = \pi(n - l)\frac{\eta}{t_b}$$

$$\text{and} \quad I_2 = t_b \delta(n - l) - I_1$$

Note that the result is only a function of N, M, the consecutive cyclic shifts order (upward or downward), and the carrier complex weights.

(b) Use the expression above to show that the ACF is zero at integer multiples of the bit duration. Give an intuitive explanation.

11.6 MCPC diverse pulse trains PACF and aperiodic ACF recurrent lobes PSL

(a) Use the results of Problem 11.2 to write simple expressions for the PACF and aperiodic ACF of the MCPC pulse train at integer multiples of the bit duration.

(b) Use the results in part (a) to argue how to lower the recurrent lobe peak sidelobe level of COCS MCPC pulse train and ICS MCPC pulse train. What can be said on lowering the recurrent lobes? What can be said on the recurrent lobe value exactly at multiples of T_r?

11.7 MOCS and ICS MCPC pulse trains frequency spectra

(a) Use the result of Problem 11.4(a) and (b) to write simple approximation to the frequency spectra of a MOCS MCPC diverse pulse train and an ICS MCPC diverse pulse train.

(b) Explain the results intuitively by observing the frequency spectra at integer multiples of $1/t_b$.

(c) Check the approximation by comparing the frequency spectra of some $8 \times 8 \times 8$ diverse MCPC pulse trains calculated directly and by using the simplified expression. Explain the results.

11.8 PACF of a MCPC CW based on cyclic shifts of an ideal sequence

(a) Show that for N odd or M even, the PACF of a MCPC CW signal based on modulating all carriers with cyclic shifts of the same ideal sequence is given by

$$\left| \sum_{n=1}^{N} \sum_{l=1}^{N} w_n w_l^* \exp\left(j2\pi l \frac{\eta}{t_b}\right) \right.$$

$$\left. \times [I_1 \delta([\phi_n - \phi_l + i + 1]_M) + I_2 \delta([\phi_n - \phi_l + i]_M)] \right|$$

where i is an integer, $0 \leq \eta < t_b$, $\delta(x)$ is Kronecker delta function and I_1 and I_2 are given by

$$I_1 = \eta \operatorname{sinc}(\beta) \exp(j\beta) \quad \text{where} \quad \beta = \pi(n-l)\frac{\eta}{t_b}$$

$$\text{and} \quad I_2 = t_b \, \delta(n-l) - I_1$$

Notice that the PACF is independent of the type of sequence used for the construction as long as the sequence exhibits ideal periodic autocorrelation. Explain why this should not come as a surprise.

(b) Find the PACF value at integer multiples of the bit duration. In what way is this result different from the one obtained for a single MCPC pulse?

11.9 PACF of MCPC CW using the Zak transform

The Zak transform of a signal $g(t)$ is defined as

$$Z_g(x, y) = \sum_{k=-\infty}^{\infty} g(x+k) \exp(-j2\pi ky)$$

(a) Show that the Zak transform satisfies the following periodicity relations:

$$Z_f(x+1, y) = \exp(j2\pi y) Z_f(x, y), \qquad -\infty < x, y < \infty$$
$$Z_f(x, y+1) = Z_f(x, y), \qquad -\infty < x, y < \infty$$

Hence, it is enough to define $Z_f(x, y)$ for $0 \leq x, y < 1$.

(b) Show that for $f(t)$ defined by

$$f(t) = \sum_{n=1}^{N} \sum_{m=1}^{M} a_{n,m} s(t-m) \exp(j2\pi nt)$$

where $s(t) \equiv 1$ for $0 \leq t < 1$, the Zak transform is given by

$$Z_f(x, y) = \sum_{n=1}^{N} \sum_{m=1}^{M} a_{n,m}$$
$$\times \exp[j2\pi(nx + my)], \qquad 0 \leq x, y < 1$$

(c) Use the Zak transform to show that the PACF of the MCPC CW can also be written as

$$\left| \frac{t_b}{M} \sum_{m=0}^{M-1} \exp\left(-2\pi j \frac{mi}{M}\right) \left[R_m\left(\frac{\eta}{t_b}\right) + (-1)^{N+1} \right. \right.$$
$$\left. \left. \times \exp\left(-2\pi j \frac{m}{M}\right) R_m^*\left(1 - \frac{\eta}{t_b}\right) \right] \right|$$

where

$$R_m\left(\frac{\eta}{t_b}\right) = M \int_0^{1-\eta/t_b} V\left(x, \frac{m}{M} + \frac{N+1}{2}\right)$$
$$V^*\left(x + \frac{\eta}{t_b}, \frac{m}{M} + \frac{N+1}{2}\right) dx$$

and $V(x, y)$ is periodic in x and y with period 1 (in x and in y) and is given by the inverse Fourier series (with respect to the pair n and x) of $w_n \exp(j2\pi\phi_n y)$.

11.10 PACF of an IIS MCPC CW signal
Use the expressions derived in Problem 11.9 to show that the PACF of an IIS MCPC CW is given by

$$|R(\tau)|$$
$$= \begin{cases} Mt_b \left| \int_0^{1-\tau/t_b} W(x) W^* \left(x + \dfrac{\tau}{t_b} \right) dx \right|, & 0 \le |\tau| < t_b \\ & (\bmod\ Mt_b) \\ 0, & t_b \le |\tau| < (M-1)t_b \\ & (\bmod\ Mt_b) \end{cases}$$

where $W(x)$ is the inverse Fourier series of w_n and is periodic with period 1.

REFERENCES

Boyd, S., Multitone signals with low crest factor, *IEEE Transactions on Circuits and Systems*, vol. CAS-33, no. 10, October 1986, pp. 1018–1022.

Friese, M., Multitone signals with low crest factor, *IEEE Transactions on Communications*, vol. 45, no. 10, October 1997, pp. 1338–1344.

Levanon, N., Multifrequency radar signals, *Records of the IEEE 2000 International Radar Conference*, Alexandria, VA, 2000a, pp. 683–688.

Levanon, N., Multifrequency complementary phase-coded radar signal, *IEE Proceedings: Radar, Sonar and Navigation*, vol. 147, no. 6, December 2000b, pp. 276–284.

Levanon, N., Train of diverse multifrequency radar pulses, *Proceedings of the IEEE 2001 International Radar Conference*, Atlanta, GA, 2001, pp. 93–98.

Levanon, N., and E. Mozeson, Multicarrier radar signal: pulse train and CW, *IEEE Transactions on Aerospace and Electronic Systems*, vol. 38, no. 2, April 2002, pp. 707–720.

Mozeson, E., and N. Levanon, Multicarrier radar signals with low peak-to-mean envelope power ratio, *IEE Proceedings: Radar, Sonar and Navigation*, vol. 150, no. 2, April 2003, pp. 71–77.

Narahashi, S., and T. Nojima, New phasing scheme of N multiple carriers for reducing peak-to-average power ratio, *Electronics Letters*, vol. 30, no. 17, August 1994, pp. 1382–1383.

Newman, D. J., An L^1 extremal problem for polynomials, *Proceedings of the American Mathematical Society*, vol. 16, December 1965, pp. 1287–1290.

Schroeder, M. R., Synthesis of low-peak factor signals and binary sequences with low autocorrelation, *IEEE Transactions on Information Theory*, vol. IT-16, no. 1, January 1970a, pp. 85–89.

Schroeder, M. R., *Number Theory in Science and Communication*, 3rd ed., Springer-Verlag, New York, 1970b, pp. 289–300.

Tseng, C. C., and C. L. Liu, Complementary sets of sequences, *IEEE Transactions on Information Theory*, vol. IT-18, no. 5, September 1972, pp. 644–652.

Van-Der Ouderaa, E., J. Schoukens, and J. Renneboog, Peak factor minimization, using time-frequency domain swapping algorithm, *IEEE Transactions on Instrumentation and Measurement*, vol. 37, no. 1, March 1988, pp. 145–147.

APPENDIX

ADVANCED MATLAB PROGRAMS

This appendix contains more elaborate MATLAB programs that supplement the programs in the text. Most of the programs utilize a graphic user interface (GUI), which simplifies parameter changes and allows quick repeated use of the program.

Note: When a continuation ellipsis (...) appears within a single quote, such as a string, it indicates that the string and what follows it (up to a continuation ellipsis that is not within a string) must be completed in the line on which it was started.

A.1 AMBIGUITY FUNCTION PLOT WITH A GUI

This is a GUI version of ambfn1.m, which appeared in Chapter 3. Running this AF plotting program requires five m-files: ambfn7.m; calplotsig7.m; cal_and_plot_amb_fn7.m; cal_and_plot_pamb7.m; cal_and_plot_acf _and_spec7.m. Calling ambfn7.m displays the GUI in Fig. A.1. The only parameter that is not trivial is r, which is explained in Appendix 3A.

A.1.1 Hints on Using ambfn7.m

The signal is defined by the vectors in the three slots in the GUI: Amplitude, Phase/pi, and frequency·t_b. When you choose a preset signal the slots are filled automatically. You can define a signal externally (in the MATLAB command window) by entering three vectors, respectively: u_amp, u_phase, and f_basic. (They must all be row vectors of the same length.) The u_amp vector must be defined. One or two of the other vectors can be avoided, but then you need to toggle off the dot next to the corresponding slot in the GUI (by clicking on it).

Radar Signals, By Nadav Levanon and Eli Mozeson
ISBN 0-471-47378-2 Copyright © 2004 John Wiley & Sons, Inc.

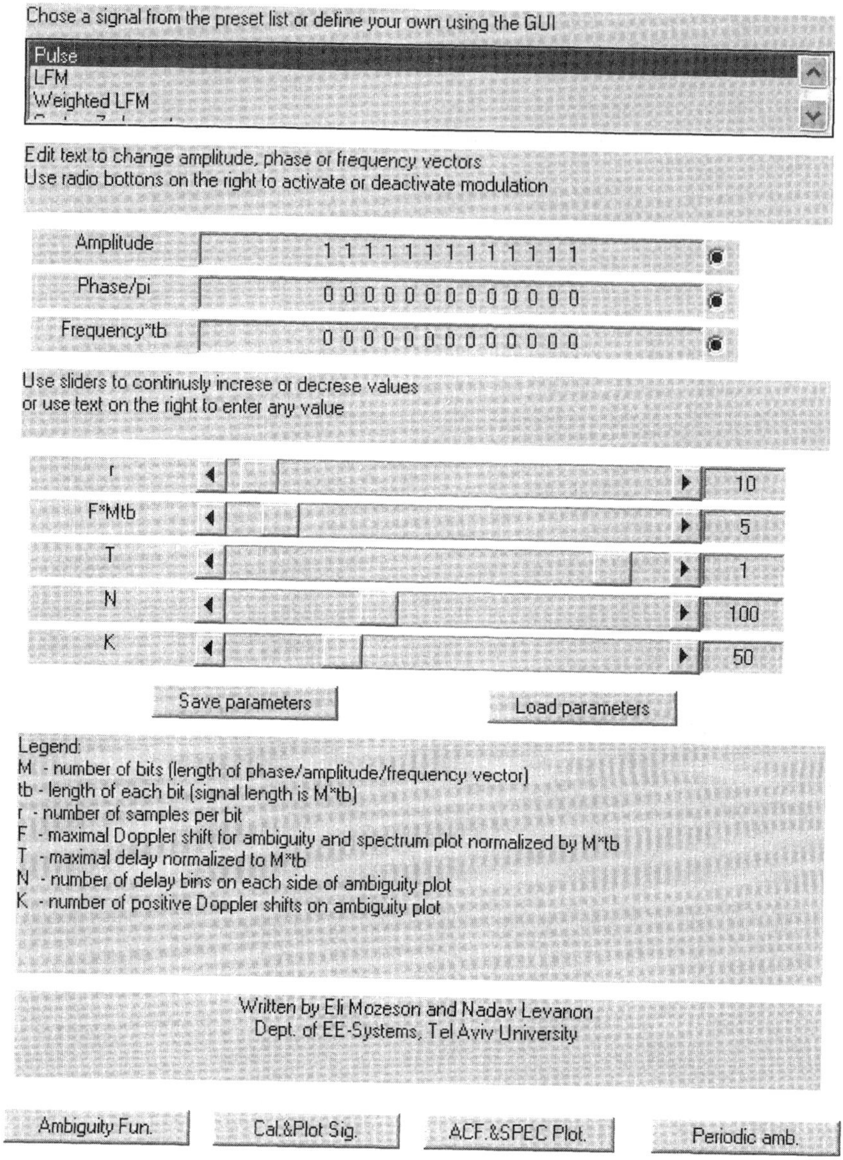

FIGURE A.1 GUI of ambfn7.m.

There are five parameters in the GUI: $\{r, F \cdot Mt_b, T, N, K\}$. Of those, the only parameter that affects the signal and not only the plots is r. Since the signal is defined by a vector, with a well-defined length (number of elements, referred to as M), it is often necessary to increase the number of samples (repeats) during each of these elements (bits), in order to meet the Nyquist criterion. This is the function of r.

A.1.2 Suggested Values for r

In a Costas signal of M elements, the signal bandwidth is approximately M/t_b. Therefore, the sampling interval should be $t_s < t_b/(2M)$. Hence, $t_b/t_s = r > 2M$. In a phase-coded signal, the main spectral lobe ends at $f = 1/t_b$. However, the spectral sidelobes extend much further at a rate of approximately 6 dB per octave. A typical spectral skirt crosses the -30 dB level at $f = 10/t_b$. Hence, choosing $r = 2$ is the minimum setting, but using $r > 10$ is recommended.

Any time that you change r, you need to click on the "Cal&Plot Sig." button to recalculate the signal. Only then can you click on any of the other buttons to get the various plots. When you change any of the other GUI parameters $\{F \cdot Mt_b, T, N, K\}$, you can usually click on the various plot buttons without recalculating the signal. The GUI parameters $\{F \cdot Mt_b, T, N, K\}$ are associated only with the plots.

- $F \cdot Mt_b$ defines the extent of the plotted Doppler axis in normalized Doppler [Doppler multiplied by the entire duration of the signal (Mt_b)]. $F \cdot Mt_b$ also defines the extent of the frequency axis in the spectrum plot (the lower subplot created when you click on "ACF. & SPEC Plot."). Although the maximum value in the $F \cdot Mt_b$ ruler is 60, you can type in a higher value.
- T is the extent of the (positive) delay axis in units of the entire duration of the signal, so it is actually T/Mt_b. For example, if you choose the signal "pulse train, 6 pulses" and choose $T = 1$, the ambiguity plot will include all five recurrent lobes. If you choose $T = 1/6$, you will get exactly one repetition period.
- N is the number of grid points on the (positive) delay axis of the plot.
- K is the number of grid points on the Doppler axis of the plot.

A.1.3 Regarding the Periodic Ambiguity Function

If you want to choose one of the preset signals but wish to plot the PAF when the processor is matched to several periods of the signal (e.g., four periods), you can do the following:
In the GUI, choose the desired signal and click on the "Cal&Plot Sig." button. Next, in the MATLAB command window, enter three commands:

```
u_amp=[u_amp u_amp u_amp u_amp];
u_phase=[u_phase u_phase u_phase u_phase];
f_basic=[f_basic f_basic f_basic f_basic];
```

Next, click again on the "Cal&Plot Sig." button.

In this way you have modified the signal to four repeats of the original signal. If you now plot the periodic ambiguity function, you will see the difference. (The volume is concentrated near Doppler values of n/period.) If you wish to display exactly two periods of the PAF, you can change T from 1 to $T = 0.25$.

ambfn7.m

```
% "ambfn7.m" - builds an MMI and calls dedicated m-files to calculate
%    the signal
% complex envelope, plot signals phase, amplitude and frequency structure, and
% ambiguity function and periodic ambiguity function
%
% Written by Nadav Levanon and Eli Mozeson, Dept. of EE-Systems, Tel Aviv
%    University
%
% ambfn7.m includes a "User defined" line to the preset values list. The line
% is activated each time the user changes one of the signal parameters
% ambfn7.m also allows the user to save/load the signal and plot parameters
%
clear all
close all

% create the parameters input figure
inputfig=figure;
set(inputfig,'Position',[6 51 520 690],'MenuBar','none','visible','off');

% create results figure (ambiguity function plot)
ambfig=figure;
set(ambfig,'Position',[274 52 749 670],'Name','ambiguity function ...
    plot','visible','off');

% create results figure (signal parameters plot)
sigfig=figure;
set(sigfig,'Position',[274 52 749 670],'Name','signal parameters ...
    plot','visible','off');

% create results figure (Autocorrelation and Spectrum plot)
acffig=figure;
set(acffig,'Position',[274 52 749 670],'Name','Autocorrelation and Spectrum ...
    plot','visible','off');

% create results figure (Periodic ambiguity plot)
pambfig=figure;
set(pambfig,'Position',[274 52 749 670],'Name','Periodic ambiguity ...
    plot','visible','off');

% set default values for the signal and plot - single carrier parameters
acode=1;                             % signal amplitude modulation flag
pcode=1;                             % signal phase modulation flag
u_amp=ones(1,13);                    % signal amplitude vector
u_phase=zeros(1,13);                 % signal phase vector
u_basic=u_amp.*exp(j*u_phase*pi);    % signal complex envelope (no
                                     %   frequency modulation)

fcode=1;                             % signal frequency modulation flag
f_basic=zeros(1,13);                 % signal frequency modulation vector

% set default values for the signal and plot - ambiguity grid parameters
F=5;                    % maximal displayed Doppler is F/Mtb
r=10;                   % signal is sampled r times in each tb (r/tb
                        %   sampling rate)
N=100;                  % number of points in the delay axis for ambiguity
                        %   plot is 2N
K=50;df=F/K/length(u_amp);  % number of points in the Doppler axis for
                        %   ambiguity plot is K
```

```
T=1;                         % maximal delay displayed for the ambiguity plot is
                             %    T*Mtb

% initializes flag of signal calculation
sigflag=0;                   %    signal was not calculated yet

% initializes strings for title of the plots
titlest='';

% create amplitude modulation input row
y_u_amp=530;

u_amp_legend = uicontrol(inputfig,'Style','text',...
    'String','Amplitude','Pos',[25 y_u_amp 100 20 ]);

u_amp_text = uicontrol(inputfig,'Style','edit',...
    'String',num2str(u_basic),'Position',[125 y_u_amp 300 20],...
    'Callback',[...
        'sigflag=0;'...
        'set(preset,''value'',12);',...
        'u_amp=str2num(get(u_amp_text,''String''));'...
        'set(u_amp_text,''String'',num2str(u_amp));']);

u_amp_on = uicontrol(inputfig,'Style','radio',...
    'Pos',[425 y_u_amp 20 20 ],'Value',1,'Callback',[...
        'sigflag=0;'...
        'set(preset,''value'',12);',...
        'set(u_amp_on,''value'',get(u_amp_on,''value'')),'...
        'acode=get(u_amp_on,''value'');']);

% create phase modulation input row
y_u_phase=505;

u_phase_legend = uicontrol(inputfig,'Style','text',...
    'String','Phase/pi','Pos',[25 y_u_phase 100 20 ]);

u_phase_text = uicontrol(inputfig,'Style','edit',...
    'String',num2str(u_phase),'Position',[125 y_u_phase 300 20],...
    'Callback',[...
        'sigflag=0;'...
        'set(preset,''value'',12);',...
        'u_phase=str2num(get(u_phase_text,''String''));'...
        'set(u_phase_text,''String'',num2str(u_phase));']);

u_phase_on = uicontrol(inputfig,'Style','radio',...
    'Pos',[ 425 y_u_phase 20 20 ],'Value',1,'Callback',[...
        'sigflag=0;'...
        'set(preset,''value'',12);',...
            'set(u_phase_on,''value'',get(u_phase_on,''value'')),'...
        'pcode=get(u_phase_on,''value'');']);

% create frequency modulation input row
y_freq=480;

freq_legend = uicontrol(inputfig,'Style','text',...
    'String','Frequency*tb','Pos',[25 y_freq 100 20 ]);

freq_text = uicontrol(inputfig,'Style','edit',...
    'String',num2str(f_basic),'Position',[125 y_freq 300 20],...
    'Callback',[...
```

```
            'sigflag=0;'...
            'set(preset,''value'',12);',...
            'f_basic=str2num(get(freq_text,''String''));'...
            'set(freq_text,''String'',num2str(f_basic));']);

freq_on = uicontrol(inputfig,'Style','radio',...
    'Pos',[ 425 y_freq 20 20 ],'Value',1,'Callback',[...
            'sigflag=0;'...
            'set(preset,''value'',12);',...
            'set(freq_on,''value'',get(freq_on,''value'')),'...
            'fcode=get(freq_on,''value'');']);

% create F input slider
y_F = 375;

sli_F = uicontrol(inputfig,'Style','Slider','sliderstep',[0.01 0.1],...
    'Position',[125 y_F 300 20],'Min',0,'Max',60,'Value',F,'Callback',[...
            'set(F_cur,''String'','',...
            'num2str(get(sli_F,''Val''))),',...
            'F=get(sli_F,''Val'');',...
            'df=F/K/length(u_amp);']);

F_cur = uicontrol(inputfig,'style','edit','Pos',[ 425 y_F 50 20 ],...
    'String',num2str(get(sli_F,'Value')),'Callback',[...
            'set(sli_F,''Val'','',...
            'str2num(get(F_cur,''String''))),',...
            'F=get(sli_F,''Val'');',...
            'df=F/K/length(u_amp);']);

F_lbl = uicontrol(inputfig,'style','Text',...
    'Pos',[ 25 y_F 100 20 ],'String','F*Mtb');

% create r input slider
y_r = 400;

sli_r = uicontrol(inputfig,'Style','Slider','Position',[125 y_r 300 20],...
    'Min',1,'Max',300,'Value',r,'Callback',[...
            'sigflag=0;'...
            'set(r_cur,''String'',num2str(floor(get(sli_r,''Val'')))),',...
            'r=floor(get(sli_r,''Val''));']);

r_cur = uicontrol(inputfig,'style','edit','Pos',[425 y_r 50 20],...
    'String',num2str(floor(get(sli_r,'Value'))),'Callback',[...
            'sigflag=0;'...
            'set(sli_r,''Val'',str2num(get(r_cur,''String''))),',...
            'r=floor(get(sli_r,''Val''));']);

r_lbl = uicontrol(inputfig,'style','Text',...
    'Pos',[25 y_r 100 20],'String','r');

% create N input slider
y_N = 325;

sli_N = uicontrol(inputfig,'Style','Slider','Position',[125 y_N 300 20],...
    'Min',1,'Max',300,'Value',N,'Callback',[...
            'set(N_cur,''String'','',...
            'num2str(floor(get(sli_N,''Val'')))),',...
            'N=floor(get(sli_N,''Val''));']);

N_cur = uicontrol(inputfig,'style','edit',...
```

```
        'Pos',[425 y_N 50 20],'String',num2str(floor(get(sli_N,'Value'))),...
            'Callback',['set(sli_N,''Val'',',...
            'str2num(get(N_cur,''String''))),',...
            'N=floor(get(sli_N,''Val''));']);

N_lbl = uicontrol(inputfig,'style','Text',...
    'Pos',[25 y_N 100 20],'String','N');

% create T input slider
y_T = 350;

sli_T = uicontrol(inputfig,'Style','Slider','Position',[125 y_T 300 20],...
    'Min',0.01,'Max',1.1,'Value',T,'sliderstep',[0.01 0.1],'Callback',[...
        'set(T_cur,''String'',',...
        'num2str(get(sli_T,''Val''))),',...
        'T=get(sli_T,''Val'');']);

T_cur = uicontrol(inputfig,'style','edit',...
    'Pos',[425 y_T 50 20],'String',num2str(get(sli_T,'Value')),'Callback',[...
        'set(sli_T,''Val'',',...
        'str2num(get(T_cur,''String''))),',...
        'T=get(sli_T,''Val'');']);

T_lbl = uicontrol(inputfig,'style','Text',...
    'Pos',[25 y_T 100 20],'String','T');

% create K input slider
y_K = 300;

sli_K = uicontrol(inputfig,'Style','Slider','Position',[125 y_K 300 20],...
    'Min',1,'Max',200,'Value',K,'Callback',[...
        'set(K_cur,''String'',',...
        'num2str(floor(get(sli_K,''Val'')))),',...
        'K=floor(get(sli_K,''Val''));',...
        'df=F/K/length(u_amp);']);

K_cur = uicontrol(inputfig,'style','edit',...
    'Pos',[425 y_K 50 20],'String',num2str(floor(get(sli_K,'Value'))),...
        'Callback',[...
        'set(sli_K,''Val'',',...
        'str2num(get(K_cur,''String''))),',...
        'K=floor(get(sli_K,''Val''));',...
        'df=F/K/length(u_amp);']);

K_lbl = uicontrol(inputfig,'style','Text',...
    'Pos',[25 y_K 100 20],'String','K');

% create calculate and draw ambiguity function push button
pushtocalculate1=uicontrol(inputfig,'Style','Push','Position',...
    [14 23 112 20 ],...
    'String','Ambiguity Fun.','Callback',[...
        'if sigflag==0,'...
        ' calplotsig7;'...
        'end;',...
        'sigflag=1;',...
        'cal_and_plot_amb_fn7;']);

% create calculate and draw signal parameters push button
pushtocalculate2=uicontrol(inputfig,'Style','Push','Position',...
```

```
    [140 23 112 20 ],...
    'String','Cal.&Plot Sig.','Callback',[...
        'sigflag=1;'...
        'calplotsig7']);

% create calculate and draw acfun and spectrum push button
pushtocalculate3=uicontrol(inputfig,'Style','Push','Position',...
    [266 23 112 20 ],...
    'String','ACF.&SPEC Plot.','Callback',[...
        'if sigflag==0,'...
        ' calplotsig7;'...
        'end;',...
        'sigflag=1;',...
        'cal_and_plot_acf_and_spec7;']);

% create calculate and draw acfun and spectrum push button
pushtocalculate3=uicontrol(inputfig,'Style','Push','Position',...
    [400 23 112 20 ],...
    'String','Periodic amb.','Callback',[...
        'if sigflag==0,'...
        ' calplotsig7;'...
        'end;',...
        'sigflag=1;',...
        'cal_and_plot_pamb7;']);

% create save signal and grid definition push button
pushtocalculate4=uicontrol(inputfig,'Style','Push','Position',...
    [100 270 112 20 ],...
    'String','Save parameters','Callback',[...
        '[newmatfile,newpath] = uiputfile(''*.mat'', ''Save As'');',...
        'if newpath~=0,',...
        ' wd=cd;',...
        '          ' cd(newpath);',...
        ' eval([''save '' newmatfile '' acode pcode fcode u_amp...
            u_phase f_basic F r N K T'']);',...
        ' cd(wd);',...
        'end']);

% create load signal and grid definition push button
pushtocalculate5=uicontrol(inputfig,'Style','Push','Position',...
    [300 270 112 20 ],...
    'String','Load parameters','Callback',[...
        '[newmatfile,newpath] = uigetfile(''*.mat'', ''Load'');',...
        'if newpath =0,',...
        ' wd=cd;',...
        '          ' cd(newpath);',...
        ' eval([''load '' newmatfile '' acode pcode fcode u_amp u_phase...
            f_basic F r N K T'']);',...
        ' cd(wd);',...
        ' set(u_amp_text,''String'',num2str(u_amp));',...
        ' set(u_amp_on,''Value'',acode);',...
        ' set(u_phase_text,''String'',num2str(u_phase));',...
        ' set(u_phase_on,''Value'',pcode);',...
        ' set(freq_text,''String'',num2str(f_basic));',...
        ' set(freq_on,''Value'',fcode);',...
        ' set(sli_F,''Value'',F);',...
        ' set(F_cur,''String'',num2str(F));',...
        ' set(sli_r,''Value'',r);',...
        ' set(r_cur,''String'',num2str(r));',...
        ' set(sli_N,''Value'',N);',...
```

```
            '   set(N_cur,''String'',num2str(N));',...
            '   set(sli_K,''Value'',K);',...
            '   set(K_cur,''String'',num2str(K));',...
            '   set(sli_T,''Value'',T);',...
            '   set(T_cur,''String'',num2str(T));',...
            '   set(preset,''value'',12);',...
            '   df=F/K/length(u_amp);',...
       'end']);

% create listbox for selection of preset type of signals
presetnames={'Pulse','LFM ','Weighted LFM','Costas, 7 elements ','Barker,...
      13 elements','Frank, 16 elements','Complementary pair',...
      'Pulse train, 6 pulses','Stepped frequency pulse train, 6 pulses...
          (zoom)',...
      'Weighted stepped frequency, 8 pulses (zoom)','P4, 25 elements',...
          'User Defined'};

z=[0 0 0 0];
presetamp={ones(1,13),ones(1,51),ones(1,51).*sqrt(chebwin(51,50))',...
       ones(1,7),ones(1,13),ones(1,16),[ones(1,3) zeros(1,7) ones(1,3)],...
       [0 0 1 z 1 z 1 z 1 z 1 z 1 0 0], [0 0 1 z 1 z 1 z 1 z 1 z 1 0 0],...
       [0 0 1 z 1 z 1 z 1 z 1 z 1 z 1 0 0].*sqrt(chebwin(40,50))',...
       ones(1,25),1};

presetphase={zeros(1,13),zeros(1,51),zeros(1,51),zeros(1,7),...
       [0 0 0 0 1 1 0 0 1 0 1 0],...
       [0 0 0 0 0 1/2 1 -1/2 0 1 0 1 0 -1/2 1 1 1/2],...
       [0 0 1 zeros(1,7) 0 1/2 0],zeros(1,30),zeros(1,30),...
       zeros(1,40),1/25*(0:24).^2-(0:24),0};

presetfreq={zeros(1,13),.0031*[-25:25],.0031*[-25:25],[4 7 1 6 5 2 3]-4,...
       zeros(1,13),zeros(1,16),zeros(1,13),zeros(1,30),...
       .78*[0 0 -2.5 z -1.5 z -.5 z .5 z 1.5 z 2.5 0 0],...
       .7*[0 0 -3.5 z -2.5 z -1.5 z -.5 z .5 z 1.5 z 2.5 z 3.5 0 0],...
       zeros(1,25),0};

presetdf={.007,.0032,.0032,.095,.0308,.0312,.03,.01,.0075,.0075,.0266,.01};

presetr={10,1,1,20,10,10,10,10,40,40,5,100};

presetN={130,50,50,140,130,160,130,285,45,45,117,100};

presetK={50,60,60,60,50,50,60,60,60,60,90,100};

presetvalues=struct('Name',presetnames,'amp',presetamp,'phase',presetphase,...
     'freq',presetfreq,'df',presetdf,'r',presetr,'N',presetN,'K',presetK);

preset = uicontrol(inputfig,'Style','ListBox','position',[20 610 480 50],...
     'String',presetnames,'callback',[...
            '  sigflag=0;',...
            '  df=presetvalues(get(preset,''value'')).df;',...
            '  r=presetvalues(get(preset,''value'')).r;',...
            '  set(r_cur,''String'',num2str(r));',...
            '  set(sli_r,''val'',r);',...
            '  N=presetvalues(get(preset,''value'')).N;',...
            '  set(N_cur,''String'',num2str(N));',...
            '  set(sli_N,''val'',N);',...
            '  K=presetvalues(get(preset,''value'')).K;',...
            '  set(K_cur,''String'',num2str(K));',...
            '  set(sli_K,''val'',K);',...
```

```
'          u_amp=presetvalues(get(preset,''value'')).amp;',...
' set(u_amp_text,''String'',...
    num2str(presetvalues(get(preset,''value'')).amp));',...
' set(u_amp_on,''val'',1);',...
' T=N/r/length(u_amp);',...
' set(T_cur,''String'',num2str(T));',...
' set(sli_T,''val'',T);',...
' F=df*K*length(u_amp);',...
' set(F_cur,''String'',num2str(F));',...
' set(sli_F,''val'',F);',...
' u_phase=presetvalues(get(preset,''value'')).phase;',...
' set(u_phase_text,''String'',...
    num2str(presetvalues(get(preset,''value'')).phase));',...
' set(u_phase_on,''val'',1);',...
' set(freq_text,''String'',...
    num2str(presetvalues(get(preset,''value'')).freq));',...
' set(freq_on,''val'',1);',...
' f_basic=presetvalues(get(preset,''value'')).freq;']);

% add credits and legend

credits=uicontrol(inputfig,'style','text','position',...
   [20 60 480 50],'String',...
   ['Written by Eli Mozeson and Nadav Levanon' sprintf('\n') ...
      'Dept. of EE-Systems, Tel Aviv University']);

legend=uicontrol(inputfig,'style','text','position',[20 120 480 140],...
   'HorizontalAlignment','left',...
   'String',...
   [  'Legend:',sprintf('\n') ...
      'M  - number of bits (length of phase/amplitude/frequency vector)',...
         sprintf('\n') ...
      'tb - length of each bit (signal length is M*tb)',sprintf('\n') ...
      'r  - number of samples per bit',sprintf('\n') ...
      'F  - maximal Doppler shift for ambiguity and spectrum plot...
         normalized by M*tb',sprintf('\n') ...
      'T  - maximal delay normalized to M*tb',sprintf('\n') ...
      'N  - number of delay bins on each side of ambiguity plot',...
         sprintf('\n') ...
      'K  - number of positive Doppler shifts on ambiguity plot']);

legend0=uicontrol(inputfig,'style','text','position',[20 660 480 20],...
   'HorizontalAlignment','left',...
   'String','Chose a signal from the preset list or define your own using...
      the GUI');

legend1=uicontrol(inputfig,'style','text','position',[20 560 480 40],...
   'HorizontalAlignment','left',...
   'String',...
   [  'Edit text to change amplitude, phase or frequency vectors',...
         sprintf('\n') ...
      'Use radio buttons on the right to activate or deactivate modulation']);

legend2=uicontrol(inputfig,'style','text','position',[20 430 480 40],...
   'HorizontalAlignment','left',...
   'String',...
   ['Use sliders to continuously increase or decrease values',sprintf('\n') ...
      'or use text on the right to enter any value']);

figure(inputfig);
```

calplotsig7.m

```
% calplotsig7.m - Written by Eli Mozeson and Nadav Levanon
% used by ambfn7 for calculation and plot of the signal when the signal is
% defined by u_amp, u_phase and u_freq (single carrier signal)
% output variables include u(t) amd t on which u is defined

tb=1;

if (acode==1)*(pcode==1),
    u_basic=u_amp.*exp(j*u_phase*pi);
elseif (acode==1)*(pcode==0),
    u_basic=u_amp;
elseif (acode==0)*(pcode==1),
    u_basic=exp(j*u_phase*pi);
else
    u_basic=ones(size(u_amp));
end

m_basic=length(u_basic);

if r==1
    dt=tb;
    m=m_basic;
    uamp=abs(u_basic);
    phas=uamp*0;
    phas=angle(u_basic);
    if fcode==1
        phas=phas+2*pi*dt*cumsum(f_basic);
    end
    uexp=exp(j*phas);
    u=uamp.*uexp;
else                            % i.e., several samples within a bit
    dt=tb/r;                    % interval between samples
    ud=diag(u_basic);
    ao=ones(r,m_basic);
    m=m_basic*r;
    u_basic=reshape(ao*ud,1,m);  % u_basic with each element repeated r times
    uamp=abs(u_basic);
    phas=angle(u_basic);
    u=u_basic;
    if fcode==1
        ff=diag(f_basic);
        phas=2*pi*dt*cumsum(reshape(ao*ff,1,m))+phas;
        uexp=exp(j*phas);
        u=uamp.*uexp;
    end
end

tscale=[0:length(uamp)-1]/r;
tscale1=[0 0:length(uamp)-1 length(uamp)-1]/r;
dphas=[NaN diff(phas)]*r/2/pi;

figure(sigfig), clf, hold off % plot the signal parameters
subplot(3,1,1)
plot(tscale1,[0 abs(uamp) 0],'linewidth',1.5)
ylabel(' Amplitude ')
titlest=presetvalues(get(preset,'value')).Name;
title(titlest);
axis([-inf inf 0 1.2*max(abs(uamp))])
```

```
subplot(3,1,2)
plot(tscale, phas,'linewidth',1.5)
axis([-inf inf -inf inf])
ylabel(' phase [rad] ')

subplot(3,1,3)
plot(tscale,dphas*ceil(max(tscale)),'linewidth',1.5)
axis([-inf inf -inf inf])
xlabel(' \itt / t_b ')
ylabel(' \itf * Mt_b ')

% variables for ambigity calculations
t=tscale;
u=u;
```

cal_and_plot_amb_fn7.m

```
% cal_and_plot_amb_fn7.m - Written by Eli Mozeson and Nadav Levanon
% calculates and plots the ambiguity function of a signal

% assumes that the work space includes a row vector u (signal complex envelope)
% and row vector t (time vector) where t is the equally spaced time points on
% which u is defined

% define the delay vector on which the ambiguity plot is calculated
dt=t(2)-t(1);           % dt is the sampling period of u(t)
m =length(t);           % total number of samples is u(t)
%T                      % normalized maximal delay (defined externally)
%N                      % number of grid points on each side of the delay
                        % axis (defined externally)

% calculate a delay vector with N+1 points that spans from zero delay to
% ceil(T*t(m)) notice that the delay vector does not have to be equally spaced
% but must have all entries as integer multiples of dt

% two cases are possible
%              a) T*m>=N - the signal is oversampled relative to the...
%                  delay axis definition
%              b) T*m<N - the signal is undersampled (decrease N,...
%                  increase T or increase r)
if T*m<N,
   msgbox(['N is too large, or r is too low for current definition of T.' ...
        'Using N=' sprintf('%d',ceil(T*m)) ' instead of...
            N=' sprintf('%d',N)],'Warnning !!!');
   Nused=ceil(T*m);
else
   Nused=N;
end

dtau=ceil(T*m)*dt/Nused;
tau=round([0:1:Nused]*dtau/dt)*dt;

% df              % spacing between adjacent grid points on the Doppler axis
                  %    (defined externally)
% calculate K+1 equally spaced grid points of Doppler axis with df spacing
f=[0:1:K]*df;

% duplicate Doppler axis to show also negative dopplers (0 Doppler
```

```
%     is calculated twice)
f=[-fliplr(f) f];

% calculate ambiguity function using sparse matrix manipulations (no loops)

% define a sparse matrix based on the signal samples u1 u2 u3 ... um
% with size m+ceil(T*m) by m (notice that u' is the conjugate transpoze of u)
% where the top part is diagonal (u*) on the diagonal and the bottom part is
%     a zero matrix
%
%                         [u1*  0    0   0 ...  0 ]
%                         [ 0   u2*  0   0 ...  0 ]
%                         [ 0   0    u3* 0 ...  0 ]   m rows
%                         [ .              .    ]
%                         [ .         .    .    ]
%                         [ .   0    0 . ...   um*]
%                         [ 0             0     ]
%                         [ .             .    ]     Nused rows
%                         [ 0   0    0   0 ...  0 ]
%
mat1=spdiags(u',0,m+ceil(T*m),m);

% define a convolution sparse matrix based on the signal samples u1 u2 u3...um
% where each row is a time(index) shifted versions of u.
% each row is shifted tau/dt places from the first row
% the minimal shift (first row) is zero
% the maximal shift (last row) is ceil(T*m) places
% the total number of rows is Nused+1
% number of columns is m+ceil(T*m)

% for example, when tau/dt=[0 2 3 5 6] and Nused=4
%
%                [u1 u2 u3 u4 ...            ... um 0  0  0  0  0]
%                [ 0  0 u1 u2 u3 u4 ...         ... um 0  0  0  0]
%                [ 0  0  0 u1 u2 u3 u4 ...         ... um 0  0  0]
%                [ 0  0  0  0  0 u1 u2 u3 u4 ...         ... um 0]
%                [ 0  0  0  0  0  0 u1 u2 u3 u4 ...         ... um]
%

% define a row vector with ceil(T*m)+m+ceil(T*m) places by padding u with
%     zeros on both sides
u_padded=[zeros(1,ceil(T*m)),u,zeros(1,ceil(T*m))];

% define column indexing and row indexing vectors
cidx=[1:m+ceil(T*m)];
ridx=round(tau/dt)';

% define indexing matrix with Nused+1 rows and m+ceil(T*m) columns
% where each element is the index of the correct place in the padded
%     version of u
index = cidx(ones(Nused+1,1),:) + ridx(:,ones(1,m+ceil(T*m)));

% calculate matrix
mat2 = sparse(u_padded(index));

% calculate the ambiguity matrix for positive delays given by
%
% [u1 u2 u3 u4 ...            ... um 0  0  0  0  0] [u1*  0    0   0 ...  0 ]
% [ 0  0 u1 u2 u3 u4 ...         ... um 0  0  0  0] [ 0   u2*  0   0 ...  0 ]
% [ 0  0  0 u1 u2 u3 u4 ...         ... um 0  0  0]*[ 0   0    u3* 0 ...  0 ]
```

```
% [ 0  0  0  0  0 u1 u2 u3 u4 ...          ... um  0] [ .                    . ]
% [ 0  0  0  0  0  0 u1 u2 u3 u4 ...       ... um]   [ .            .       . ]
%                                                    [ .  0  0  . ...      um*]
%                                                    [ 0                    0 ]
%                                                    [ .                    . ]
%                                                    [ 0  0  0  0 ...       0 ]
%
% where there are m columns and Nused+1 rows and each element gives an element
% of multiplication between u and a time shifted version of u*. each row gives
% a different time shift of u* and each column gives a different entry in u.
%
%
uu_pos=mat2*mat1;

clear mat2 mat1

% calculate exponent matrix for full calculation of ambiguity function.
% The exponent matrix is 2*(K+1) rows by m columns where each row represents a
% possible Doppler and each column stands for a different place in u.
e=exp(-j*2*pi*f'*t);

% calculate ambiguity function for positive delays by calculating the integral
% for each possible delay and Doppler over all entries in u.
% a_pos has 2*(K+1) rows (Doppler) and Nused+1 columns (Delay)
a_pos=abs(e*uu_pos');

% normalize ambiguity function to have a maximal value of 1
a_pos=a_pos/max(max(a_pos));

% use the symmetry properties of the ambiguity function to transform the
% negative Doppler positive delay part to negative delay, positive Doppler
a=[flipud(conj(a_pos(1:K+1,:))) fliplr(a_pos(K+2:2*K+2,:))];

% define new delay and Doppler vectors
delay=[-fliplr(tau) tau];
freq=f(K+2:2*K+2)*ceil(max(t));

% exclude the zero Delay that was taken twice
delay=[delay(1:Nused) delay((Nused+2):2*(Nused+1))];
a=a(:,[1:Nused (Nused+2):2*(Nused+1)]);


% plot the ambiguity function and autocorrelation cut
[amf amt]=size(a);

% create an all blue color map
cm=zeros(64,3);
cm(:,3)=ones(64,1);

figure(ambfig), clf, hold off
mesh(delay, [0 freq], [zeros(1,amt);a])

hold on
surface(delay, [0 0], [zeros(1,amt);a(1,:)])

colormap(cm)
view(-40,50)
axis([-inf inf -inf inf 0 1])
xlabel(' {\it\tau}/{\itt_b}','Fontsize',12);
ylabel(' {\it\nu}*{\itMt_b}','Fontsize',12);
```

```
zlabel(' |{\it\chi}({\it\tau},{\it\nu})| ','Fontsize',12);
title(titlest);
hold off
```

cal_and_plot_acf_and_spec7.m

```
% cal_and_plot_acf_and_spec7.m - Written by Eli Mozeson and Nadav Levanon
% plot correlation function and spectrum of a signal defined by
% u(t), t, and F (maximal Doppler normalized by signal length)

% calculate normalized ACF
acfun=20*log10(abs(xcorr(u))+eps);
acfun=acfun-max(acfun);
acfun=max(acfun,-60);
acfun=acfun(1:(length(acfun)+1)/2);
acfun=fliplr(acfun);
scalet=[0:length(acfun)-1]/(length(acfun)-1)*t(length(t));
% calculate spectrum
fftlength=max(1024*8,length(u)*32);
spec=20*log10(max(abs(fft(u,fftlength)),eps));
spec=spec-max(spec);
spec=max(spec,-60);
spec=spec(1:fftlength/2);
scales=[0:fftlength/2-1]/(fftlength/2-1)*r/2;

% calculate normalized PACF
zeru=zeros(size(u));
pacfun=20*log10(abs(xcorr([u u u],[zeru u zeru]))+eps);
clear zeru;
pacfun=pacfun-max(pacfun);
pacfun=max(pacfun,-60);
pacfun=pacfun(3*length(u):4*length(u)-1);

figure(acffig);
set(acffig,'Visible','on');
subplot(311);
plot(scalet,acfun);
xlabel('{\it\tau}/\itt_b');
ylabel('Autocorrelation [dB]');
title(titlest);
axis([0 max(scalet) -60 0]);

subplot(312);
plot(scalet,pacfun);
xlabel('{\it\tau}/\itt_b');
ylabel('Periodic Autocorrelation [dB]');
axis([0 max(scalet) -60 0]);

subplot(313);
plot(scales*length(u)/r,spec);
xlabel('{\itf*Mt_b}');
ylabel('|{\itS}({\itf})|');
axis([0 F -60 0]);
```

cal_and_plot_pamb7.m

```
% cal_and_plot_pamb7.m - written by Eli Mozeson and Nadav Levanon
% calculates and plots the periodic ambiguity function of a signal
```

```
% assumes that the work space includes a row vector u (signal complex envelope)
% and row vector t (time vector) where t is the equally spaced time points on
% which u is defined

% define the delay vector on which the ambiguity plot is calculated
dt=t(2)-t(1);           % dt is the sampling period of u(t)
m =length(t);           % total number of samples is u(t)
%T                      % normalized maximal delay (defined externally)
%N                      % number of grid points on each side of the delay axis
                        %    (defined externally)

% calculate a delay vector with N+1 points that spans from zero delay to
% ceil(T*t(m)) notice that the delay vector does not have to be equally spaced
% but must have all entries as integer multiples of dt

% two cases are possible
%               a) T*m>=N - the signal is oversampled relative to the delay
%                           axis definition
%               b) T*m<N  - the signal is undersampled (decrease N, increase
%                           T or increase r)
if ceil(T*m)>m,
    msgbox(['T is larger then 1 using 1 (single period) instead'],...
        'Warning !!!');
    Tused=1;
else
    Tused=T;
end

if Tused*m<N,
    msgbox(['N is too large, or r is too low for current definition of T.' ...
          'Using N=' sprintf('%d',ceil(Tused*m)) ' instead of N='
             sprintf('%d',N)],'Warnning !!!');
    Nused=ceil(Tused*m);
else
    Nused=N;
end

dtau=ceil(Tused*m)*dt/Nused;
tau=round([0:1:Nused]*dtau/dt)*dt;

% df      spacing between adjacent grid points on the Doppler axis
%         (defined externally) calculate K+1 equally spaced grid points
%         of Doppler axis with df spacing
f=[0:1:K]*df;

% duplicate Doppler axis to show also negative Dopplers (0 Doppler is...
%    calculated twice)
f=[-fliplr(f) f];

% calculate ambiguity function using sparse matrix manipulations (no loops)

% define a sparse matrix based on the signal samples u1 u2 u3 ... um
% with size m+ceil(T*m) by m (notice that u' is the conjugate transpose of u)
% where the top part is diagonal (u*) on the diagonal and the bottom part
% is a zero matrix
%
%                 [u1*  0    0   0  ...   0  ]
%                 [ 0  u2*   0   0  ...   0  ]
%                 [ 0   0   u3*  0  ...   0  ]  m rows
%                 [ .    .         .          ]
```

```
%                [ .     .        .        ]
%                [ .   0 0     . ... um*]
%                [ 0                 0  ]
%                [ .              .  ]    Nused rows
%                [ 0   0 0 0 ... 0  ]
%
mat1=spdiags(u',0,m+ceil(Tused*m),m);

% define a cyclic convolution sparse matrix based on the signal samples
% u1 u2 u3 ... um where each row is a cyclic time(index) shifted versions of u.
% each row is shifted tau/dt places from the first row
% the minimal shift (first row) is zero
% the maximal shift (last row) is ceil(T*m) places
% the total number of rows is Nused+1
% number of columns is m+ceil(T*m)

% for example, when tau/dt=[0 2 3 5 6] and Nused=4
%
%              [u1    u2   u3  u4 ...              ... um u1 u2 u3 u4 u5 u6]
%              [um-1  um   u1  u2 u3 u4 ...        ... um u1 u2 u3 u4]
%              [um-2 ...   um  u1 u2 u3 u4 ...     ... um u1 u2 u3]
%              [um-4        ... um u1 u2 u3 u4 ...             ... um u1]
%              [um-5           ... um u1 u2 u3 u4 ...              ... um]
%

% define a row vector with ceil(T*m)+m+ceil(T*m) places by padding...
%    cyclically padding u
u_padded=[u(m-ceil(Tused*m)+1:m),u,u(1:ceil(Tused*m))];

% define column indexing and row indexing vectors
cidx=[1:m+ceil(Tused*m)];
ridx=round(tau/dt)';

% define indexing matrix with Nused+1 rows and m+ceil(T*m) columns
% where each element is the index of the correct place in the padded...
% version of u
index = cidx(ones(Nused+1,1),:) + ridx(:,ones(1,m+ceil(Tused*m)));

% calculate matrix
mat2 = sparse(u_padded(index));

% calculate the periodic ambiguity matrix for positive delays given by
%
% where there are m columns and Nused+1 rows and each element gives an element
% of multiplication between u and a cyclically time shifted version of u*. Each
% row gives a different time shift of u* and each column gives a different
% entry in u.
%
uu_pos=mat2*mat1;

clear mat2 mat1

% calculate exponent matrix for full calculation of ambiguity function. The
% exponent matrix is 2*(K+1) rows by m columns where each row represents a
% possible Doppler and each column stands for a different place in u.
e=exp(-j*2*pi*f'*t);

% calculate ambiguity function for positive delays by calculating the integral
% for each possible delay and Doppler over all entries in u.
% a_pos has 2*(K+1) rows (Doppler) and Nused+1 columns (Delay)
```

```
a_pos=abs(e*uu_pos');

% normalize ambiguity function to have a maximal value of 1
a_pos=a_pos/max(max(a_pos));

% use the symmetry properties of the ambiguity function to transform the
% negative Doppler positive delay part to negative delay, positive Doppler
a=[flipud(conj(a_pos(1:K+1,:))) fliplr(a_pos(K+2:2*K+2,:))];

% define new delay and Doppler vectors
delay=[-fliplr(tau) tau];
freq=f(K+2:2*K+2)*ceil(max(t));

% excludes the zero Delay that was taken twice
delay=[delay(1:Nused) delay((Nused+2):2*(Nused+1))];
a=a(:,[1:Nused (Nused+2):2*(Nused+1)]);

% plot the ambiguity function and autocorrelation cut
[amf amt]=size(a);

% create an all blue color map
cm=zeros(64,3);
cm(:,3)=ones(64,1);

figure(pambfig), clf, hold off
mesh(delay, [0 freq], [zeros(1,amt);a])

hold on
surface(delay, [0 0], [zeros(1,amt);a(1,:)])

colormap(cm)
view(-40,50)
axis([-inf inf -inf inf 0 1])
xlabel(' {\it\tau}/{\itt_b}','Fontsize',12);
ylabel(' {\it\nu}*{\itMt_b}','Fontsize',12);
zlabel(' |{\it\chi}({\it\tau},{\it\nu})| ','Fontsize',12);
title(titlest);
hold off
```

A.2 CREATING COMPLEX SIGNALS FOR USE WITH AMBFN1.M OR AMBFN7.M

The u_amp, u_phase and/or f_basic vectors of complex signals needed for the plotting programs may require considerable preparation. Some examples were given in Appendixes 6B and 6C and in Appendix 9A. The latter one, steplfm.m, generates a stepped-frequency train of LFM pulses. By changing its parameters, many different signals can be created. To allow easy parameter changes without rerunning the program, a GUI version is presented. The GUI version involves two m-files, steplfm_gui.m and steplfm_cal.m. Note that steplfm_gui.m is called when the GUI of ambfn7.m is already present. When steplfm_gui.m is entered at the MATLAB workspace, the GUI shown in Fig. A.2 appears. The parameters are the number of pulses n, the ratio of pulse repetition period to pulse duration T_r/T, the pulse time–bandwidth product $T \cdot B$, and the time frequency–step product $T \cdot \Delta f$. The first toggle determines if the LFM slopes are all positive or alternate. The second toggle determines if the frequency steps are ordered linearly or according to Costas. The "Cal. Sig. Amplitude and Frequency" bar should be clicked on before the corresponding "Cal. & Plot Sig." bar in the ambfn7.m GUI is clicked on.

steplfm_gui.m

```
% steplfm_gui.m - stepped train of LFM pulses with GUI
% written by Eli Mozeson and Nadav Levanon
% creates signal for use with ambiguity function plotting program

% create the signal parameters input figure
steplfmfig=figure;
set(steplfmfig,'Position',[460 51 520 390],'MenuBar','none','visible','on');

% set default values for the steplfm signal
tpdf=3;
tpb=18;
trovertp=10;
nn=8;
rv=-1;
cost=0;

% create tpdf input slider
y_tpdf = 175;
```

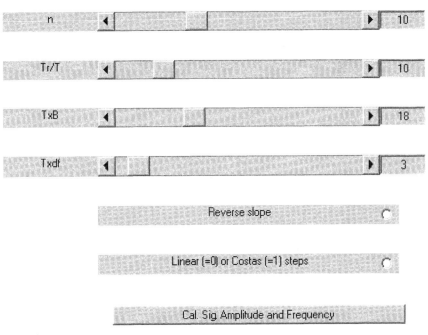

FIGURE A.2 GUI of steplfm_gui.m.

```
sli_tpdf = uicontrol(steplfmfig,'Style','Slider',...
    'sliderstep',[0.01 0.1],...
    'Position',[125 y_tpdf 300 20],'Min',0,'Max',60,'Value',tpdf,...
        'Callback',[...
        'set(tpdf_cur,''String'','',...
        'num2str(get(sli_tpdf,''Val''))),',...
        'tpdf=get(sli_tpdf,''Val'');']);

tpdf_cur = uicontrol(steplfmfig,'style','edit','Pos',[ 425 y_tpdf 50 20 ],...
    'String',num2str(get(sli_tpdf,'Value')),'Callback',[...
        'set(sli_tpdf,''Val'','',...
        'str2num(get(tpdf_cur,''String''))),',...
        'tpdf=get(sli_tpdf,''Val'');']);

tpdf_lbl = uicontrol(steplfmfig,'style','Text',...
    'Pos',[ 25 y_tpdf 100 20 ],'String','Txdf');

% create tpB input slider
y_tpB = 225;

sli_tpB = uicontrol(steplfmfig,'Style','Slider','sliderstep',[0.01 0.1],...
    'Position',[125 y_tpB 300 20],'Min',0,'Max',60,'Value',tpb,'Callback',[...
        'set(tpB_cur,''String'','',...
        'num2str(get(sli_tpB,''Val''))),',...
        'tpb=get(sli_tpB,''Val'');']);

tpB_cur = uicontrol(steplfmfig,'style','edit','Pos',[ 425 y_tpB 50 20 ],...
    'String',num2str(get(sli_tpB,'Value')),'Callback',[...
        'set(sli_tpB,''Val'','',...
        'str2num(get(tpB_cur,''String''))),',...
        'tpb=get(sli_tpB,''Val'');']);

tpB_lbl = uicontrol(steplfmfig,'style','Text',...
    'Pos',[ 25 y_tpB 100 20 ],'String','TxB');

% create trovertp input slider
y_trovertp = 275;

sli_trovertp = uicontrol(steplfmfig,'Style','Slider','sliderstep',...
    [0.01 0.1],...
    'Position',[125 y_trovertp 300 20],'Min',0,'Max',60,'Value',trovertp,...
    'Callback',['set(trovertp_cur,''String'','',...
        'num2str(get(sli_trovertp,''Val''))),',...
        'trovertp=get(sli_trovertp,''Val'');']);

trovertp_cur = uicontrol(steplfmfig,'style','edit','Pos',...
    [ 425 y_trovertp 50 20 ],...
    'String',num2str(get(sli_trovertp,'Value')),'Callback',[...
        'set(sli_trovertp,''Val'','',...
        'str2num(get(trovertp_cur,''String''))),',...
        'trovertp=get(sli_trovertp,''Val'');']);

trovertp_lbl = uicontrol(steplfmfig,'style','Text',...
    'Pos',[ 25 y_trovertp 100 20 ],'String','Tr/T');

% create nn input slider
y_nn = 325;

sli_nn = uicontrol(steplfmfig,'Style','Slider','sliderstep',...
    [0.01 0.1],...
```

```
    'Position',[125 y_nn 300 20],'Min',1,'Max',30,'Value',trovertp,...
       'Callback',...
       ['set(nn_cur,''String'',',...
       'num2str(floor(get(sli_nn,''Val'')))),',...
       'nn=floor(get(sli_nn,''Val''));']);

nn_cur = uicontrol(steplfmfig,'style','edit','Pos',[ 425 y_nn 50 20 ],...
    'String',num2str(get(sli_nn,'Value')),'Callback',[...
       'set(sli_nn,''Val'',',...
       'floor(str2num(get(nn_cur,''String'')))),',...
       'nn=floor(get(sli_nn,''Val''));']);

trovertp_lbl = uicontrol(steplfmfig,'style','Text',...
    'Pos',[ 25 y_nn 100 20 ],'String','n');

% create reverse slope push botton
y_rv=125;

rv_on = uicontrol(steplfmfig,'Style','radio',...
    'Pos',[425 y_rv 20 20 ],'Value',(rv+1)/2,'Callback',[...
       'set(rv_on,''value'',get(rv_on,''value'')),'...
       'rv=2*get(rv_on,''value'')-1;']);

rv_text = uicontrol(steplfmfig,'Style','text',...
    'String','Reverse slope','Position',[125 y_rv 300 20]);

% create linear / costas push botton
y_cost=75;

cost_on = uicontrol(steplfmfig,'Style','radio',...
    'Pos',[425 y_cost 20 20 ],'Value',cost,'Callback',[...
       'set(cost_on,''value'',get(cost_on,''value'')),'...
       'cost=get(cost_on,''value'');']);

cost_text = uicontrol(steplfmfig,'Style','text',...
    'String','Linear (=0) or Costas (=1) steps','Position',[125 y_cost 300 20]);

% create calculate and draw signal parameters push button
pushtocalculatestplfm=uicontrol(steplfmfig,'Style','Push','Position',...
    [140 23 312 20 ],...
    'String','Cal. Sig. Amplitude and Frequency ','Callback',[...
       'sigflag=0;'...
       'steplfm_cal;',...
       'set(u_amp_text,''String'',num2str(u_amp));',...
       'set(u_amp_on,''Value'',acode);',...
       'set(u_phase_text,''String'',num2str(u_phase));',...
       'set(u_phase_on,''Value'',pcode);',...
       'set(freq_text,''String'',num2str(f_basic));',...
       'set(freq_on,''Value'',fcode);']);
```

steplfm_cal.m

```
% steplfm_cal.m - stepped train of LFM pulses calculation
% written by Eli Mozeson and Nadav Levanon
% used with steplfm_gui.m

mm=50;
mm2=mm/2-1;
ufm=ones(1,mm);
```

```
ddf=tpb/mm^2;
ffm=ddf*(-mm2:mm2+1)-ddf/2;
mtr=round((trovertp-1)*mm);

space1=zeros(1,mtr);
u_step1=[ufm space1];
f_step1=[ffm space1];
f1=-(nn-1)/2;
u_step=u_step1;
f_step=f_step1;
f_add=[f1*u_step1];

if cost ==1;
    if nn==16
        qq=[6 14 2 3 10 8 11 15 9 1 13 12 5 7 4] ;
        % qq= [2 8 9 12 4 14 10 15 13 7 6 3 11 1 5]
        % qq= [13 8 6 12 11 14 5 15 2 7 9 3 4 1 10]
    elseif nn==8
            qq=[7 2 5 1 6 4 3];
    elseif nn==4
                qq=[2 3 1];
    elseif nn==3
                qq=[2 1];
    end
end
if cost==0  % linear steps
    if nn==16
        qq=[1:15] ;
    elseif nn==8
            qq=[1:7];
    elseif nn==4
                qq=[1:3];
    else
                qq=[1:nn-1];
    end
end

for q= qq
            u_step=[u_step u_step1];
            f_step=[f_step (rv)^q*f_step1];
            f_add=[f_add   (f1+q)*u_step1];
end
f_total=f_step+(tpdf/mm)*f_add;
u_amp = u_step;
f_basic = f_total;
u_phase = zeros(size(u_step));

acode=1;
pcode=0;
fcode=1;
```

A.3 CROSS-AMBIGUITY FUNCTION PLOT

When the reference signal in the receiver is not matched to the signal transmitted, the delay–Doppler response of the receiver output is different from the ambiguity function. The MATLAB code cross_ambfn2.m calculates that response.

This program is different from the ambiguity function plot programs in several respects. For example, because the cross-ambiguity does not obey the symmetry rule that the AF obeys, it was not possible to switch quadrants as was done in ambfn1.m or ambfn7.m.

cross_ambfn2.m calls for two complex vectors representing the amplitude and phase of the complex envelopes of two signals: the transmitted signal and the reference signal. The frequency vector is identical in both signals. When preparing the reference signal it is possible to introduce Doppler compensation, making the receiver matched to nonzero Doppler. An example of a code that prepares both a fixed-amplitude CW periodic signal and a corresponding weighted reference signal is cw_sgnl.m, which is listed in the next section.

cross_ambfn2.m

```
% cross_ambfn2.m - a modification of "ambfn1.m" for plotting cross ambiguity
%                  between two signals OF THE SAME LENGTH
%                  designed to allow mismatch caused by FFT Doppler
%                  processing a pulse train.
%                  the two signals differ only in phases (if there is
%                  frequency mode it is the same in both signals.
% ambfn1.m - plots ambiguity function of a signal u_basic (row vector)
%
% The m-file returns a plot of quadrants 1 and 2 of the cross ambiguity
%    function of a signal
% The ambiguity function is defined as:
%
% a(t,f) = abs ( sumi( u(k)*u'(i-t)*exp(j*2*pi*f*i) ) )
%
% The user is prompted for the signal data:
% u_basic is a row complex vector representing amplitude and phase
% f_basic is a corresponding frequency coding sequence
%
% The duration of each element is tb (total duration of the signal is
%    tb*(m_basic-1))
%
% F is the maximal Doppler shift
% T is the maximal Delay
% K is the number of positive Doppler shifts (grid points)
% N is the number of delay shifts on each side (for a total of 2N+1 points)
% The code allows r samples within each bit
%
% Written by Eli Mozeson and Nadav Levanon, Dept. of EE-Systems, Tel Aviv
%    University

% clear all

% prompt for signal data
u_basic=input(' Signal elements (row complex vector, each element last tb...
   sec) = ? ');
m_basic=length(u_basic);
v_basic=input(' 2nd Signal elements (row complex vector, each element last...
   tb sec) = ? ');

fcode=input(' Allow frequency coding (yes=1, no=0) = ? ');
if fcode==1
```

```
      f_basic=input(' Frequency coding in units of 1/tb (row vector of same...
          length) = ? ');
end

F=input(' Maximal Doppler shift for ambiguity plot [in units of 1/Mtb]...
    (e.g., 1)= ? ');
K=input(' Number of Doppler grid points for calculation (e.g., 100) = ? ');
df=F/K/m_basic;
T=input(' Maximal Delay for ambiguity plot [in units of Mtb] (e.g., 1)= ? ');
N=input(' Number of delay grid points on each side (e.g. 100) = ? ');

sr=input(' Over sampling ratio (>=1) (e.g. 10)= ? ');
r=ceil(sr*(N+1)/T/m_basic);

if r==1
   dt=1;
   m=m_basic;
   uamp=abs(u_basic);
                   vamp=abs(v_basic);
   phas=uamp*0;
   phas=angle(u_basic);
                   phasv=angle(v_basic);
   if fcode==1
      phas=phas+2*pi*cumsum(f_basic);
                   phasv=phas+2*pi*cumsum(f_basic);

   end
   uexp=exp(j*phas);
   u=uamp.*uexp;
                   vexp=exp(j*phasv);
                   v=vamp.*vexp;


else                        % i.e., several samples within a bit
   dt=1/r;                  % interval between samples
   ud=diag(u_basic);
                   vd=diag(v_basic);
   ao=ones(r,m_basic);
   m=m_basic*r;
   u_basic=reshape(ao*ud,1,m); % u_basic with each element repeated r times
   uamp=abs(u_basic);
   phas=angle(u_basic);
   u=u_basic;

   v_basic=reshape(ao*vd,1,m); % v_basic with each element repeated r times
   vamp=abs(v_basic);
   phasv=angle(v_basic);
   v=v_basic;
   if fcode==1
       ff=diag(f_basic);
       phas=2*pi*dt*cumsum(reshape(ao*ff,1,m))+phas;
       uexp=exp(j*phas);
       u=uamp.*uexp;

       phasv=2*pi*dt*cumsum(reshape(ao*ff,1,m))+phasv;
       vexp=exp(j*phasv);
       v=vamp.*vexp;
   end
end
```

```
t=[0:r*m_basic-1]/r;
tscale1=[0 0:r*m_basic-1 r*m_basic-1]/r;
dphas=[NaN diff(phas)]*r/2/pi;
dphasv=[NaN diff(phasv)]*r/2/pi;

% plot the signal parameters
figure(1), clf, hold off
subplot(3,1,1)
plot(tscale1,[0 abs(uamp) 0],'linewidth',1.5)
ylabel(' Amplitude ')
axis([-inf inf 0 1.2*max(abs(uamp))])

subplot(3,1,2)
plot(t, phas,'linewidth',1.5)
axis([-inf inf -inf inf])
ylabel(' Phase [rad] ')

subplot(3,1,3)
plot(t,dphas*ceil(max(t)),'linewidth',1.5)
axis([-inf inf -inf inf])
xlabel(' \itt / t_b ')
ylabel(' \itf * Mt_b ')


% plot the 2nd signal parameters
figure(2), clf, hold off
subplot(3,1,1)
plot(tscale1,[0 abs(vamp) 0],'linewidth',1.5)
ylabel(' Amplitude ')
axis([-inf inf 0 1.2*max(abs(vamp))])

subplot(3,1,2)
plot(t, phasv,'linewidth',1.5)
axis([-inf inf -inf inf])
ylabel(' Phase [rad] ')

subplot(3,1,3)
plot(t,dphasv*ceil(max(t)),'linewidth',1.5)
axis([-inf inf -inf inf])
xlabel(' \itt / t_b ')
ylabel(' \itf * Mt_b ')

% calculate a delay vector with N+1 points that spans from zero delay to
% ceil(T*t(m)) notice that the delay vector does not have to be equally spaced
% but must have all entries as integer multiples of dt

dtau=ceil(T*m)*dt/N;
% tau=round([0:1:N]*dtau/dt)*dt;
tau=round([0:1:2*N]*dtau/dt)*dt;

% calculate K+1 equally spaced grid points of Doppler axis with df spacing
f=[0:1:K]*df;
ff=f;

% duplicate Doppler axis to show also negative Doppler (0 Doppler is calculated
%     twice)
f=[-fliplr(f) f];

% calculate ambiguity function using sparse matrix manipulations (no loops)
```

```
% define a sparse matrix based on the signal samples u1 u2 u3 ... um
% with size m+ceil(T*m) by m (notice that u' is the conjugate transpose of u)
% where the top part is diagonal (u*) on the diagonal and the bottom part is a
% zero matrix
%
%                   [u1*  0    0   0 ...  0  ]
%                   [ 0   u2*  0   0 ...  0  ]
%                   [ 0   0    u3* 0 ...  0  ]   m rows
%                   [ .       .                ]
%                   [ .         .              ]
%                   [ .   0    0   ....  um* ]
%                   [ 0                    0  ]
%                   [ .   .                   ]   N rows
%                   [ 0   0    0   0 ...  0  ]
%
%   mat1=spdiags(u',0,m+ceil(T*m),m);   <====== replaced by the 2nd signal
mat1=spdiags(v',0,m+ceil(T*m),m);


% define a convolution sparse matrix based on the signal samples
% u1 u2 u3 ... um where each row is a time(index) shifted versions of u.
% each row is shifted tau/dt places from the first row
% the minimal shift (first row) is zero
% the maximal shift (last row) is ceil(T*m) places
% the total number of rows is N+1
% number of columns is m+ceil(T*m)

% for example, when tau/dt=[0 2 3 5 6] and N=4
%
%              [u1 u2 u3 u4 ...              ... um 0  0  0  0  0]
%              [ 0  0 u1 u2 u3 u4 ...        ... um 0  0  0  0]
%              [ 0  0  0 u1 u2 u3 u4 ...     ... um 0  0  0]
%              [ 0  0  0  0  0 u1 u2 u3 u4 ...        ... um 0]
%              [ 0  0  0  0  0  0 u1 u2 u3 u4 ...        ... um]

% define a row vector with ceil(T*m)+m+ceil(T*m) places by padding u with zeros
% on both sides u_padded=[zeros(1,ceil(T*m)),u,zeros(1,ceil(T*m))];
u_padded=[zeros(1,ceil(T*m)),u,zeros(1,2*ceil(T*m))];

% define column indexing and row indexing vectors
cidx=[1:m+ceil(T*m)];
ridx=round(tau/dt)';

% define indexing matrix with Nused+1 rows and m+ceil(T*m) columns
% where each element is the index of the correct place in the padded version
% of u index = cidx(ones(N+1,1),:) + ridx(:,ones(1,m+ceil(T*m)));
index = cidx(ones(2*N+1,1),:) + ridx(:,ones(1,m+ceil(T*m)));

[mmm,nnn]=size(index);
% calculate matrix
mat2 = sparse(u_padded(index));

% calculate the ambiguity matrix for positive delays given by
%
% [u1 u2 u3 u4 ...           ... um 0 0 0 0 0 0] [u1*  0   0  0 ...  0 ]
% [ 0  0 u1 u2 u3 u4 ...     ... um 0 0 0 0]    [ 0   u2* 0  0 ...  0 ]
% [ 0  0  0 u1 u2 u3 u4 ...  ... um 0 0 0]*     [ 0   0   u3* 0 ...  0 ]
% [ 0  0  0  0  0 u1 u2 u3 u4 ...   ... um 0]    [ .              .    ]
% [ 0  0  0  0  0  0 u1 u2 u3 u4 ... ... um]    [ .              .    ]
%                                                [ .   0   0  ....  um*]
```

```
%                                                    [ 0          0 ]
%                                                    [ .          . ]
%                                                    [ 0  0  0 0 ... 0 ]

% where there are m columns and N+1 rows and each element gives an element
% of multiplication between u and a time shifted version of u*. each row gives
% a different time shift of u* and each column gives a different entry in u.
%
uu_pos=mat2*mat1;

% clear mat2 mat1

% calculate exponent matrix for full calculation of ambiguity function.
% the exponent matrix is 2*(K+1) rows by m columns where each row represents
% a possible Doppler and each column stands for a different place in u.
% e=exp(-j*2*pi*f'*t);
e=exp(-j*2*pi*ff'*t);

% calculate ambiguity function for positive delays by calculating the integral
% for each possible delay and Doppler over all entries in u.
% a_pos has 2*(K+1) rows (Doppler) and N+1 columns (Delay)
a_pos=abs(e*uu_pos');

% normalize ambiguity function to have a maximal value of 1
a_pos=a_pos/max(max(a_pos));

% use the symmetry properties of the ambiguity function to transform the
% negative Doppler positive delay part to negative delay, positive Doppler
% a=[flipud(conj(a_pos(1:K+1,:))) fliplr(a_pos(K+2:2*K+2,:))];
a=a_pos;

% define new delay and Doppler vectors
delay0=[-fliplr(tau) tau];
% freq=f(K+2:2*K+2)*ceil(max(t));
% freq=f*ceil(max(t));
freq=ff*ceil(max(t));

% excludes the zero Delay that was taken twice
% delay=[delay0(1:N) delay0((N+2):2*(N+1))];
delay=[delay0(N+1:2*N+1) delay0(2*N+3:(3*N+2))];

% a=a(:,[1:N (N+2):2*(N+1)]);

% plot the ambiguity function and autocorrelation cut
[amf amt]=size(a);

% create an all blue color map
cm=zeros(64,3);
cm(:,3)=ones(64,1);

figure(3), clf, hold off
mesh(delay, [0 freq], [zeros(1,amt);a])

hold on
surface(delay, [0 0], [zeros(1,amt);a(1,:)])

colormap(cm)
view(-40,50)
axis([-inf inf -inf inf 0 1])
xlabel(' {\it\tau}/{\itt_b}','Fontsize',12);
ylabel(' {\it\nu}*{\itMt_b}','Fontsize',12);
```

```
zlabel(' |{\it\chi}({\it\tau},{\it\nu})| ','Fontsize',12);
hold off
```

A.4 GENERATING A CW PERIODIC SIGNAL WITH WEIGHTING ON RECEIVE

The m-file cw_sgnl.m prepares a fixed-amplitude CW periodic signal and a corresponding amplitude-weighted reference signal. Both phase-coded and frequency-modulated signals can be generated. The reference signal can have both inter- and intraperiod amplitude weighting. The reference signal can also include Doppler compensation, creating a receiver matched to nonzero Doppler. The transmitted signal contains 48 identical periods. In the reference signal only the middle 16 periods are not of zero amplitude. This allows plotting a *periodic* delay–Doppler response (*periodic* cross-ambiguity function) of a mismatched receiver coherently processing 16 periods of the transmitted signal. The delay scale of the plot can extend as far as ± 16 periods, although plotting ± 1 period will be representative enough. Once the transmitted and reference signals are generated, the plotting can be done using cross_ambfn2.m.

cw_sgnl.m

```
% cw_sgnl.m - prepares a 16 period CW signal
% written by Nadav Levanon
% for use with "cross_ambfn2.m" plotting code
sig_flag=input('Signal:Frank16=0,P4=1,Golomb15=2,sinusoidal-FM=3,...
   triangular-FM=4,sawtooth-FM=5; =? ');

if sig_flag==0
    f1=[0 0 0 0 0 .5 1 -.5 0 1 0 1 0 -.5 1 .5];
    elseif sig_flag==1
        mp4=input(' code length = ? ');
        k=1:mp4;
        f1=-1/mp4*(k-1).^2+(k-1);
    elseif sig_flag==2
        alfa=acos(-7/8);
        f1=[0 0 0 0 alfa alfa alfa 0 alfa alfa 0 0 alfa 0 alfa]./pi;
        mp4=15;
    elseif sig_flag==3
        df_fm=input('Df/fm= ? ');
        mp4=input('frequency vector length = ? ');
        k=1:mp4;
        f1= 0.5/mp4*df_fm*sin(2*pi/mp4*(k-1));  % this is frequency not phase
    elseif sig_flag==4
        df_tr=input('Df*Tr= ? ');
        mp4=input('frequency vector length (even) = ? ');
        k1=1:mp4/2;  k2=mp4/2+1:mp4;
        f1_a=2/mp4^2*df_tr*(k1-mp4/4);
        f1_b=-2/mp4^2*df_tr*(k2-3*mp4/4);
        f1=[f1_a  f1_b];
    elseif sig_flag==5
        df_tr=input('Df*Tr= ? ');
        mp4=input('frequency vector length (odd) = ? ');
```

```
        k1=1:mp4;
        f1=-1/mp4/(mp4-1)*df_tr*(-k1+(mp4+1)/2);
end

f4=[f1 f1 f1 f1];
f4=[f4 f4 f4 f4];   %  N=16
lf=length(f4);
lfsingle=length(f1);
f4_3=[f4 f4 f4];

u_amp0=zeros(size(f4));
u_amp1=ones(size(f4_3));

weight_flag=input(' Inter-period weight: uniform=0, Hamming=1,...
   Hann=2; = ? ');
if weight_flag==0
    u_amp_mid=ones(1,lf);
elseif weight_flag==1
    u_amp_mid=hamming(lf)';
elseif weight_flag==2
    u_amp_mid=hann(lf)';
end
int_weight_flag=input(' Intra-period weight: uniform=0,...
   Hamming=1,  Hann=2;  = ? ');
if int_weight_flag==1
      rw=hamming(lfsingle)';
      rwn=[rw rw rw rw rw rw rw rw rw rw rw rw rw rw rw rw];
      u_amp_mid=u_amp_mid.*rwn;
elseif int_weight_flag==2
      rw=hann(lfsingle)';
      rwn=[rw rw rw rw rw rw rw rw rw rw rw rw rw rw rw rw];
      u_amp_mid=u_amp_mid.*rwn;
end

u_amp_ref=[ u_amp0    u_amp_mid   u_amp0];

sig_trans=u_amp1.*exp(j*pi*f4_3);
sig_ref=u_amp_ref.*exp(j*pi*f4_3);

t_freq=0:length(f4_3)-1;
if sig_flag==3
   sig_trans=u_amp1;
   sig_ref=u_amp_ref;
end

% Adding Doppler equivalent phase ramp in the reference phase slope

dop_flag=input('Add Doppler compensation  (no = 0, yes = 1)  = ? ');
if dop_flag==1
   mult=input(' Dphi*M/pi = ? ');
   dphi=mult*pi/length(f1);
   num_phase_steps=0:length(f4_3)-1;
   phase_ramp=dphi*num_phase_steps;
   sig_ref=sig_ref.*exp(j*phase_ramp);
end

disp(' ')
disp('  Call "cross_ambfn2" ')
disp('  For "signal elements" use "sig_trans" and for "2nd signal elements"...
   use "sig_ref"  ')
```

```
disp('  If an FM signal: for freq coding use "f4_3" ')
disp(' ')
disp('  To plot two periods of the Periodic Cross Amb. Func.  ')
disp('  set maximal delay ... = 0.024 ')
disp(' ')
```

INDEX

A/D, *see* Analog-to-Digital converter
Acceleration, 6
Accuracy, 8
ACF, *see* Autocorrelation function
Acoustic propagation, 4
AF, *see* Ambiguity function
Air traffic control radar, 191
Airborne radar, 14, 168
Almost complementary quadriphase set, 290
Altitude line, 15, 212
Ambiguity, 12
Ambiguity function
  cuts through, 40
  definition, 8, 31, 34, 40, 43, 53, 181
  LFM property, 38
  plotting MATLAB code, 47
  properties, 8–9, 15, 34–38, 42, 45
  volume distribution, 42, 272
Ambiguity function of
  Barker phase-coded pulse, 114
  Biphase pulse, 10
  coherent pulse train, 11, 68–70, 230
  coherent train of LFM pulses, 170–172 177, 181
  Costas frequency stepped pulse, 75, 77, 78, 84, 86
  Costas frequency stepped pulse train, 86
  Frank phase-coded pulse, 118–119, 122
  Huffman coded pulse, 144
  interpulse weighted pulse train, 176
  LFM pulse, 10, 12, 40, 58, 60–62, 64, 72, 170, 229
  minimum peak sidelobe phase-coded pulse, 10
  modified Costas pulse, 261
  multicarrier phase-coded pulse, 328, 330, 349
  NLFM pulse, 90–91, 93
  $P(n, k)$ phase-coded pulse, 131
  P1 phase-coded pulse, 122
  P2 phase-coded pulse, 122
  Px phase-coded pulse, 122
  pulse train, 11
  quadriphase code, 150, 152
  stepped frequency LFM pulse train, 231, 248, 256
  stepped-frequency pulse train, 233
  subcomplementary pulse train, 271
  weighted LFM pulse, 64
  Zadoff–Chu phase coded pulse, 126–127
  unmodulated pulse, 9, 53–55, 58, 68
Amplitude weighting, 16, 57
  coherent pulse train, 177, 226, 304
  LFM pulse, 61, 63, 86, 176
  multicarrier pulse, 335–337, 340–341
  on recieve, 179
  windows, *see* Weight window
Analog-to-digital converter, 146, 209, 238
Angle measurement, 7
A-periodic ACF, *see* Autocorrelation function

INDEX

Asymptotically perfect codes, 132–133
ATC, see Air traffic control radar
Autocorrelation function
   definition, 27, 40
   properties, 41, 103, 105, 122, 161
   sidelobe reduction, 176
Autocorrelation function of
   amplitude weighted LFM pulse, 63
   Barker phase-coded pulse, 106, 142, 145–146, 148, 343
   chirplike phase code, 115
   clutter, 204
   coherent pulse train, 69, 226
   complementary code, 263
   Costas frequency stepped pulse, 78–79, 80, 84–85
   diverse pulse train, 227
   Frank phase-coded pulse, 116
   Huffman-coded pulse, 144–145
   LFM pulse, 59–61, 72, 86
   m-sequence, 133
   modified Barker phase-coded pulse, 152
   modified Costas pulse, 258, 260–262
   modified P4 coded pulse, 16, 152–155
   multicarrier phase-coded pulse, 328, 339, 341–348
   nested codes, 109
   NLFM pulse, 89–92, 94–95
   orthogonal coded LFM pulse train, 279
   orthogonal coded LFM-LFM pulse train, 280–282
   orthogonal coded NLFM-LFM pulse train, 282
   orthogonal coded pulse train, 274–277
   $P(n, k)$ phase-coded pulse, 130
   P4 phase-coded pulse, 130, 155
   phase-coded pulse, 103
   polyphase Barker phase-coded pulse, 113
   Px phase-coded pulse, 120
   quadriphase coded pulse, 149–150, 152
   stepped frequency LFM pulse train, 245–248, 250–254, 256, 258
   stepped frequency pulse train, 231–233, 235, 254, 286
   subcomplementary coded pulse train, 270–271
   unmodulated pulse, 54
   Zadoff–Chu phase-coded pulse, 123, 125–126
Automotive radar, 237, 239

Bandpass filter, 321–322
Bandwidth efficiency, 57
Bandwidth limitations of
   $P(n, k)$ phase-coded pulse, 132
   P4 phase-coded pulse, 132
Bank of filters, 7, 321
Barker code
   ambiguity function, 114
   autocorrelation function, 106
   binary, 106
   definition, 105
   Doppler tolerance, 113
   Gaussian-windowed sinc, 152
   polyphase, see Polyphase Barker code
   properties, 106
   quaternary, see Welti code
   sextic, 113
   table of, 106
   ternary, 113
Baseband signals, 21
Bed of nails AF, 12, 69, 193, 195, 258, 299
Bessel function, 313–314
Binary alexis sequences, 156
Binary complementary sequences, 263
Binary integration, 191, 220
Binomial weights, 205–206
Biphase Golomb code, 134–135
Biphase-to-quadriphase, 147
Blind speed, 195–197, 226
Blind zone, 210–211, 213, 215
BPF, see Bandpass filter
BTQ, see Biphase to quadripahse transform
Burst, 14, 210, 214

Canonical form, 20
Carrier frequency, 2, 5, 18, 21, 23, 43, 210, 333
CDS, see Cyclic difference set
Chaff, 13–14
Chebyshev weight window, 92, 235, 240, 243, 349
Chinese remainder theorem, 213
Chirplike phase code, 113–114, 333
   ambiguity function, 126
   autocorrelation function, 115
   Doppler tolerance, 113
   periodic autocorrelation function, 115
   properties, 265, 333, 340
Chirplike sequence, 265, 339, 349, 365
Chu phase code, 114–115, 122–123, 128
   autocorrelation function, 123, 127
   definition, 122
   properties, 122
CLS, see Chirplike sequence
Clutter, 7, 8, 13, 14, 191, 195, 208, 215, 223
   attenuation, 202, 204

INDEX

autocorrelation function of, 203
average power of, 207
spectral density of, 202–204
spectral width of, 207
visibility, 202
COCS, *see* Consecutive ordered cyclic shifts
COCS of a CLS, 345
Coherent on receive, 7
Coherent pulse train, 6, 42–43, 45, 67, 70, 168, 173, 181, 191, 226
 ambiguity function, 68–70
 autocorrelation function, 69
 definition, 6
 periodic ambiguity function, 12
 resolution, 6, 56, 70
Coherent train of diverse pulses, 226
Coherent train of LFM pulses, 168
Coincidence
 algorithm, 213
 detection, 220
Complementary
 pair, 263–268
  generation of, 268
 pulses, 262–263
 sequence, 263–264, 267, 269–270
 set, 263–267, 269–270, 272–274, 277, 339, 347, 358–359, 361–362, 364
Complex envelope, 20–24, 27, 30–31, 34, 98
Complex envelope of
 coherent LFM pulse train, 169
 coherent pulse train, 43, 68, 169
 Costas frequency stepped pulse, 77
 LFM pulse, 57–58, 181
 multicarrier bit, 333
 multicarrier phase-coded pulse, 327–328
 multicarrier phase-coded pulse train, 358
 narrow bandpass signal, 30
 phase-coded pulse, 100
 sawtooth FM signal, 309
 sinusoidal FM signal, 311
 unmodulated pulse, 53–54
Complex signal, 23
Compression ratio, 11, 56, 59, 94, 181, 261, 343, 354
 of a NLFM pulse, 93–94
Consecutive ordered cyclic shifts, 333, 339, 341, 349, 353, 360, 365
Constant-frequency pulse, 53
Continuous wave, 14, 133, 168, 176, 294, 372
 multicarrier phase-coded signal, 350
Continuous correlation function, 101
Convolution, 23, 27, 182–183, 301–302
 properties, 182, 301

Costas, 11, 47, 74–75, 244, 258, 260–262, 284, 391
 ambiguity function, 11, 75, 77–78, 84–85
 autocorrelation function, 78–80, 84–85
 complex envelope, 77
 definition, 75
 Doppler tolerance, 86
 frequency coding, 74
 properties, 75, 77, 80
 pulse train AF, 86
 resolution, 78, 84
 spectrum, 80, 84–85
 longer array, 83
 pulse compression of, 78
Costas arrays
 construction algorithms, 75
 number of, 81
Countermeasure, 13
CPT, *see* Coherent pulse train
Crest-factor, 333, 338
Cross ambiguity function, 140, 174, 176, 179, 185, 394
 of LFM pulse train, 178, 186
 of LFM pulses with Hamming weigthing, 189
Cross-correlation, 66, 140
 properties, 104
Cumulative probability, 220
Cusping, 146
CW, *see* Continuous wave
CW multicarrier signal, 350
CW radar, 168, 294, 297, 305
 block diagram, 321
 mixer implementation, 318
Cyclic difference set, 134–136, 139
 definition, 134
Cyclic Hadamard difference sets, 134

Decimation, 105, 122
Delay and add receiver, 290
Delay–Doppler response, 176, 179–181, 183, 186, 242, 244, 301, 305–306, 394
 definition, 182, 301
 of a coherent train of weighted LFM pulses, 180
 of a sawtooth FM signal, 309–310, 312
 of a sinusoidal FM signal, 313–314
 of a triangular FM signal, 315
 of coherent LFM pulse train, 180
 of Golomb biphase signal, 305
 properties, 184
Difference matrix, 76–77
Discrete correlation function, 101
Discrete Fourier transform, 137
Difference set, 134–136, 139

Diverse pulse train, 67
Diversity, 67, 168, 170, 226, 233
  pulse-to-pulse, 227
Doppler, 32, 35, 54
  bandwidth limitations, 5
  compensation, 174
  effect, *see* Doppler shift
  recurrent lobes, 171
  filter, 173–174
Doppler navigation radars, 294, 314
Doppler resolution, 6, 9, 13, 44–45, 84, 168, 295, 297, 301
  of a coherent pulse train, 70
  of a constant frequency pulse, 54
Doppler shift, 2–5, 22, 31, 34, 60–61, 113, 128, 131, 137, 173–174, 176
  definition, 4
  development of, 3
Doppler tolerance, 61, 113, 128, 131, 137, 176
Doppler tolerance of
  Barker phase-coded pulse, 113
  chirplike phase-coded pulse, 113
  Costas frequency stepped pulse, 86
  Frank phase-coded pulse, 128
  Golomb codes, 137
  LFM pulse, 60–61, 92, 173, 176
  NLFM pulse, 91–92
  $P(n, k)$ phase-coded pulse, 131
  P1 phase-coded pulse, 128
  P2 phase-coded pulse, 128
  P3 phase-coded pulse, 128
  P4 phase-coded pulse, 128
  Px phase-coded pulse, 128
Double canceller, 193, 196, 198–199
  frequency response, 195, 197
  improvement factor, 203, 206
Duty cycle, 12, 69, 84, 99, 176, 215, 359, 361
Dwell, 43, 210–211, 213, 216, 296
Dwell-to-dwell PRF stagger, 211
Dynamic range, 192, 209

Eclipsing, 168, 211, 214–215
Eigenvalue, 207
Eigenvector, 207
Electromagnetic propagation, 4
Environmental diagram, 13–14
Equivalent complementary sets, 265–266
Equivalent phase codes, 103
Evolutionary algorithms, 215

Fast Fourier transform, 7, 63, 173–174, 176, 191, 240, 242, 331–332
FFT, *see* Fast Fourier transform

Finite fields, 157
Finite impulse response filter, 66, 192–193, 195, 208
FIR, *see* Finite impulse response filter
Fixed-PRF radar, 202
Flat-top pulses, 183
Fourier transform, 20, 24–25, 30, 36, 41, 52, 61, 63, 86, 254, 330
  matrix, 115
  of weight windows, 183
  properties, 182, 301
Frank matrix, 115–117
Frank phase code, 12, 114–115, 118, 121, 123, 297
  spectrum, 132
  ambiguity function, 118–119, 122
  autocorrelation function, 116, 118, 120, 126, 130
  bandwidth limitations, 122, 128
  definition, 115, 117
  Doppler tolerance, 128
  matrix, *see* Frank matrix
  periodic autocorrelation function, 122, 124
  properties, 116–117, 291
Frequency deviation, 58, 61, 245
Frequency modulated CW signals, 306
Frequency modulation, 35, 41, 61, 80, 306
Frequency response, 24–25
  of a Double canceller, 195, 198
  of a single canceller, 193
Frequency weighting, 61, 244
Fresnel integral, 64

Galois fields, 81, 133, 139, 156
Gaussian-windowed sinc, 16–17, 151–153, 155, 159
Generalized Barker, *see* Polyphase Barker codes
GF, *see* Galois field
Ghosts, 212–213, 216
GMW sequence, 134
Golomb
  biphase code, 297, 299, 305
  code, 134
  construction of, 134
  table of, 135
  phase evolution of, 128
  polyphase code, 126
    definition, 128
    periodic autocorrelation function, 128
  signal, 304, 306
Grating lobes, 231–232, 235, 245–246, 248, 250, 252–253, 258, 260, 263
  nullifying, 245–246
Group time delay, 88

INDEX 407

Hadamard matrix, 134, 136, 269–270, 277
Hamming window, 61, 67, 86, 176–177,
    179–180, 183, 186, 189, 209, 304, 310,
    345
  expression of, 183
  Fourier transform, 183, 303
  square root of, 63–64
Hann window, 61, 86, 180, 183, 209
  expression of, 183
  Fourier transform, 183, 303
Heimiller code, 114
High power RF amplifier, 179
High PRF, 71, 211
High-order cancellers, 201
Hilbert transform, 23
Huffman code, 16, 142–143
  autocorrelation of, 144
  real envelope of, 144
Hybrid FM, 96

I/Q detector, 21–22
Ideal periodic autocorrelation code, see
    Perfect code
Improvement factor, 202–206, 222
  of a filter with binomial coefficients, 208
Impulse response, 20, 24–25, 27
Inner-outer code, 107
Index codes, 156
Instantaneous bandwidth, 226, 228, 236
Instantaneous frequency, 58, 61, 86–90, 129,
    310, 335
  of LFM pulse, 86
  of NLFM pulse, 90
Instantaneous peak power, 330
Integrated sidelobe ratio, 8, 141, 347
  of a Frank phase-coded pulse, 118
  of a Px phase-coded pulse, 118
Interpulse amplitude weighting, 168, 176,
    179–182, 235, 305
Interrupted CW, 43
Intrapulse amplitude weighting, 179–181,
    183–184
Intrapulse modulation, 173
Ipatov code, 137, 302
Ipatov signal, 302–303
ISLL, 141–142
ISLR, see Integrated sidelobe ratio

Jacobian, 37

Kaiser window, 92, 349
Kernels, 264–265
Kronecker product, 107, 270

Legendre sequences, 134
Legendre symbol, 162
LFM property of the AF, 35, 38
LFM pulse, 169
  ambiguity function, 10, 12, 58, 60, 62, 64,
    72, 170
  autocorrelation function, 59–60, 63, 72, 86
  complex envelope, 57–58
  compression ratio, 59
  Doppler tolerance, 60–61, 92, 176
  instantaneous frequency, 58
  resolution, 39, 59–60
  spectrum, 41, 59, 63
LFM pulse train, 168
  ambiguity function, 170–171, 181
  complex envelope, 169
  resolution, 169
  with intrapulse and interpulse weighting,
    180
Linear frequency modulation, see LFM
Linear power amplifiers, 2, 16, 86, 152, 328
Low-pass filter, 21
Low PRF, 211
Low probability of intercept, 294
LPF, see Low-pass filter
LPI, see Low probability of intercept

M out-of N, 191, 214, 216, 220
Mainlobe clutter, 14, 215
Major-minor PRF set selection, 215
Matched filter, 8, 13, 16–17, 20, 23–24, 26,
    30, 40, 146, 173, 180, 301, 317
  frequency response, 25
  impulse response, 24–25, 27
  of a narrow bandpass signal, 30
  of an unmodulated pulse, 27
  of a rectangular pulse, 27
  response to Doppler shift, 31
MATLAB, 18, 46, 338, 373
  ambiguity function plotting program,
    373
MATLAB code, 47
  for Gaussian-windowed sinc, 159
  for generating train of LFM pulses, 284
  for quadriphase Barker 13, 158
  for Welch construction of Costas arrays,
    82, 96
Matched receiver, see Matched filter
Matched weighting, 63, 66, 179
Maximum-length linear feedback sequences,
    see m-sequences
MCPC, see Multicarrier phase-coded signal
Medium PRF, 191, 211, 214
Minimal separable distance, 8

Minimum peak sidelobe phase code, 106, 108
  ambiguity function, 11
  autocorrelation function, 133
  properties, 107
Minimum shift keying, 147
Mismatch loss, 63, 66, 136, 140, 179
Mismatched filter, 62, 66
MLC, *see* Mainlobe clutter
MOCS, *see* Mutually orthogonal complementary sets
Modified Costas pulse, 260
Monopulse, 7
Moving target, 8, 31, 191
MPS, *see* Minimum peak sidelobe code
MSD, *see* Minimal separable distance
m-sequence, 133–134, 139, 158
  autocorrelation function, 133
  generation, 133
  periodic autocorrelation function, 133
Moving target indicator, 191–192, 199, 205, 222
  filter, 207
  frequency response of, 209
MTI, *see* Moving target indicator
MTI radar, 191, 195
MTI weights, 205
Multicarrier 16, 349
Multicarrier bit, 333–334, 337
  phasing, 335
Multicarrier phase-coded signal, 327, 345
  COCS of CLS diverse pulse train, 360
  CW periodic signal, 350
  MOCS pulse train, 361
  pulse diversity, 360
  single pulse, 341
  train of diverse pulses, 358
Mutually orthogonal complementary sets, 358, 361–362, 366
Muzzle velocity radar, 294

Narahashi and Nojima phasing, 335–337
Narrow bandpass
  filter, 318
  matched filter, 29
  signal, 5, 20–23, 29, 31
Natural envelope, 20, 23
NBP, *see* Narrow bandpass
Nested codes, 107
  autocorrelation function, 113
Newman phases, 334–337
NLFM pulse, 74, 86–88
  ambiguity function, 90–91, 93
  autocorrelation function, 89–92, 94–95
  compression ratio, 93–94
  Doppler tolerance, 91–92
  instantaneous frequency, 90
  spectrum, 89–90, 94–95
Noise radars, 18
Normalized form, 103, 109
North filter, 24
Nyquist criterion, 47, 186, 374

OFDM, *see* Orthogonal frequency division multiplexing
Optimal MTI weights, 202
On receive, 66, 179, 186, 240, 295, 315
Optimal filters for sidelobe suppression, 140
Orthogonal, 17, 78, 269, 274, 330
Orthogonality, 274
  of a Costas pulse, 78
Orthogonal coded LFM pulse train, 277, 284
  frequency spectrum, 284
Orthogonal coding, 274
Orthogonal frequency division multiplexing, 327, 330
Orthogonal matrix, 116, 264, 269
Orthogonal pulses, 273
Outer code, 270

$P(n, k)$ phase code, 129
  ambiguity function, 131
  autocorrelation function, 130–131
  bandwidth limitations, 132
  definition, 129
  resolution, 130
  spectrum, 130, 132
P1 phase code, 115, 118, 121, 123, 268
  autocorrelation function, 130
  bandwidth limitations, 122
  definition, 121
  Doppler tolerance, 128
  periodic autocorrelation function, 122, 124
P2 phase code, 115, 118, 121–122, 268, 269
  ambiguity function, 122
  bandwidth limitations, 122
  definition, 121, 123
  Doppler tolerance, 128
  properties, 121
P3 phase code, 115, 126, 269, 272, 299
  autocorrelation function, 130
  bandwidth limitations, 128
  definition, 127
  Doppler tolerance, 128

P4 phase code, 12, 16–17, 73, 115, 126, 152, 161, 272, 277, 297, 299, 303–305, 309, 340, 346, 349, 353
  ambiguity function, 298
  autocorrelation function, 130
  bandwidth limitations, 128, 132
  definition, 127, 297
  Doppler tolerance, 128
  palindromic, 115, 128
  periodic ambiguity function, 298, 304
  periodic autocorrelation function, 128, 299
  properties, 299
  spectrum, 17, 132
  using Gaussian-windowed sinc, 154
PACF, see Periodic autocorrelation function
PAF, see Periodic ambiguity function
Palindromic, 115, 121, 123, 128
  P2, 121
  P4, 115, 128
PAPR, see Peak-to-average power ratio
Parseval's theorem, 36
Peak sidelobe level, 8, 109, 350
Peak sidelobe ratio, 8
Peak-to-average power ratio, 334
Peak-to-mean envelope power ratio, 144, 328, 330, 332–334, 336, 345, 347, 359
Perfect codes, 105, 115, 122, 132, 134, 155, 227, 264, 297, 350
  definition, 132
  test of, 323
Perfect periodic autocorrelation function, 297
Periodic ambiguity function, 42–43, 227, 296
  definition, 43–44, 296
  properties, 44–45, 170, 296–297
  variants, 43–44
Periodic ambiguity function of
  coherent pulse train, 12, 69
  Golomb biphase coded CW, 300
  multicarrier phase-coded CW, 357
  multicarrier phase-coded pulse train, 360–361, 363
  P4 phase-coded CW, 298
  pulse train, 12, 170
  triangular FM, 316
Periodic autocorrelation function, 227, 296
  properties, 105, 160
Periodic autocorrelation function of
  chirplike phase code, 115
  Golomb polyphase code, 128
  m-sequence, 133
  multicarrier phase-coded CW, 350, 354

multicarrier phase-coded pulse train, 359, 362
  P1 phase code, 122
  P4 phase code, 128, 299
  two-valued Golomb code, 134
  Zadoff–Chu phase code, 124
Periodic correlation, 105, 128, 134
  of phase-coded pulse, 101
Periodic cross ambiguity function, 400
Periodic complementary binary sequences, 290
Permutation codes, 105
Phase-coded pulse, 12, 16, 100
  correlation function, 101, 103–104
  resolution, 103
Phase-coded CW signal, 304
Phase-coded signal, 16, 47, 155, 375
PMEPR, see Peak-to-mean envelope power ratio
Polarity reversals, 226
Polarization, 18
Police radar, 294
Polyphase Barker code, 109–110, 113, 131
  autocorrelation function, 113
  definition, 109
  properties, 113
  table of, 110
Polyphase code, 103, 272
PONS
  complementary set, 277
  construction, 267–269
  matrix, 268–270, 272
Power spectral density, 41, 61
Power amplifier, 62
PRF, see Pulse repetition frequency
PRI, see Pulse repetition interval
Primitive, 139, 155, 158
Primitive element, 81–82, 157
Primitive polynomial, 157–158
Primitive root code, 162
Probability of detection, 24, 220
Proximity fuze, 294
Pseudo-random, 18
PSLL, see Peak sidelobe level
PSLR, see Peak sidelobe ratio
Pulse compression, 1, 11, 35, 46, 56–57, 78, 83, 100, 168, 270, 327
Pulse Doppler radar, 211
Pulse repetition frequency, 71, 195
Pulse repetition interval, 12, 45, 68, 70, 181, 191, 226, 358
  diversity, 191, 210
  staggering, 191

Pulse width, 9
Px phase code, 118, 121–123
  ambiguity function, 122
  autocorrelation function, 120, 130
  bandwidth limitations, 122
  definition, 118–120
  Doppler tolerance, 128
  matrix, 119
  phase evolution, 120
  phase matrix, 119

Quadratic residue codes, 155
Quadrature components, 21
Quadriphase code, 147
  ACF of, 149
Quaternary residue code, 162

Radar altimeters, 294
Rain, 13
Radar cross section, 7–8, 212
Raised cosine window, 89, 92
Random waveforms, 18
Range
  Doppler ambiguities, 212
  Doppler map, 215
  error, 60, 295
  folding, 12
  gate, 215
  rate, 2
  resolve, 215
  window, 40–41, 199
RC filter, 29
RCS, see Radar cross section
Rectangular pulse, 57
Recurrent lobes, 12, 45, 69, 71, 168–169, 226–227, 233, 240, 246, 359, 363–364
  reduction of, 226
Recursive filters, 195
Resolution, 6–7, 11, 15, 42, 46, 52, 56, 61, 191, 228, 233, 236, 245, 258, 295, 301, 314
Resolution of
  coherent pulse train, 45, 70
  coherent train of LFM pulses, 169
  Costas frequency stepped pulse, 84
  CW signal, 295
  LFM pulse, 12, 39, 40, 59–60
  multicarrier phase-coded pulse, 328
  $P(n, k)$ phase-coded pulse, 130
  P4 phase-coded pulse, 130
  unmodulated pulse, 54
Ridge shaped ambiguity function, 12, 114

Saturated power amplifier, 87, 328
Sawtooth FM, 309–310

Schroeder code, 114
Schroeder phases, 334–336, 343, 339, 358–359
Schwarz inequality, 25, 36
SCV, see Subclutter visibility
Second time around, 12, 240
Sextic residue sequence, 134
Sidelobe clutter, 14, 212
Sidelobe matrix, 77–79, 84
Sidelobes, 8–9, 11, 74, 109, 161, 168, 264, 297, 311, 347, 349
  of a Costas AF, 84
  of AF, 31
Signal-to-clutter ratio, 209
Signal-to-noise ratio, 7–8, 16, 20, 24–29, 67, 87, 141, 179–180, 301, 311
  highest attainable, 24–25, 27
  loss, 67–87, 141, 179–180, 311
Single canceller, 192–193, 199
  frequency response, 193
  improvement factor, 203
Sinusoidal FM, 311, 314
SLC, see Sidelobe clutter
Smooth window, 176
SNR, see Signal-to-noise ratio
Spectral density, 24, 63, 87, 256
  of a Costas signal, 80
  of a train of LFM pulses, 254
  of MCPC, 343
Spectral shaping, 61
Spectrum, 16, 23, 25, 30, 41, 61, 63, 86, 146, 148, 152, 256
  properties, 41
  sidelobes, 16, 57
Spectrum of
  Barker phase-coded pulse, 145
  clutter, 202
  Costas frequency stepped pulse, 80, 84–85
  Frank phase-coded pulse, 132
  LFM pulse, 41, 59, 63
  multicarrier phase-coded CW, 356
  multicarrier phase-coded pulse, 328, 343, 348
  multicarrier phase-coded pulse train, 366
  NLFM pulse, 89–90, 94–95
  $P(n, k)$ phase-coded pulse, 130, 132
  P4 phase-coded pulse, 132
  unmodulated pulse, 41, 57
Stagger ratio, 196–199
Staggered PRF, 195, 199, 202, 205
  with single canceler, 199
Stationary phase principle, 87–88, 92, 129
Stepped-frequency, 73, 226, 228, 231–232, 236, 244–246, 248, 284, 390

radar, 237
train of LFM pulses, 229, 245, 247–248, 250–251, 254
train of unmodulated pulses, 228, 231, 233, 236
pulse, 116
Stepwise window, 176
Straddle, 214–215
  loss, 146, 176
Stretch
  processing, 236–237, 238
  processor, 226, 239, 242, 244
Subclutter visibility, 202
Subcomplementary sequences, 270, 274, 277, 280
Supercomplementary set, 265, 267, 269
Surveillance radar, 13

Target space, 13
Thumbtack AF, 9, 11, 67, 78, 84, 114, 327–328
Toeplitz matrix, 212
Taylor weighting, 185
Ternary Barker codes, 113
Time-bandwidth product, 11–12, 58–61, 89, 93, 181, 252, 261
  of a phase-coded pulse, 11
  of an unmodulated pulse, 11
Time-varying MTI weights, 202
Transfer function, 20, 199, 333
  of a single canceller, 200
Triangular FM, 315, 321–323
Twin prime sequence, 134
Two-valued asymptotically perfect sequences, 134
Two-valued Golomb code, 134, 136
Two-valued PACF, 133–134

Ultra-wide bandwidth, 5
Unambiguous
  range, 210–211, 213, 215
  velocity, 210
Unmodulated pulse, 9
  ambiguity function, 9, 54–55, 58, 68
  autocorrelation function, 54
  complex envelope, 53–54
  matched filter, 27
  resolution, 9, 54
  spectrum, 41, 57

Variable amplitude, 2, 16, 41, 56, 86, 152, 176, 179
  of a multicarrier phase-coded pulse, 328
Variable pulse amplitudes, 237
Variable-top pulses, 183
VCO, *see* Voltage controlled oscillator
Velocity measuring radars, 294
Velocity of propagation, 2
Voltage controlled oscillator, 237
Voltage spectral density, 57

Wavelength, 5
Weight window, 16, 62, 66, 176, 179–180, 182–184, 237, 258, 295, 301, 303–304
  Fourier transform of, 183, 302
Welch algorithm, 81–82, 96
Welti codes, 113, 270

Zadoff–Chu phase code, 114–115, 122, 124–125, 128, 132, 272, 297
  ambiguity function, 127
  autocorrelation function, 115, 123, 125, 130
  definition, 122–123, 126
  periodic autocorrelation function, 124, 128
Zak transform, 370
Zero correlation zone, 270, 273
Zero cross-correlation complementary code, 291
Zero-padding, 176, 240

Printed in Poland
by Amazon Fulfillment
Poland Sp. z o.o., Wrocław
04 November 2020

ad00fbb2-e41c-4090-afe1-eb64e48665b9R01